Digital Signal Processing using MATLAB®

Digital Signal Processing using MATLAB®

Edited by
André Quinquis

iSTE

WILEY

First published in France in 2007 by Hermes Science/Lavoisier entitled "Le traitement du signal sous Matlab®: pratique et applications", 2nd edition
First published in Great Britain and the United States in 2008 by ISTE Ltd and John Wiley & Sons, Inc.

ISTE Ltd
6 Fitzroy Square
London W1T 5DX
UK

www.iste.co.uk

John Wiley & Sons, Inc.
111 River Street
Hoboken, NJ 07030
USA

www.wiley.com

Library of Congress Cataloging-in-Publication Data

Quinquis, André.
[Traitement du signal sous MATLAB. English]
Digital signal processing using MATLAB / André Quinquis.
p. cm.
Includes bibliographical references and index.
ISBN 978-1-84821-011-0
1. Signal processing--Digital techniques. I. Title.
TK5102.9.Q853 2008
621.382'2--dc22

2007043209

British Library Cataloguing-in-Publication Data
A CIP record for this book is available from the British Library
ISBN: 978-1-84821-011-0

Printed and bound in Great Britain by Antony Rowe Ltd, Chippenham, Wiltshire.

Table of Contents

Preface

Why and How this Book was Written

Sometimes it is easier to say what a book is not than what it exactly represents. It may be also better to resume the authors' motivations than to explain the book content itself.

From this point of view, our book is certainly not a traditional course, although it recalls many theoretical signal processing concepts. Indeed, we emphasize a limited number of important ideas instead of making a detailed description of the involved concepts. Intuitive manners have been used to link these concepts to physical aspects. Hence, we hope that reading this book will be much more exciting than studying a traditional signal processing course.

This book is also not a physics course, although a major purpose of most proposed exercises is to link abstract signal processing concepts to real-life problems. These connections are illustrated in a simple and comprehensive manner through MATLAB® simulations.

The main topics of this book cover the usual program of an undergraduate signal processing course. It is especially written for language and computer science students, but also for a much larger scientific community who may wish to have a comprehensive signal processing overview. Students will certainly find here what they are looking for, while others will probably find new and interesting knowledge.

This book is also intended to illustrate our pedagogical approach, which is based on three major reasons:

1. Students need to know how the teaching provided can be useful for them; it is their customer attitude.

2. Students have good potential for doing independent work; their interest and curiosity should be continuously stimulated by:

– using a diversified pedagogical approach that combines the two sides of a complete presentation methodology: from components to the system and vice versa;

– encouraging them to take advantage of their creativity through interactive educational tools; they should be allowed to make changes and even contribute to their development.

3. Students have to improve and validate their knowledge through written work; writing is still the best way to focus someone's concentration.

The role of simulations is becoming more and more important in the framework of a scientific education because it is an effective way to understand many physical phenomena, some of them less known or mastered, and to take into account their complexity. Simulations may be thus very useful for:

– *understanding* working principles and deriving behavior laws;

– *learning* about processing methods and systems running using algorithms to reproduce them off-line;

– *evaluating* the performance and robustness of various algorithms and estimating the influence of different parameters.

Simulations in signal processing education enable students to learn faster and facilitate the comprehension of the involved physical principles. From a teaching point of view, simulation tools lead to lower costs and time efficiency.

This book is based on a signal processing course, which has been successfully given for many years in several universities. According to our experience, signal theory abstract concepts and signal processing practical potentialities can be linked only through tutorial classes and simulation projects. In this framework, simulations appear to be the necessary complement for the classical tripod theory – modeling – experimentation.

This book brings together into a clear and concise presentation the main signal processing methods and the related results, using MATLAB software as a simulation tool. Why MATLAB? Because it is:

– simple to learn and to use;

– powerful and flexible;

– accurate, robust and fast;

– widespread in both academic and industrial environments;

– continuously updated by professionals.

The word "signal" stands for a physical entity, most often of an electrical nature, like that observed at a microphone output. It is submitted to various transformations when it goes through a system. Thus, in a communication chain, the signal is subject to some changes (distortion, attenuation, etc.), which can make it unrecognizable. The aim is to understand this evolution in order to properly recover the initial message.

In other words, a signal is a physical support of information. It may carry the orders in a control and command equipment or multimedia (speech and image) over a network. It is generally very weak and it has to be handled with much caution in order to reach the signal processing final goal, i.e. information extraction and understanding.

Signal processing is widely used in many industrial applications such as: telecommunications, audio and speech signal processing, radar, sonar, non-destructive control, vibrations, biomedicine, imagery, etc. Standard signal processing functions include signal analysis, improvement, synthesis, compression, detection, classification, etc., which depend on and interact with each other in an integrated information processing chain.

The digital signal processing methods provide noteworthy capabilities: accurate system design, excellent equipment reproducibility, high stability of their exploitation characteristics and an outstanding supervision facility.

The digital signal processing boom is related to the development of fast algorithms to calculate the discrete Fourier transform. Indeed, this is the equivalent of the Fourier transform in the discrete domain and so it is a basic tool to study discrete systems. However, related concepts are generally considered highly theoretical and accessible to scientific researchers rather than to most engineers. This book aims to overcome this difficulty by putting the most useful results of this domain within the understanding of the engineer.

Chapter 1 briefly describes essential concepts of MATLAB software, which is an interactive software tailored for digital signal processing. Language rules, elementary operations as well as basic functions are presented. Chapter 2 illustrates the generation of 1D or 2D (image) digital signals as data vectors and matrices respectively.

Finding the solution of a signal processing problem involves several distinct phases. The first phase is the modeling: the designer chooses a representation model for an observed data. When it can be done very accurately the signals are said to be deterministic. A powerful tool for analyzing them is provided by the Fourier transform, also called frequency representation, which is presented in Chapter 5. Its equivalent in the discrete domain is represented by the z-transform, which is developed in Chapter 6.

There are many other processes, which give different and apparently unpredictable results, although they are observed using identical experimental conditions. They are known as random processes, such as the receiver's thermal noise. The wide sense stationary random processes, which form a particularly interesting class of these signals, are presented in Chapter 3. Some useful statistical tools for testing different hypothesis about their parameters behavior are provided in Chapter 4.

From a very general point of view, digital signal processing covers all the operations, arithmetical calculations and number handling performed on the signal to be processed, defined by a number series, in order to obtain a transformed number series representing the processed signal. Very different functions can be carried out in this way, such as classical spectral analysis (Chapter 10), time-frequency analysis (Chapters 11 and 12), linear filtering (Chapters 7 and 8), detection and estimation (Chapter 9), and feature extraction for information classification or compression (Chapters 13 and 14).

Theoretical developments have been reduced to the necessary elements for a good understanding and an appropriate application of provided results. A lot of MATLAB programs, solved examples and proposed exercises make it possible to directly approach many practical applications. The reader interested in some more complementary information will find this in the references cited at the end of this book.

Finally, I would like to acknowledge all the members of my team, Emanuel Radoi, Cornel Ioana, Ali Mansour and Hélène Thomas, for their contributions to this book.

André QUINQUIS

Chapter 1

Introduction

1.1. Brief introduction to MATLAB

1.1.1. *MATLAB software presentation*

MATLAB (MATrix LABoratory) is an interactive software, developed by Math Works Inc. and intended especially for digital signal processing. It is particularly effective when the data format is vector or matrix.

MATLAB integrates digital calculus, data visualization and open environment programming. MATLAB exists under both Windows and UNIX. Many demonstrations are available using the command `demo`.

This digital simulation software enables a fast and simple visualization of the obtained results.

MATLAB was primarily written in FORTRAN and C. However, MATLAB knows to interpret commands, while a compilation of the source code is required by FORTRAN and C.

MATLAB is especially designed for digital signal processing and for complex digital system modeling and simulation. It is also suitable for processing data series, images or multidimensional data fields.

MATLAB software general structure is provided in Figure 1.1.

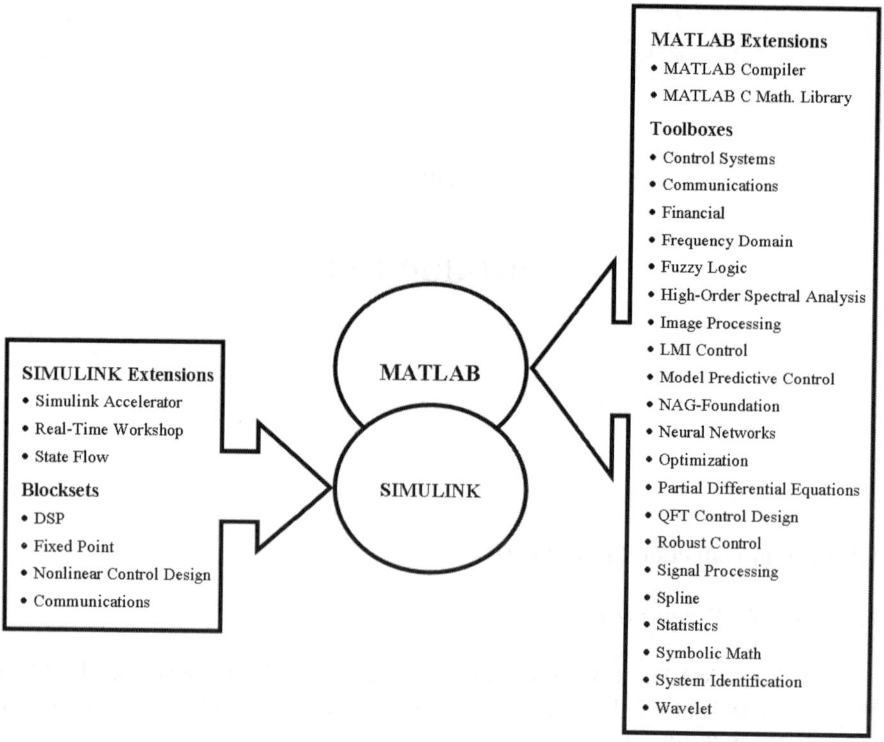

Figure 1.1. *MATLAB software general structure*

The toolboxes extend the basic MATLAB functions and perform specific tasks corresponding to different digital processing fields, such as image processing, optimization, statistics, system control and identification, neural networks, fuzzy systems, etc.

SIMULINK is an interactive software designed for modeling and simulating continuous-time or discrete-time dynamical systems or hybrid structures containing both analog and digital systems. It makes use of a mathematical equation set and provides a large variety of predefined or user-defined functional blocks.

MATLAB has been developed for several years, especially as a consequence of its use in the academic environment as an excellent education tool in mathematics, engineering and science. In addition, MATLAB has already proven its utility for scientific research and technological development.

In order to run MATLAB, type the command `matlab` with UNIX shell (if a MATLAB license under UNIX is available) or double click on the MATLAB icon if the operating system is Windows. To exit MATLAB, type `exit` or `quit`. If MATLAB is running under UNIX, you may have access to all UNIX commands using just before the symbol! (example: `!ls -l`).

1.1.2. *Important MATLAB commands and functions*

`who`	lists the variables in the current workspace
`whos`	the same as previous, but lists more information about each variable
`what`	lists MATLAB-specific files in directory
`size`	provides the size of a data array
`length`	provides the size of a data vector
`help`	displays help text in Command Window
`exit, quit`	exits from MATLAB

Table 1.1. *General commands*

`dir, chdir,`
`delete, load,`
`save, type`
	similar to the corresponding DOS commands

`pack`	consolidates workspace memory

Table 1.2. *Commands related to the workspace*

`+, -, *, /, ^`	usual arithmetical operators
`.`	followed by an arithmetical operator for applying it to each array element
`'`	Hermitian operator
`.'`	transpose operator

Table 1.3. *Arithmetical operators*

<table>
<tr><td><, <=, >, >=</td><td>usual relational operators</td></tr>
<tr><td>==</td><td>equality operator</td></tr>
<tr><td>~=</td><td>inequality operator</td></tr>
<tr><td>&</td><td>element-wise logical AND</td></tr>
<tr><td>|</td><td>element-wise logical OR</td></tr>
<tr><td>~</td><td>logical complement (NOT)</td></tr>
</table>

Table 1.4. *Relational and logical operators*

=	variable assignment operator
,	used to separate the arguments of a function or the elements of a data array
[]	used to build data arrays
()	used in arithmetical expressions
:	used for indexing variables
;	used at the end of a statement to cancel displaying any output
...	used to continue a command on the next line
%	used to enter a comment

Table 1.5. *Special characters*

ans	default name of a variable or a result
eps	spacing of floating point numbers
pi	value of $\pi = 3.14159...$
i,j	value of $\sqrt{-1}$
Inf	IEEE arithmetic representation for positive infinity (1/0)
NaN	IEEE arithmetic representation for Not-a-Number (0/0)
nargin	returns the number of function input arguments
nargout	returns the number of function output arguments

Table 1.6. *Special variables and constants*

abs	absolute value function
sqrt	square root function
real	real part of a complex variable
imag	imaginary part of a complex variable
angle	returns the phase angles, in radians, of a complex variable
conj	complex conjugate operator
sign	signum function
rem	returns the remainder after division
exp	exponential function
log	natural logarithm function
log10	base 10 logarithm function

Table 1.7. *Elementary mathematical functions*

sin, cos, tan, cot, sec	usual trigonometric functions
asin, acos, atan, acot, asec	inverse trigonometric functions
sinh, cosh, tanh, coth, sech	hyperbolic functions
asinh, acosh, atanh, acoth, asech	inverse hyperbolic functions

Table 1.8. *Trigonometric functions*

max	largest component
min	smallest component
mean	average or mean value
std	standard deviation
sum	sum of elements
cumsum	cumulative sum of elements
prod	product of elements
cumprod	cumulative product of elements

Table 1.9. *Data analysis functions*

`conv`	convolution and polynomial multiplication
`deconv`	deconvolution and polynomial division
`roots`	finds polynomial roots
`poly`	converts roots to polynomial
`polyval`	evaluates polynomial
`residue`	partial-fraction expansion (residues)

Table 1.10. *Polynomial related functions*

`zeros`	enables generation of zero arrays
`ones`	enables generation of ones arrays
`rand`	enables generation of uniformly distributed random numbers
`randn`	enables generation of normally distributed random numbers
`linspace`	enables generation of linearly spaced vectors
`logspace`	enables generation of logarithmically spaced vector
`det`	calculates the determinant of a square matrix
`norm`	calculates matrix or vector norm
`inv`	calculates matrix inverse
`eig`	calculates matrix eigenvalues and eigenvectors

Table 1.11. *Vector or matrix related functions*

`input`	gives the user the prompt and then waits for input from the keyboard
`ginput`	gets an unlimited or a predefined number of points from the current axes and returns their coordinates

Table 1.12. *Input functions*

plot	plot vectors or matrices
subplot	create axes in tiled positions
bar	draws a bar graph
hist	draws a histogram graph
polar	makes a plot using polar coordinates
stairs	draws a stairstep graph
stem	plots the data sequence as stems
semilogx, semilogy	semi-log scale plot: a logarithmic (base 10) scale is used for the x-axis or y-axis
loglog	log-log scale plot: a logarithmic (base 10) scale is used for both the x-axis and y-axis
xlabel, ylabel	adds text beside the x-axis or y-axis
title	adds text at the top of the current axes
grid	adds grid lines to the current axes
figure	creates a new figure window
clf	clears current figure
close all	closes all the open figure windows
hold on/off	holds/discards the current plot and all axis properties
axis	controls axis scaling and appearance
legend	puts a legend on the current plot using the specified strings as labels
gtext	allows placing text with mouse
image	displays a matrix as an image

Table 1.13. *1D and 2D graphical commands*

plot3	plot lines and points in 3-D space
mesh/surf	plots a 3-D mesh/colored surface
contour	plots a contour plot of a matrix treating its values as heights above a plane

Table 1.14. *3D graphical commands*

`if`	conditionally executes statements
`else, elseif`	used with `if` command
`end`	terminates scope of `for`, `while`, `switch`, `try` and `if` statements
`for`	repeats statements a specific number of times
`while`	repeats statements an indefinite number of times
`switch`	switches among several cases based on expression
`break`	terminates execution of `while` or `for` loop
`return`	causes a return to the invoking function or to the keyboard
`pause`	pauses and waits for the user response

Table 1.15. *Control commands*

1.1.3. *Operating modes and programming with MATLAB*

The "online command" default operating mode is available after MATLAB gets started. It displays the prompt >> and then waits for an input command. Running a command usually results in creating one or several variables in the current workspace, displaying a message or plotting a graph. For instance, the following command:

```
v = 0:10
```

creates the variable v and displays its elements on screen. A semicolon has to be added at the end of the statement if it is not necessary to display the result.

The previously typed commands can be recalled with the key ↑, while a statement can be modified using the keys ← and →. You may also analyze the effects on the command lines of the following keys: ↓, *home, end, esc, del, backspace* and of the following key combinations: ctrl + →, ctrl + ←, ctrl + *k*.

Besides the "online command" operating mode, MATLAB can also create script files and function files. Both of these are saved with the extension .m, but the function files accept input arguments and return output arguments and operate on variables within their own workspace.

In order to create a script file you have to select the menu *File/New/M-file*, while to edit an existing file you have to first select *File/Open M-file* etc., and then choose the appropriate file. After these commands, an edition session will be open using the chosen editor from *Edit/View/Edit Preference*. The edited file can be saved with the menu *File/Save As* etc., followed by the file name (with the extension .m).

In MATLAB, many functions are predefined and saved as m-files. Some of them are intrinsic, the others being provided by external libraries (*toolbox*): they cover specific domains such as mathematics, data analysis, signal processing, image processing, statistics, etc.

A function may use none, one or several input arguments and return none, one or several output values. These different cases for a MATLAB function are called:

– one output value and no input argument:

variable_name = function_name

– no output value and one input argument:

function_name (argument_name)

– several output values and several input arguments:

[var_1, var_2, ...,var_n] = function_name (arg_1, arg_2, ..., arg_m)

For the last case, the first line of the file `function_name.m` has the following form:

– function [var_1, var_2, ...,var_n] = function_name(arg_1, arg_2, ..., arg_m)

Usually, the input arguments are not modified, even if their values change during the function execution. In fact, all the variables are local by default. Nevertheless, this rule can be changed using the command: `global variable_name`.

In a MATLAB file, the comment lines have to begin with the symbol %.

The on-line help can be obtained using: `help` *<function_name>*. The first lines of the file *<function_name>.m* beginning with % are then displayed. It is also possible to search all the files containing a given keyword in their help using the command: `lookfor` *<keyword>*.

NOTE.– The user-defined MATLAB files are recognized only in the current directory, unlike the original MATLAB functions (*toolbox,* etc.). In order to make available a user-defined file *<file_name.m>* outside the current directory you have to type the command:

```
path(path,'<file_acces_path>/file_name>')
```

(see `help path, help addpath`).

The data from the current workspace can be saved in a *.mat file using the command `save`. They can be reloaded using the command `load`. (Type `help save` and `help load` for more information).

Another possibility is to use the same procedure to manage the files as in the C language:

```
fid = fopen('x.dat','wb'); fwrite(fid,x,'double'); fclose(fd);
```

MATLAB is also able to manage other file formats, such as postscript.

1.1.4. *Example of work session with MATLAB*

Format

All the calculations are performed in MATLAB using the format double, but the display format can be controlled using the function `format` (type `help format`). Some examples are provided here after:

– `format short`: scaled fixed point format with 5 digits (default);

– `format long`: scaled fixed point format with 15 (7) digits for double (simple);

– `format short e`: floating point format with 5 digits;

– `format long e`: floating point format with 15 (7) digits for double (simple).

Scalars, vectors, matrices

MATLAB handles only one data type, because all the variables are considered as floating point complex matrices. It is not necessary to declare or to size these matrices before using them. In fact, when a variable is assigned a value, MATLAB replaces the previous value if this variable exists in the work space; otherwise the variable is created and sized properly.

A vector is a one row or a one column matrix, while a scalar is a 1×1 matrix. MATLAB is optimized for matrix calculations. You should try to use matrix operation as much as possible instead of loops in order to save execution time and memory space.

The effectiveness of an algorithm can be measured using the functions `flops` (*number of floating point operations*) and `etime` (*elapsed time*). Thus, the couple of commands `flops(0)` and `flops` inserted just before and after an algorithm code line returns the number of operations required. The function `clock` yields the present time, while `etime(t1,t2)` provides the time elapsed between t1 and t2.

EXAMPLE

```
t = clock;
%Algorithm;
time = etime(clock,t)
```

`etime` is not an accurate measure of the algorithm effectiveness because the execution speed depends on the CPU.

EXERCISE 1.1.

Type `a = 3` and then `a = 3;`

What is the signification of the symbol ";"?

There are some predefined variables. For instance `pi` = π, while *i* and *j* are defined as the square root of –1. Type `a = 1+2*i`.

Pay attention to the use of these keywords for defining new variables: any assignment replaces the predefined value by the new input (for instance the assignment `pi = 3` replaces the value π). Type `clear pi` to recover the initial value of this variable.

You should avoid assigning *i* and *j* other values in a MATLAB program which handles complex numbers.

EXERCISE 1.2.

Type `i = 2`, then `a = 1+2*i` and finally `clear i`.

`clear` command allows one or several variables to be removed.

Elementary operations

An operation involving 2 variables is possible only if the corresponding matrix sizes match.

EXERCISE 1.3.

Type `v = [1 2 3]` then `v = [1; 2; 3]` and `v(1)`.

As opposed to the case of C language, where the array index begins with 0, in MATLAB it begins with 1: see the effect of `v(0)`.

A vector filled with equally spaced values is defined in the following manner: initial_value:increment:final value (for example `v = 4:-0.1:3.2`).

A matrix can be defined as indicated below:

– `M = [1 2; 3 4];`

– `N(1,:) = [1 2]` and `N(2,:) = [3 4]`.

Type `M(:,1)`,`M(:,2)`,`N(:,1)` and `M(:,2)`.

The pointwise operators: `".*", "./"` or `".^"` are useful for performing matrix operations.

EXERCISE 1.4.

Define the following matrix: `A = [exp(j) 1; 2 j]` and see `A', A.', A^2, A.^2`.

The relational operators: `<, <=, >, >=, ~=` and `==` compare couples of elements belonging to equal size matrices and return a matrix filled with 1 (true) and 0 (false).

The logical operators such as: `&, |, ~, any` or `all` consider all the non-zero elements as true and return a matrix filled with 0 and 1, according to the logical operation result.

MATLAB has no pointer structures, but it automatically allocates (when using =) and recovers (when using `clear`) memory space. For example, for solving `A*x=y`, MATLAB automatically creates a vector for `x`.

Notice the difference between matrix right division and matrix left division: `X=A\B` (equivalent to A^{-1}`*B`) is the solution to `A*X=B` while `X=A/B` (equivalent to `A*`B^{-1}) is the solution to `X*B=A`.

EXERCISE 1.5.

```
A = [1 2 1; 2 1 3; 4 0 5];
y = [3; 2; 1];
x = A\y
z = A/y
```

The matrices can be concatenated either line by line or column by column.

`N = [1 2]; P = [M; N];` then `Q = [M'; N'];`

The inverse submatrix extraction can be performed using brackets as indicated below:

Type `B=A(1:3,:)` and `C=A([1 3],:)`.

1.1.5. *MATLAB language*

MATLAB is a true programming language. However, it is an uncompiled language and thus is not particularly suitable for developing very complex applications. However, it is provided with all the necessary algorithmic structures for rigorous programming.

The "for" loops

```
for (expression)
     code lines;
end
```

The "while" loops

```
while (condition)
     code lines;
end
```

The "if ... then" loops

```
if (condition1)
     code lines;
else if (condition2)
     code lines;
else
     code lines;
end
```

1.2. Solved exercises

EXERCISE 1.6.

Define a 4×3 matrix zero everywhere excepting the first line that is filled with 1.

```
b = ones (1,3); m = zeros (4,3); m(1,:) = b

m =
      1     1     1
      0     0     0
      0     0     0
      0     0     0
```

EXERCISE 1.7.

Consider the couples of vectors (x_1, y_1) and (x_2, y_2). Define the vector x so that:

$x(j) = 0$ if $y_1(j) < y_2(j)$;

$x(j) = x_1(j)$ if $y_1(j) = y_2(j)$;

$x(j) = x_2(j)$ if $y_1(j) > y_2(j)$.

```
function x = vectors(x1,y1,x2,y2)
x = x1.*[y1 == y2] + x2.*[y1 > y2];

vectors([0 1],[4 3],[-2 4],[2 0])

ans =
   -2     4
```

EXERCISE 1.8.

Generate and plot the signal: $y(t) = \sin(2\pi t)$ for $0 \leq t \leq 2$, with an increment of 0.01, then undersample it (using the function `decimate`) with the factors 2 and 16.

```
t = 0:0.01:2;
y = sin(2*pi*t);
subplot(311)
plot(t,y) ;
ylabel('sin(2.pi.t)');
title('Original signal');
t2 = decimate(t,2);
t16 = decimate(t2,8);
y2 = decimate(y,2);
y16 = decimate(y2,8);
subplot(312)
plot(t2,y2);
ylabel('sin(2.pi.t)')
title('Undersampled signal with a factor 2');
subplot(313);
plot(t16,y16);
ylabel('sin(2.pi.t)');
xlabel('Time t');
title('Undersampled signal with a factor 16');
```

You can save the figures in *eps* (Encapsulated PostScript) format, which is recognized by many software programs. The command `print -eps file_name` creates the file *file_name.eps*.

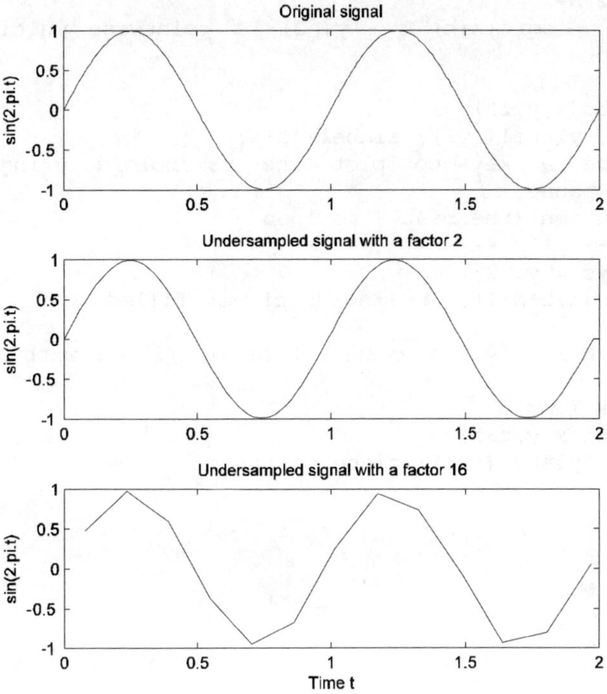

Figure 1.2. *Sinusoid waveform corresponding to different sample frequencies*

EXERCISE 1.9.

Plot the paraboloid defined by the equation: $z^2 = x^2 + y^2$ for $-50 \le x, y \le 50$.

```
N = 50; x = -N:N; y = -N:N;
% first solution (to avoid): two nested loops
%-------------------------------------------------------
for k = 1: 2*N+1
    for l = 1: 2*N+1
        z1(k,l) = sqrt(x(k)^2 + y(l)^2);
    end;
end;
figure; meshc(x,y,z1);
xlabel('x'); ylabel('y'); zlabel('z');

fprintf('Type a key to plot the paraboloid using another
method\n'); pause;
% second solution: one loop
%----------------------------------------
z2 = zeros(2*N+1,2*N+1);
```

```
for k = 1: 2*N+1
    z2(k,:) = sqrt(x(k)^2 + y.^2); % pointwise multiplication
for y
end;
figure; meshc(x,y,z2);
xlabel('x'); ylabel('y'); zlabel('z');
fprintf('Type a key to plot the paraboloid using another
method\n'); pause;
% third solution (the best): no loop
%----------------------------------------------
xc = x.^2; yc = y.^2;
mx=xc.'*ones(1,2*N+1); % line k of mx filled with the value
xc[k]
my=ones(1,2*N+1).'*yc; % column l of my filled with the value
yc[l]
z3 = sqrt(mx + my);
figure; meshc(x,y,z3);
xlabel('x'); ylabel('y'); zlabel('z');
```

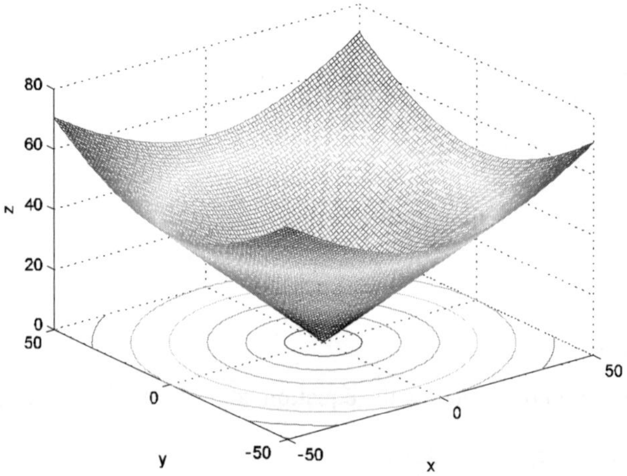

Figure 1.3. *Paraboloid plot*

EXERCISE 1.10.

1. Generate 1,000 independent values $x_1,...,x_{1,000}$ of a zero-mean random Gaussian variable with variance 4 using the function `randn`.

Plot the corresponding histogram and calculate the mean and the standard deviation of the generated series using the functions `hist`, `mean` and `std`.

Find out the mean and the standard deviation of the random series $x^2_1, \ldots, x^2_{1,000}$. Then compare the obtained results with the theoretical results.

```
clear all
n = 1000;

% If X ~ N(m,sigma^2) then Y = (X-m)/sigma ~ N(0,1)
Y=randn(1,n); X=2*Y;
[histoX,bins]=hist(X);
plot(bins,histoX);
xlabel('Bins');
ylabel('Number of values belonging to each bin');
title('Histogram of X using 10 bins');

% Find below 2 ways for displaying the results:

% 1) Character chain concatenation:
moyX=num2str(mean(X));
ecartX=num2str(std(X));
varX=num2str(var(X))
fprintf(strcat('\nMean of X            = ',moyX, '\n'));
fprintf(strcat('St. dev. of X        = ',ecartX, '\n'));
fprintf(strcat('Variance of X        = ',varX, '\n\n'));

% 2) Use of formats, just like in C:
% (type "help format" for more explanations)
fprintf('Mean of X               = %2.5f\n',mean(X));
fprintf('St. dev. of X           = %2.5f\n',std(X));
fprintf('Variance of X           = %2.5f\n',std(X)^2);
Z = X.*X;
fprintf('\nMean of Z               = %2.5f\n',mean(Z));
fprintf('St. dev. of Z           = %2.5f\n',std(Z));
fprintf('Variance of Z           = %2.5f\n',std(Z)^2);
fprintf('Var Z - 2*(Var X)^2 = %2.5f\n\n',std(Z)^2-2*std(X)^4);
```

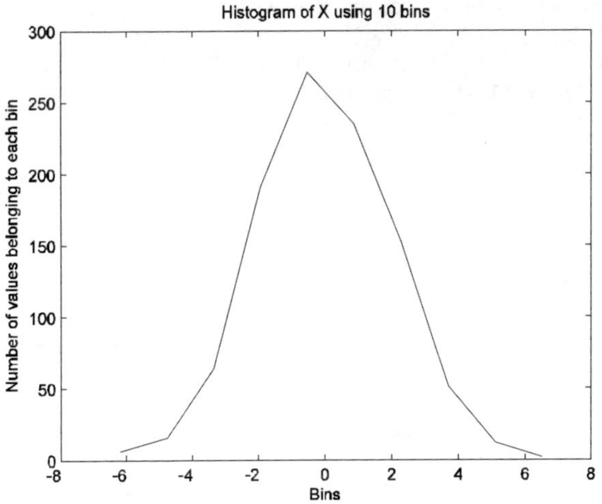

Figure 1.4. *Histogram of a Gaussian random variable*

```
Mean of X          = 0.0085986
St. dev. of X      = 1.963
Variance of X      = 3.8533

Mean of X          = 0.00860
St. dev. of X      = 1.96298
Variance of X      = 3.85328

Mean of Z          = 3.84950
St. dev. of Z      = 5.55695
Variance of Z      = 30.87972
Var Z - 2*(Var X)^2 = 1.18418
```

2. Use the function rand to generate 1,000 independent values of the random variable X defined by:

$$P(X=-1)=p_0; \quad P(X=0)=p_1; \quad P(X=1)=1-p_0-p_1;$$

where p_0 and p_1 are the probabilities to be entered by the user.

```
function va_gen(n,po,p1)
help va_gen; Y = rand(1,n);
X = -1*[Y< po*ones(1,n)] + 1*[Y>((po+p1)*ones(1,n))];
% If Y is a uniformly distributed variable between 0 and 1, then the X
current value is obtained from the combination of 2 tests, so that X = -
1*(Y<p0) + 1*(Y>p0+p1):
% - if Y < p0 (this case occurs with the probability p0) then
%   the first test is true and the second is false, so X = -1
% - if Y > p0+p1 (this case occurs with the probability 1-p0-p1)
%   then the first test is false and the second is true, so X = 1
% - if p0 < Y < p0+p1 (this case occurs with the probability p1)
%   then the two tests are false, so X = 0
prob = hist(X,3)/n;
fprintf('\np [X = -1] = %1.4f\n', prob(1));
fprintf('p [X = 0]  = %1.4f\n', prob(2));
fprintf('p [X = 1]  = %1.4f\n\n', prob(3));
```

When the function va_gen is called:

```
va_gen(1000,0.1,0.5)
```

it provides the following result:

```
If Y is a uniformly distributed variable between 0 and 1, then the X current
value is obtained from the combination of 2 tests, so that X = -1*(Y<p0) +
1*(Y>p0+p1):
   - if Y < p0 (this case occurs with the probability p0) then
     the first test is true and the second is false, so X = -1
   - if Y > p0+p1 (this case occurs with the probability 1-p0-p1)
     then the first test is false and the second is true, so X = 1
```

```
- if p0 < Y < p0+p1 (this case occurs with the probability p1)
  then the two tests are false, so X = 0

p [X = -1] = 0.1000
p [X = 0]  = 0.4840
p [X = 1]  = 0.4160
```

EXERCISE 1.11.

Plot in polar coordinates the poles of the filter having the following transfer function:

$$H(z) = \frac{1}{1 + az^{-1} + bz^{-2}}$$

The values of a and b are entered by the user and the function returns the poles. (use the commands roots and polar).

```
function c = filter_bis(a,b)
c = roots([1 a b]); % Comment: H(z) = poly(c)
fprintf('The poles are:\n'); z1 = c(1,:)
z2 = c(2,:)
if (abs(z1)> 1 | abs(z2) > 1)
    fprintf ('There is at least an instable pole\n');
else
    clf; figure; polar(angle(z1),abs(z1),'r+');
    % Second solution: use zplane
    hold on; polar(angle(z2),abs(z2),'r+');
    legend('Polar plot of H(z) poles',0);
end
```

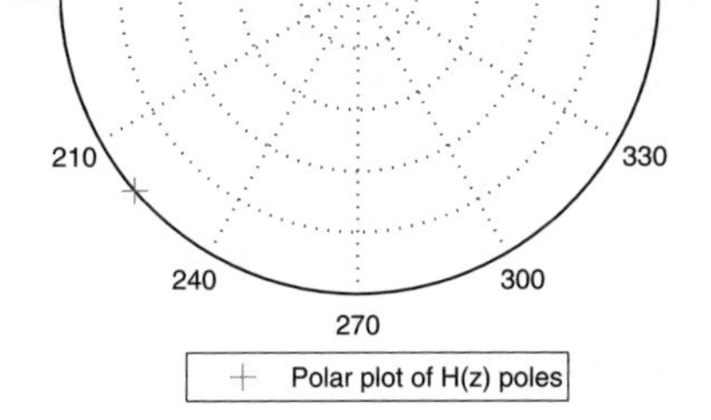

Figure 1.5. *Function call example – filter_bis (1.5,1)*

```
The poles are:

z1 =

   -0.7500 + 0.6614i

z2 =

   -0.7500 - 0.6614i
```

EXERCISE 1.12.

Generate the signal: $x(t) = A \cdot \sin(2\pi ft + \phi) + b(t)$, $t = 0..1024$, where ϕ is a uniformly distributed random variable on $[0, 2\pi]$ and $b(t)$ is a white Gaussian noise with mean zero and variance one (use `rand` and `randn`). A and f are chosen by the user.

Estimate the mean value, the autocorrelation function (`xcorr`) and the spectrum of $x(t)$ using the periodogram and the correlogram (use `fft` and `fftshift`). Compare the obtained results to the theoretical results. Change A in order to control the SNR.

```
function noisy_sin(A,f)
N = 1024;     % Number of calculated frequencies
nech = 1024; % Number of samples
t = 0:nech;
b   = randn(1,nech+1);
phi = 2*pi*rand(1);
x = A*sin(2*pi*f*t+phi)+b;
fprintf('\nMean of x(t) = %2.4f\n',mean(x));
fprintf('Mean of b(t) = %2.4f\n',mean(b));
cx = xcorr(x);
% plot(cx);
% Correlogram based spectrum estimation:
sx_correlo = (abs(fft(cx,N))).^2;
sx_correlo = sx_correlo / max(sx_correlo);
% the first N/2 values correspond to 0<=f<0.5
% the last N/2 values correspond to 0.5<=f<1 (or -0.5<=f< 1)
sx_correlo = fftshift(sx_correlo);
% The spectrum is centred around 0:
% the first N/2 values correspond to -0.5<=f<0
% the last N/2 values correspond to 0<=f<0.5

% Periodogram based spectrum estimation:
%-------------------------------------------
sx_periodo = (abs(fft(x,N))).^2;
sx_periodo = sx_periodo / max(sx_periodo);
sx_periodo = fftshift(sx_periodo);

% SNR estimation for a noisy sinusoid
%-----------------------------------------------
```

```
vector(1:N) = sx_periodo;
vector(N+1:2*N) = sx_correlo;

plot(-0.5:1/N:0.5-1/N,10*log10(sx_correlo(1:N)),'r-',-0.5:1/N:0.5-
1/N,10*log10(sx_periodo(1:N)),'b:');
legend('Correlogram','Periodogram',0);
xlabel('Normalized frequency');
ylabel('Magnitude spectrum [dB]');
axis([-0.5 0.5 min(10*log(vector)) 0]);

% 0 dB <=> 10.log10 (Psignal + Pnoise)
% background_noise <=> 10.log10 (Pnoise) < 0
% Psignal = A^2/(2.N)  (periodogram)
% Pnoise = sigma^2 = 1

SNRth = A^2/2;

fprintf('\nSignal SNR = %2.4f dB \n',10*log10(SNRth));
fprintf('\t=>  Theoretical  mean  background  noise  corresponding  to  the
periodogram estimated spectrum\n');
fprintf('\t    in the range [-0.5:%1.2f] & [%1.2f:0.5] =  %2.4f dB\n\n',-f-
0.05,0.05+f,-10*log10(N*SNRth/2+1));

background_noise1 = mean(10*log10(sx_periodo(1:round(N*(0.45-f)))));
background_noise2 = mean(10*log10(sx_periodo(round(N*(0.65+f)):N)));

mean_background_noise=mean([background_noise1,background_noise2]);

fprintf('Mean  background  noise  corresponding  to  the  periodogram  estimated
spectrum \n');
fprintf('in  the  range  [-0.5:%1.2f]  &  [%1.2f:0.5]  =  %2.4f  dB\n',-f-0.05,
0.05+f, mean_background_noise);
fprintf('\t=>   Estimated   SNR   =   %2.4f   dB   \n',10*log10((2/N)*(-1+exp(-
noise_moy*log(10)/10))));
```

Function call example: `noisy_sin(2,0.15)`

```
Mean of x(t) = 0.0529
Mean of b(t) = 0.0499

Signal SNR = 3.0103 dB
      => Theoretical mean background noise corresponding to the periodogram
estimated spectrum in the range
[-0.5:-0.20] & [0.20:0.5] =  -30.1072 dB

Mean background noise corresponding to the periodogram estimated spectrum in
the range [-0.5:-0.20] & [0.20:0.5] = -29.7643 dB
      => Estimated SNR = 2.6670 dB
```

The SNR estimation error is related to several odd spectrum values, which lead to a biased mean background noise level.

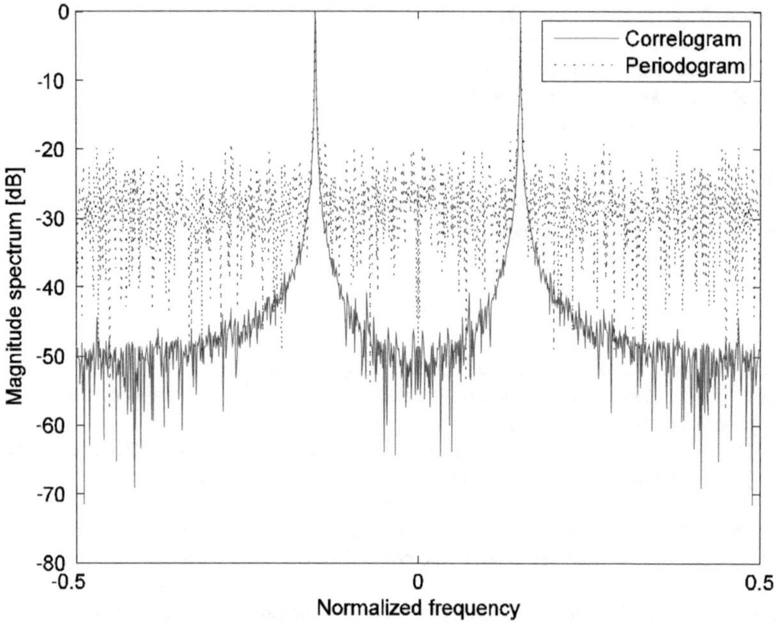

Figure 1.6. *Spectral representation of a noisy sinusoidal signal*

Chapter 2

Discrete-Time Signals

2.1. Theoretical background

A signal is a physical support for information, depending on one or several independent variables, such as: time, range, temperature, pressure, etc. The signal concept generally refers to its mathematical or physical model, chosen in the most appropriate manner for describing the complexity of real signals.

According to whether a signal depends on only one variable or two variables, it is called one-dimensional (1D) or two-dimensional (2D). As a general rule, a multidimensional signal is defined as a function of several variables.

A discrete-time signal is represented by a digital series, uniformly or non-uniformly sampled. The first case is considered in the following and corresponds to a constant time delay between each two successive samples.

Usually, $s(k)$ stands for the kth sample of the discrete-time signal $\{s(k)\}_{k=1,2,\ldots}$. However, the two notations usually overlap in order to simplify the theoretical presentation.

The signal theory is mainly related to the signal mathematical representation in the original or a transformed space, and to its algorithmic processing in order to extract the useful information.

2.1.1. *Mathematical model of 1D and 2D discrete-time signals*

The mathematical model of a discrete-time signal can be defined in terms of the function indicated below:

$$x : T \to X, \ n \to x[n] \qquad [2.1]$$

so that:

$$\forall \ n \in T \subseteq \mathbf{N} \ \text{ or } \ \mathbf{Z} \Rightarrow x[n] \in X \subseteq \mathbf{N}, \ \mathbf{R} \ \text{ or } \ \mathbf{C} \qquad [2.2]$$

while, for 2D discrete-time signals $n = (n_1, n_2) \to x(n_1, n_2)$ and $\forall \ (n_1, n_2) \subseteq \mathbf{N} \times \mathbf{N}$ or $\mathbf{Z} \times \mathbf{Z} \Rightarrow x[n_1, n_2] \in \mathbf{N}, \mathbf{R}$ or $\mathbf{C}$.

In addition to the mathematical model above, 1D discrete-time signals can be also described as:

a. a data vector:

$$x = \{x_n \ \text{ with } \ n \in \mathbf{N}, \ \mathbf{Z} \ \text{ or } \ n \in 0..N-1\} \qquad [2.3]$$

b. a polynomial depending on a real variable *z*:

$$X(z) = \sum_{n=0}^{N-1} x_n z^n = x_0 + x_1 z + x_2 z^2 + \ldots + x_{N-1} z^{N-1} \qquad [2.4]$$

In the same manner, 2D discrete-time signals can be also represented as:

a. a data matrix:

$$x = \{x(n_1, n_2) \ \text{ or } \ x_{n_1, n_2} \ \text{ with } \ n_1 = 0..N'-1 \ \text{ and } \ n_2 = 0..N''-1\} \qquad [2.5]$$

b. a set of polynomials depending on one variable:

$$X_{n_1}(z) = \sum_{n_2=0}^{N''-1} x_{n_1, n_2} z^{n_2} \ \text{ with } \ n_1 = 0..N'-1 \qquad [2.6]$$

c. a polynomial depending on two variables:

$$X(z_1, z_2) = \sum_{n_1=0}^{N'-1} \sum_{n_2=0}^{N''-1} x_{n_1, n_2} z_1^{n_1} z_2^{n_2} \qquad [2.7]$$

2.1.2. *Basic 1D and 2D discrete-time signals*

1D	2D
Dirac pulse	
$\delta[n]=\begin{cases}1, \text{ if } n=0\\0, \text{ otherwise}\end{cases}$, so that: $x[n]=\sum_{(r)} x[r]\,\delta[n-r]$	$\delta[n_1,n_2]=\begin{cases}1, \text{ if } n_1=n_2=0\\0, \text{ otherwise}\end{cases}$, so that: $x[n_1,n_2]=\sum_{(r_1)}\sum_{(r_2)} x[r_1,r_2]\,\delta[n_1-r_1,n_2-r_2]$
Step signal	
$u[n]=\begin{cases}1, \text{ if } n\geq 0\\0, \text{ otherwise}\end{cases}$ and: $u[n]=\sum_{r=0}^{\infty}\delta[n-r]$	$u[n_1,n_2]=\begin{cases}1, \text{ if } n_1,n_2\geq 0\\0, \text{ otherwise}\end{cases}$ and: $u[n_1,n_2]=\sum_{r_1=0}^{\infty}\sum_{r_2=0}^{\infty}\delta[n_1-r_1,n_2-r_2]$
Complex exponential signal	
$x[n]=e^{j\omega_0 n}=e^{j\omega_0(n+N)}$ with: $N=2\pi/\omega_0$	$x[n_1,n_2]=e^{j\omega_{01}n_1}e^{j\omega_{02}n_2}=e^{j\omega_{01}(n_1+N_1)}e^{j\omega_{02}(n_2+N_2)}$ with: $N_1=2\pi/\omega_{01}$ and $N_2=2\pi/\omega_{02}$
Set of finite orthogonal complex signals	
$\varphi_k[n]=\{e^{jk\omega_0 n}\}$, where: $k=0..N-1$ and: $\langle\varphi_i,\varphi_j\rangle=\begin{cases}N, i=j\\0, i\neq j\end{cases}$	$\varphi_{k_1,k_2}[n_1,n_2]=\{e^{jk_1\omega_{01}n_1}e^{jk_2\omega_{02}n_2}\}$ where: $k_1=0..N_1-1$ and $k_2=0..N_2-1$ and: $\langle\varphi_{i_1,i_2},\varphi_{j_1,j_2}\rangle=\begin{cases}N_1N_2, i_1=j_1=i_2=j_2\\0, i_{1,2}\neq j_{1,2}\end{cases}$
Set of orthonormal complex signals	
$\left\{\frac{1}{\sqrt{N}}\varphi_k[n]\right\}$, $k=0..N-1$	$\left\{\frac{1}{\sqrt{N_1N_2}}\varphi_{k_1,k_2}[n_1,n_2]\right\}$, $k_{1,2}=0..N_{1,2}$

Table 2.1. *Mathematical representations of discrete-time signals*

2.1.3. *Periodic 1D and 2D discrete-time signal representation using the discrete-time Fourier series*

Harmonic analysis is the most important tool in signal analysis theory. The generalized Fourier transform, which makes use of distributions, allows the spectral representation of deterministic signals to be obtained. This describes the frequency distribution of the signal amplitude, phase, energy or power.

The discrete-time Fourier series (DTFS) consists of decomposing a periodic signal as a sum of several basic functions, which are easier to generate and to observe. They may be the *sin* and *cos* functions in the case of the trigonometric Fourier series or the exponential function exp($j\omega t$) in the case of the complex Fourier series.

A periodic 1D discrete-time signal, denoted[1] by $\tilde{x}[n]$, has the following form:

$$\tilde{x}[n]=\tilde{x}[n+N]=\sum_{k=0}^{N-1} c_k \exp\left[jk(2\pi/N)n\right] \qquad [2.8]$$

where:

$$c_k=\frac{1}{N}\sum_{n=0}^{N-1}\tilde{x}[n]\exp\left[-jk(2\pi/N)n\right],\ k=0..N-1 \qquad [2.9]$$

A periodic 2D discrete-time signal, denoted by $\tilde{x}[n_1, n_2]$ can be expressed as:

$$\begin{aligned}\tilde{x}[n_1,n_2]&=\tilde{x}[n_1+N_1,n_2]=\tilde{x}[n_1,n_2+N_2]=\\&=\sum_{k_1=0}^{N_1-1}\sum_{k_2=0}^{N_2-1} c_{k_1,k_2}\exp\left(jk_1\frac{2\pi}{N_1}n_1+jk_2\frac{2\pi}{N_2}n_2\right)\end{aligned} \qquad [2.10]$$

where:

$$\begin{aligned}c_{k_1,k_2}&=\frac{1}{N_1N_2}\sum_{n_1=0}^{N_1-1}\sum_{n_2=0}^{N_2-1}\tilde{x}[n_1,n_2]\exp\left(-jk_1\frac{2\pi}{N_1}n_1-jk_2\frac{2\pi}{N_2}n_2\right)\\k_1&=0..N_1-1,\ k_2=0..N_2-1\end{aligned} \qquad [2.11]$$

Consequently, $DTFS_{1D}$ and $DTFS_{2D}$ perform the following transformations:

$$\tilde{x}[n]\xleftrightarrow{\text{DTFS}_{1D}}c_k,\ \tilde{x}[n_1,n_2]\xleftrightarrow{\text{DTFS}_{2D}}c_{k_1k_2} \qquad [2.12]$$

1 Different notations are sometimes used for periodic ($\tilde{x}[n]$ or $\tilde{x}[n_1, n_2]$) and non-periodic discrete-time signals ($x[n]$ or $x[n_1, n_2]$).

2.1.4. *Representation of non-periodic 1D and 2D discrete-time signals by discrete-time Fourier transform*

The discrete-time Fourier transform (DTFT) of a 1D discrete-time signal $x[n]$ has the following form:

$$X\left(e^{j\omega}\right)=\mathrm{DTFT}_{1D}\left\{x[n]\right\}=\sum_{n=-\infty}^{+\infty} x[n]\exp(-j\omega n) \quad [2.13]$$

while the inverse transform is defined by:

$$x[n]=\mathrm{DTFT}_{1D}^{-1}\left\{X\left(e^{j\omega}\right)\right\}=\frac{1}{2\pi}\oint_{2\pi} X\left(e^{j\omega}\right)e^{j\omega n}\,d\omega \quad [2.14]$$

Direct and inverse discrete Fourier transform of a discrete-time 2D signal are given by very similar relationships:

$$\begin{aligned} X\left(e^{j\omega_1+j\omega_2}\right) &= X(\omega_1,\omega_2)=\mathrm{DTFT}_{2D}\left\{x[n_1,n_2]\right\} \\ &= \sum_{n_1=-\infty}^{+\infty}\sum_{n_2=-\infty}^{+\infty} x[n_1,n_2]\exp[-j\omega_1 n_1 - j\omega_2 n_2] \end{aligned} \quad [2.15]$$

$$x[n_1,n_2]=\frac{1}{(2\pi)^2}\oint_{2\pi}\oint_{2\pi} X(\omega_1,\omega_2)\exp[j\omega_1 n_1 + j\omega_2 n_2]\,d\omega_1\,d\omega_2 \quad [2.16]$$

Consequently, DTFT_{1D} and DTFT_{2D} perform the following transformations:

$$x[n]\xleftrightarrow{\mathrm{DTFT}_{1D}} X\left(e^{j\omega}\right)=X(\omega),\; x[n_1,n_2]\xleftrightarrow{\mathrm{DTFT}_{2D}} X(\omega_1,\omega_2) \quad [2.17]$$

2.1.5. *Analytic signals*

It is well known that the Fourier transform of a real signal has the Hermitian symmetry property (odd absolute value and real part, even argument and imaginary part). This means that giving the Fourier transform (FT) of a real signal for the positive frequency axis is enough for its complete characterization.

It is thus possible to represent a real signal $x(t)$ by a complex one $z_x(t)$, without any information loss. The new complex signal $z_x(t)$ is called the analytic signal associated with the real signal $x(t)$.

Generally, a signal $z(t)$ is an analytic signal if its spectrum $Z(\nu)$ is zero for any negative frequency value. Bearing in mind that $Z(\nu) = Z(\nu)U(\nu)$, where $U(\nu)$ stands for the frequency step function, it can be easily shown that the real and imaginary parts of $z(t)$ are a couple of the Hilbert transform. An analytical signal is therefore completely determined if its real part is known.

Let us consider the DTFT of a real, non-periodic and causal 1D discrete-time signal $x[n]$:

$$\mathrm{DTFT_{1D}}\left\{x[n]\right\} = X\left(\mathrm{e}^{j\omega}\right) = X_{\mathrm{Re}}\left(\mathrm{e}^{j\omega}\right) + jX_{\mathrm{Im}}\left(\mathrm{e}^{j\omega}\right)$$

The real and imaginary parts of this DTFT are related by the Hilbert transform (HT):

$$X_{\mathrm{Im}}\left(\mathrm{e}^{j\omega}\right) = \mathrm{PV}\left\{\frac{1}{2\pi}\int_{-\pi}^{+\pi} X_{\mathrm{Re}}\left(\mathrm{e}^{j\theta}\right)\cot\frac{\theta-\omega}{2}\,\mathrm{d}\theta\right\} = \mathrm{HT_{1D}}\left\{X_{\mathrm{Re}}\left(\mathrm{e}^{j\omega}\right)\right\} \quad [2.18]$$

$$X_{\mathrm{Re}}\left(\mathrm{e}^{j\omega}\right) = -\mathrm{PV}\left\{\frac{1}{2\pi}\int_{-\pi}^{+\pi} X_{\mathrm{Im}}\left(\mathrm{e}^{j\theta}\right)\cot\frac{\theta-\omega}{2}\,\mathrm{d}\theta\right\} + x[0]$$
$$= -\mathrm{HT}_{1D}\left\{X_{\mathrm{Im}}\left(\mathrm{e}^{j\omega}\right)\right\} + x[0] \quad [2.19]$$

where "PV" denotes the principal value of an integral and the Hilbert transform is defined by:

$$\mathrm{HT_{1D}}\{f\}(\omega) = \mathrm{PV}\left\{\frac{1}{2\pi}\int_{-\pi}^{\pi} f(\theta)\cot\left(\frac{\theta-\omega}{2}\right)d\theta\right\} \quad [2.20]$$

If the spectrum of a non-periodical discrete-time signal $x[n]$ is constrained by the causality condition over $\omega \in [-\pi,+\pi]$, we obtain:

$$X\left(\mathrm{e}^{j\omega}\right) = 0, \ -\pi \le \omega < 0 \quad [2.21]$$

$x[n]$ is thus an analytical signal and can be written in the following form:

$$x[n] = x_{re}[n] + jx_{im}[n] \quad [2.22]$$

The relationship between its real and imaginary parts can be easily derived:

$$x_{im}[n] = \sum_{m=-\infty}^{+\infty} x_{re}[m]h[n-m] = \mathrm{DHT}_{1D}\{x_{re}[n]\} \quad [2.23]$$

$$x_{re}[n] = -\sum_{m=-\infty}^{+\infty} x_{im}[m]h[n-m] = -\mathrm{DHT}_{1D}\{x_{im}[n]\} \quad [2.24]$$

where $h[n]$ is the impulse response corresponding to an ideal discrete Hilbert transformer:

$$h[n] = \begin{cases} 0, \text{ if } n=0 \\ \dfrac{2}{\pi}\dfrac{\sin^2(\pi n/2)}{n}, \text{ if } n \neq 0 \end{cases} \quad [2.25]$$

2.2. Solved exercises

EXERCISE 2.1.

The MATLAB code below generates and plots some basic discrete-time signals.

```
subplot(3,3,1);
stem([1;zeros(49,1)]);
title('Dirac pulse')
subplot(3,3,2); stem(ones(50,1));
title('Step function')
subplot(3,3,3);
stem([ones(1,5),zeros(1,3)])
title('Rectangular pulse')
subplot(3,3,4);
stem(sin(2*pi/8*(0:15)))
title('Sinusoidal signal')
subplot(3,3,5); stem(sinc(0:0.25:8));
title('"Sinc" signal')
subplot(3,3,6); stem(exp(-(0:15)));
title('e^-^n signal')
subplot(3,3,7);
stem(pow2(-0.5*(0:15)))
title('2^-^0^.^5^n signal')
subplot(3,3,8); stem(3.^(0:15));
title('3^n signal')
subplot(3,3,9); stem(randn(1,16));
title('Gaussian random signal')
```

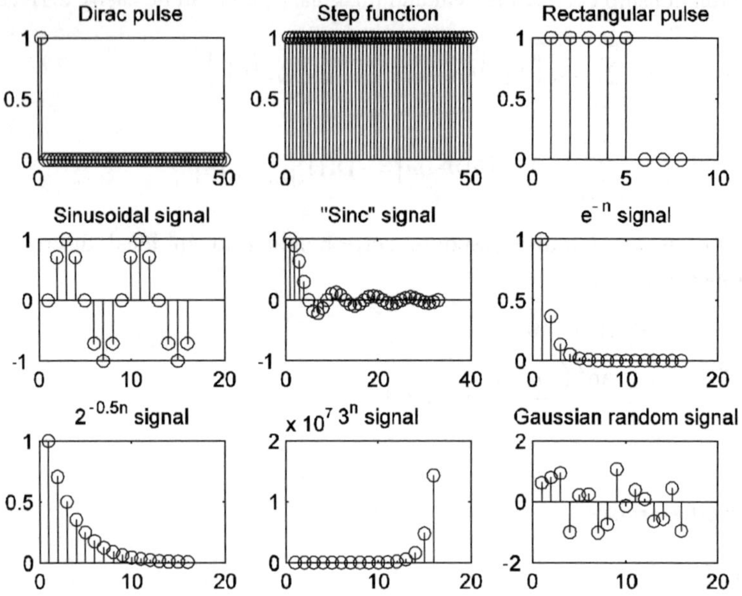

Figure 2.1. *Examples of discrete-time signals*

EXERCISE 2.2.

Generate the following signal:

$$x(n) = K \cdot \exp[c \cdot n],$$

where: $K = 2,\ c = -1/12 + j\pi/6$, $n \in \mathbb{N}$ and $n = 0..40$.

```
c = -(1/12)+(pi/6)*i;
K = 2; n = 0:40;
x = K*exp(c*n);
subplot(2,1,1); stem(n,real(x));
xlabel('Time index n');
ylabel('Amplitude');
title('Real part');
subplot(2,1,2); stem(n,imag(x));
xlabel('Time index n');
ylabel('Amplitude');
title('Imaginary part');
```

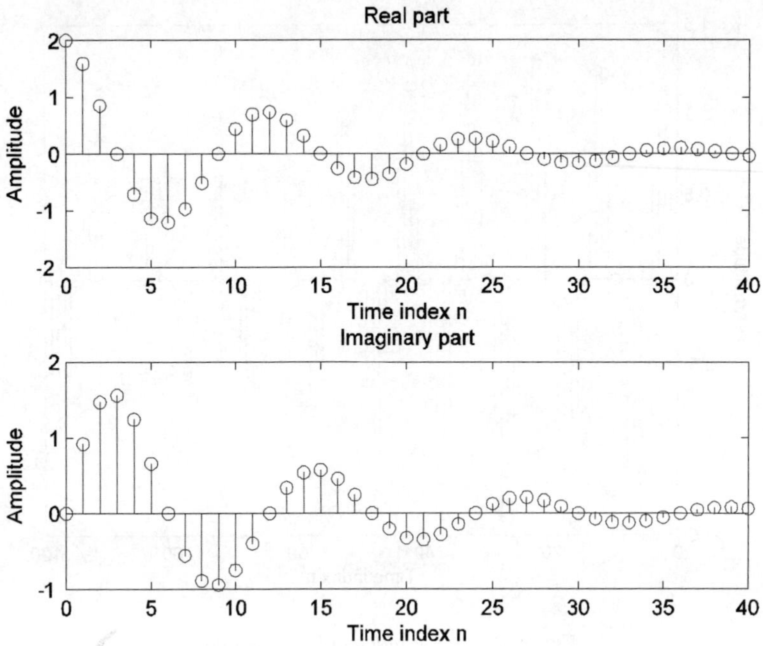

Figure 2.2. *Real and imaginary parts of a complex discrete-time signal*

K is a constant amplitude factor and $\mathrm{Re}\{c\}$ sets the attenuation, while $\mathrm{Im}\{c\}$ is related to the dumped signal period (12 points per period).

EXERCISE 2.3.

Generate the following amplitude modulated signal:

$$y(n) = (1 + m \cdot \sin(2\pi \upsilon_b n)) \cdot \sin(2\pi \upsilon_h n)$$

where $m = 0.4$, $\upsilon_b = 0.01$, $\upsilon_h = 0.1$, $n \in \mathbf{N}$ and $n = 0..100$.

```
n = 0:100; m = 0.4;
fH = 0.1 ; fL = 0.01;
xH = sin(2*pi*fH*n);
xL = sin(2*pi*fL*n);
y = (1+m*xL).*xH;
stem(n,y); grid ;
xlabel('Time index n');
ylabel('Amplitude');
```

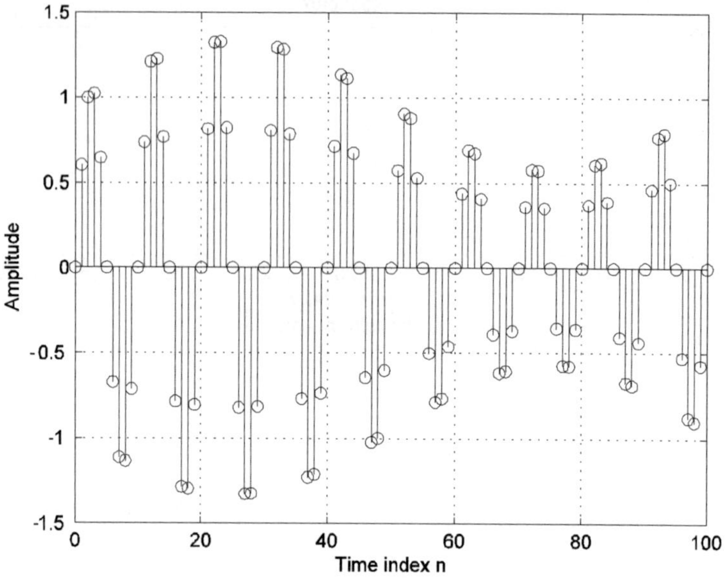

Figure 2.3. *Amplitude modulated signal*

EXERCISE 2.4.

Let us consider the continuous-time signal below:

$$x(t) = 2te^{-t}, \quad t \in [0, 10]$$

Show that sampling this signal results in spectrum aliasing if the sampling period is $T_s = 0.005$ s.

```
t = 0:0.005:10; xa = 2*t.*exp(-t);
subplot(2,2,1); plot(t,xa); grid
xlabel('Time [ms]'); ylabel('Amplitude');
title('Continuous-time signal x_{a}(t)');
wa = 0:10/511:10 ; ha = freqs(2,[1 2 1],wa);
subplot(2,2,2); plot(wa/(2*pi),abs(ha)); grid;
xlabel('Frequency [kHz]'); ylabel('Amplitude');
title('|X_{a}(j\Omega)|'); axis([0 5/pi 0 2]);
T = 1; n = 0:T:10 ; xs = 2*n.*exp(-n); k = 0:length(n)-1;
subplot(2,2,3); stem(k,xs);grid;
xlabel('Time index n'); ylabel('Amplitude');
title('Discrete-time signal x[n]');
wd = 0:pi/255:pi; hd = freqz(xs,1,wd);
subplot(2,2,4); plot(wd/(T*pi), T*abs(hd)); grid;
```

```
xlabel('Frequency [kHz]'); ylabel('Amplitude');
title('|X(e^{j\omega})|'); axis([0 1/T 0 2])
```

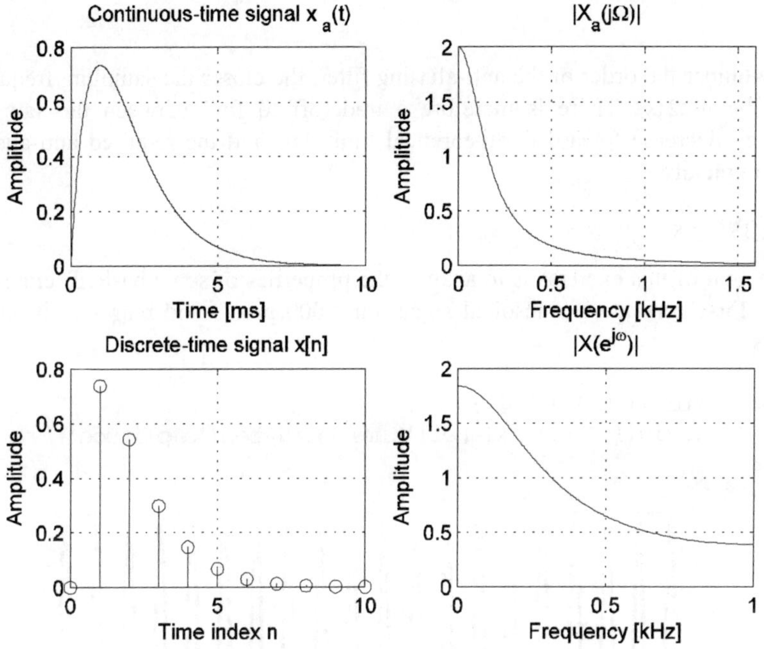

Figure 2.4. *Spectral aliasing illustration*

A spectral representation provides information about the variation rate of the corresponding signal in the time domain. The more extended the signal spectrum, the faster the signal temporal variation is and the higher the sampling frequency has to be in order to avoid information loss.

According to Nyquist's theorem, there is no information loss if the sampling frequency is at least equal to the double of the highest spectral component of the signal. The discrete-time signal obtained by sampling the continuous-time one will then account for all its variations.

It is always necessary to use an anti-aliasing filter before the sampling stage in order to avoid any spectral aliasing risk and to set an appropriate sampling frequency. In practice, a causal approximation of this ideal filter is used. Thus, depending on the chosen filter synthesis method, some imperfections are introduced, such as a passband amplitude ripple, a transition band and a stopband finite attenuation.

Butterworth filters, whose 3 dB cutoff frequency is roughly equal to υ_m, are generally considered for this task because there is no bandpass ripple in this case. The main drawback of this type of filter is the residual aliasing due to the stopband ripple.

The higher the order of the anti-aliasing filter, the closer the sampling frequency υ_s can be to $2\upsilon_m$. There is therefore a trade-off to find between the sampling frequency decrease toward the theoretical limit $2\upsilon_m$ and the required anti-aliasing filter complexity.

EXERCISE 2.5.

The goal of this exercise is to analyze the properties of some basic discrete-time signals. First generate a sinusoidal signal on 1,000 points and represent it on 200 samples.

```
xsin=sin(2*pi*[1:0.1:1000]);
plot(1:200,xsin(1:200)); xlabel('time'); ylabel('amplitude')
```

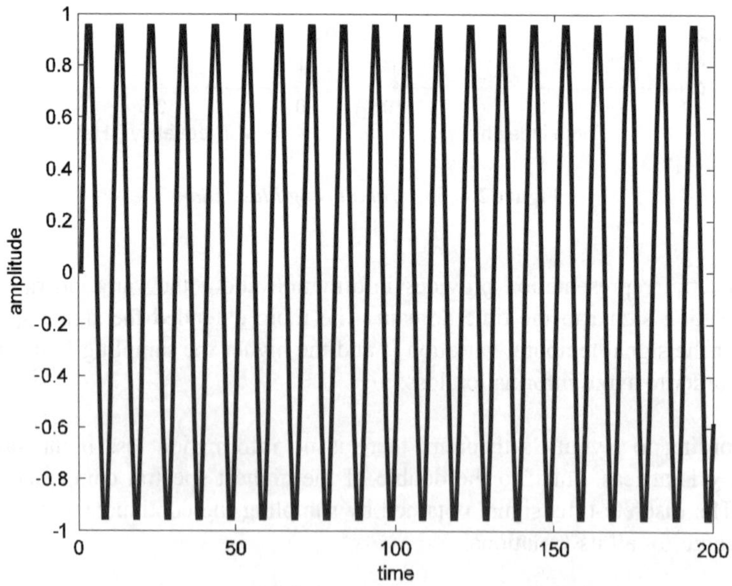

Figure 2.5. *Sine signal*

Find out its maximum value, minimum value, mean value, median value and its standard deviation.

max(xsin) ans = 1.0000	min(xsin) ans = -1.0000
mean(xsin) ans = 4.6721e-004	median(xsin) ans = 0.0026
std(xsin) ans = 0.7070	

Repeat the same exercise for the signals generated by the following MATLAB commands: `square`, `sawtooth`, `rand` and `randn`.

EXERCISE 2.6.

Plot the waveforms corresponding to a continuous-time sinusoidal signal $x(t)$ and to its sampled version $x[n]$, knowing that:

$$x(t) = A\sin(\Omega_0 t + \phi_0)$$

The sampled signal can be thus written as indicated below:

$$x(t)\,|_{t=nT} = x(nT) = A\sin(\Omega_0 nT + \phi_0) = A\sin(2\pi F_0 nT + \phi_0) =$$
$$= A\sin\left(2\pi\frac{F_0}{1/T}n + \phi_0\right) = A\sin\left(2\pi\frac{F_0}{F_s}n + \phi_0\right)$$

Consequently:

$$x[n] = A\sin(\omega_0 n + \phi_0) = A\sin(2\pi f_0 n + \phi_0)$$

where $\omega_0 = 2\pi f_0 = 2\pi F_0 / F_s$ with $F_s = 1/T$.

Because $t = nT$ and $\omega_0 = \Omega_0 T$, we obtain:

$$\omega_0 = \Omega_0 \frac{t}{n} \text{ [rad/sample]}$$

The MATLAB code below plots the waveforms corresponding to a continuous-time and a discrete-time sinusoidal signal for the following parameters:

$$F_0 = 1{,}200 \text{ Hz}, \; F_s = 16 \text{ kHz}, \; \varphi_0 = \pi/4 \text{ rad}, \; A = 10, \; t_0 = 0, \; t_f = 5 \text{ ms}$$

```
Fs=16e3; t=0:1/Fs:5e-3;
n=0:length(t)-1;
subplot(211);
plot(t,10*sin(2*pi*1200*t+pi/4));
xlabel('continuous time');
ylabel('amplitude')
title('Continuous-time sinusoidal signal')
subplot(212);
stem(10*sin(2*pi*1200/16000*n+pi/4))
xlabel('discrete time');
ylabel('amplitude')
title('Discrete-time sinusoidal signal')
```

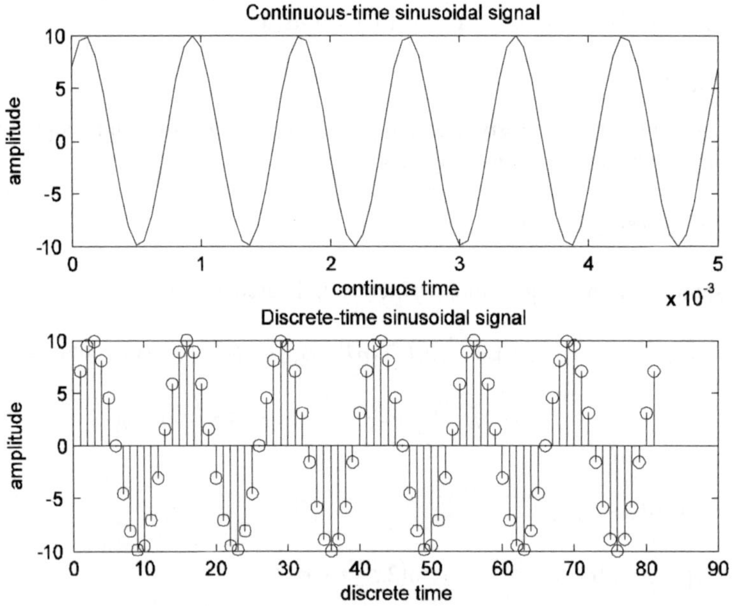

Figure 2.6. *Continuous (top) and sampled (bottom) sinusoid*

EXERCISE 2.7.

The generation of a digital signal using a computer or a digital signal processor requires some cautions concerning the choice of the sampling frequency. The aim of this exercise is to show that if the sampling frequency is not properly chosen, serious interpretation errors may occur while simulating or generating synthetic signals.

1. Generate during 0.5 s a signal obtained by the sum of two sinusoids having the same amplitude (1 V), sampled at 256 Hz and whose frequencies are 100 Hz and 156 Hz respectively. Plot this signal and conclude about its shape.

2. Generate and plot a sinusoid of 356 Hz sampled at 256 Hz. Compare this signal to another sinusoid of 100 Hz and sampled at the same frequency.

3. Generate during 0.5 s a signal obtained by the sum of two sinusoids, etc.

```
t=[1:128]; f1=100;
f2=156; fe=256;
y=sin(2*pi*f1/fe*t)+sin(2*pi*f2/fe*t);
plot(t/fe,y);
xlabel('time [s]');
ylabel('amplitude [V]')
```

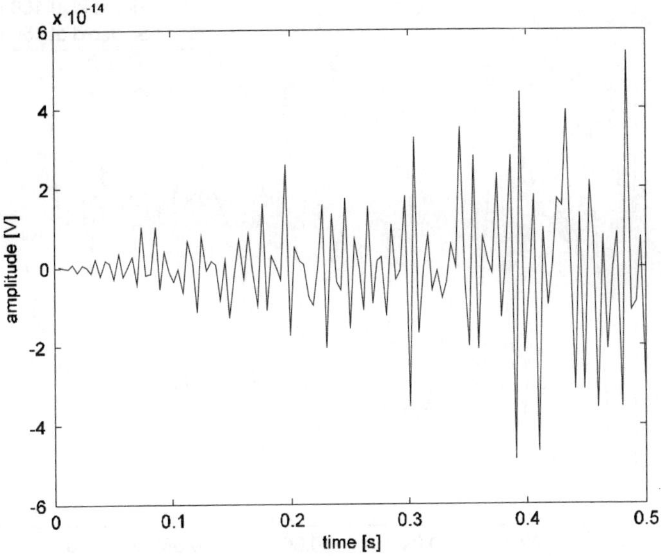

Figure 2.7. *Sum of two sampled sinusoids*

It can be readily seen that the final result does not represent the sum of two sinusoids of amplitude equal to 1 V. In fact, the signal depicted on the figure above is nothing else but the residual error related to the MATLAB computation precision.

This result can be explained by a wrong application of the sampling theorem. Indeed, the sampling frequency has to be equal at least to the double of the frequency of each one of the two sinusoids. This constraint is fulfilled only for the first sinusoid, because for the second the sampling frequency would have been at least 312 Hz.

Using some standard trigonometric relationships it is easy to see that after sampling the two sinusoids appear to have the same frequency (100 Hz) and opposite phases, as it is illustrated on the figure below. Thus, they cancel each other and their sum is theoretically zero.

```
t=[1:30]; f1=100; f2=156; fs=256;
y1=sin(2*pi*f1/fs*t); y2=sin(2*pi*f2/fs*t);
plot(t/fs,y1,'-xb',t/fs,y2,'-or');
xlabel('time [s]') ; ylabel('amplitude [V]')
legend('Sinusoid at 100 Hz','Sinusoid at 156 Hz')
axis([min(t/fs) max(t/fs) -1.1 1.5])
```

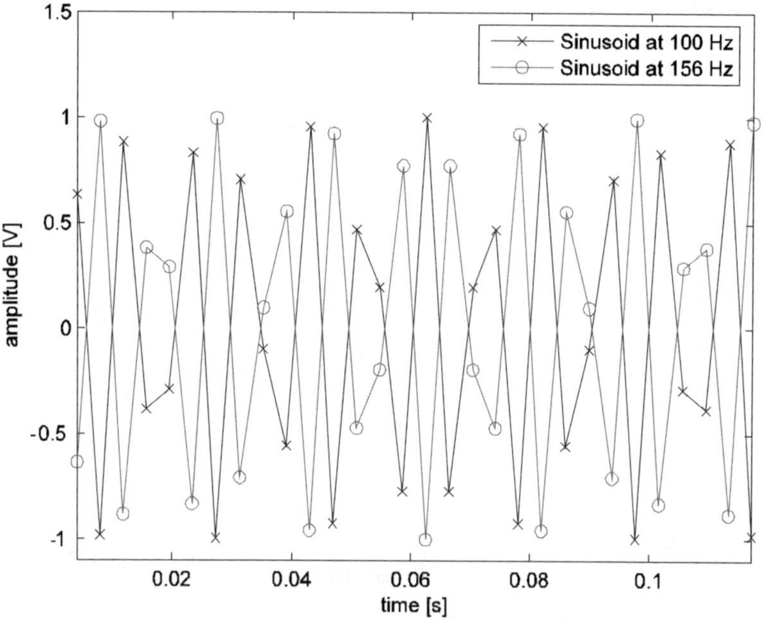

Figure 2.8. *Inappropriate choice of the sampling frequency makes two different sinusoids appearing as opposite phase signals*

2. Generate and plot a sinusoid of 356 Hz sampled at 256 Hz, etc.

```
t=(1:100); f1=100; f2=356; fs=256;
y1=sin(2*pi*f1/fs*t); y2=sin(2*pi*f2/fs*t);
subplot(211); plot(t/fs,y1); ylabel('amplitude [V]');
title('Sinusoid at 100 Hz')
subplot(212); plot(t/fs,y2); xlabel('time [s]');
ylabel('amplitude [V]'); title('Sinusoid at 356 Hz')
```

It can be readily seen from the equations below that the two sinusoids are identical:

$$\sin\left(\frac{2\pi 356n}{256}\right)=\sin\left(\frac{2\pi(256+100)n}{256}\right)$$

$$\sin\left(\frac{2\pi 356n}{256}\right)=\sin\left(\frac{2\pi 256n}{256}\right)\cos\left(\frac{2\pi 256n}{256}\right)+\sin\left(\frac{2\pi 100n}{256}\right)\cos\left(\frac{2\pi 256n}{256}\right)$$

$$\sin\left(\frac{2\pi 356n}{256}\right)=\sin\left(\frac{2\pi 100n}{256}\right)$$

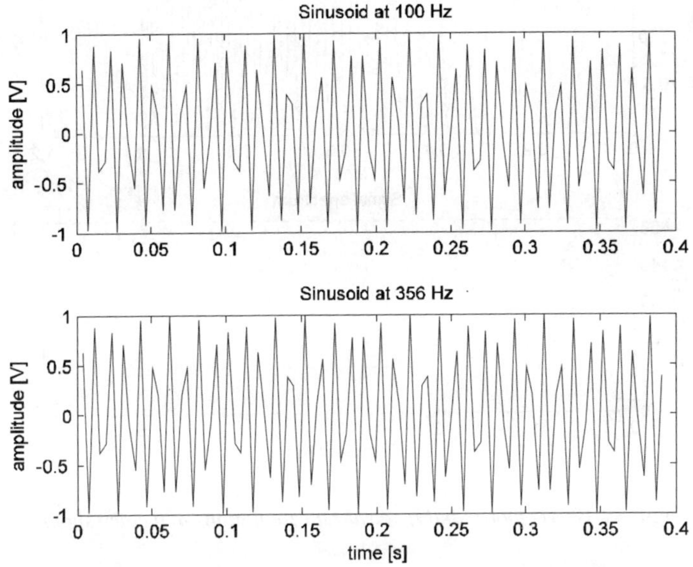

Figure 2.9. *Sinusoids at 100 Hz (top) and 356 Hz (bottom)*

EXERCISE 2.8.

A *chirp* pulse of width T can be expressed as: $x(t)=A_0\cos\phi(t)$, where the instantaneous phase is given by: $\phi(t)=\Omega_0 t+\beta t^2$. A linear variation of the instantaneous frequency $\Omega(t)$ during the time support T is then obtained according to: $\Omega(t)=\mathrm{d}\phi(t)/\mathrm{d}t=\Omega_0+2\beta t$, where $\beta=\Delta\Omega/(2T)=(\Omega_f-\Omega_0)/(2T)$.

The MATLAB code below performs the temporal and spectral analysis of a *chirp* signal, whose instantaneous frequency varies between 0 and 5 MHz, during its time support T = 10 μs. The sampling frequency is considered 50 MHz.

```
f0=0; ff=5e6; T=10e-6; beta=(ff-f0)*pi/T; Fs=5e7; t=0:1/Fs:T;
x=cos(2*pi*f0*t+beta*t.^2);
subplot(211);plot(t,x);xlabel('time [s]');ylabel('amplitude [V]');
title('Chirp signal'); N=length(x);
X=fftshift(abs(fft(x))); freqv=linspace(-Fs/2,Fs/2,length(X));
subplot(212); plot(freqv,X)
xlabel('frequency [Hz]'); ylabel('amplitude [V]');
title('Signal spectrum'); axis([0 1.2*ff 0 1.1*max(X)])
```

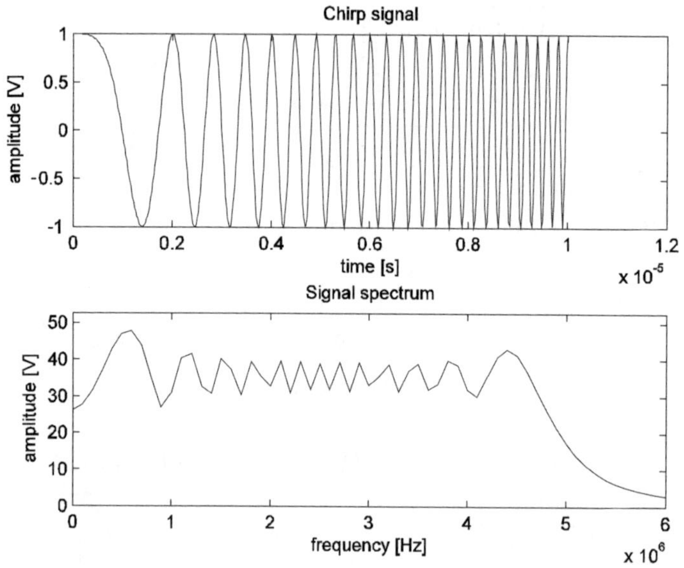

Figure 2.10. *Temporal and spectral representations of a chirp signal*

EXERCISE 2.9.

In the case of a binary information transmission, the message consists of a series of symbols, which are transmitted at a given constant rate. The time delay between two symbols is called symbol width.

Plot the pulse modulation corresponding to the binary sequence below:

1 0 1 0 1 1 0 0 1 0 1 0 0 1 1 1 0 1 0 1

using the following codes:

– unipolar with and without return to zero (RZ);

– polar with and without return to zero;

– bipolar with and without return to zero.

Plot the PSD (power spectral density) of each signal. Use a binary rate of 1,000 symbols per second and a sampling frequency of 10 KHz.

```
Rb = 1000; fs = 10000;
Ts = 1/fs; Tb = 1/Rb;
sequence=[1 0 1 0 1 1 0 0 1 0 1 0 0 1 1 1 0 1 0 1 ];
no= length(sequence);
no_ech = no* Tb/Ts;
time_t = [0:(no-1)] * Ts;
pulse=ones(1,fs/Rb);
x=(sequence'*pulse)'; x=x(:);
t_time=Ts*[0:no_ech];
figure(1);
subplot(231); plot(t_time,[x;0]);
axis([0 0.021 0 1.2]);
title('Unipolar WRZ code');
xlabel('t [s]'); ylabel('A [V]')
[y1,f]=psd(x,64,10000,64);
figure(2);
subplot(231); plot(f,y1);
xlabel('frequency [Hz]'); ylabel('PSD [W]')
title('Unipolar WRZ code');
n_middle = no /2;
pulse = ones(1,no);
pulse((n_middle + 1):(no)) = zeros(1,(no-n_middle));
x=(sequence'*pulse)';
x=x(:); t_time=Ts*[0:length(x)];
figure(1);
subplot(234); plot(t_time,[x;0]);
axis([0 0.041 0 1.2]);
title('Unipolar RZ code');
xlabel('t [s]');ylabel('A [V]')
[y1,f]=psd(x,64,10000,64);
figure(2);
subplot(234); plot(f,y1);
title('Unipolar RZ code');
xlabel('frequency [Hz]');
ylabel('PSD [W]')
pulse=ones(1,fs/Rb);
x=(2*sequence'*pulse)'-ones(size(pulse,2),size(sequence,2));
x=x(:); t_time=Ts*[0:length(x)];
figure(1);
subplot(232); plot(t_time,[x;0]);
axis([0 0.021 -1.2 1.2]);
xlabel('t [s]'); ylabel('A [V]')
title('Polar WRZ code');
[y1,f]=psd(x,64,10000,64);
figure(2);
subplot(232); plot(f,y1);
xlabel('frequency [Hz]');
ylabel('PSD [W]')
title('Polar WRZ code');
pulse = ones(1,no);
```

```
pulse((n_middle + 1):(no)) = zeros(1,(no-n_middle));
x=(2*sequence'*pulse)'-ones(size(pulse,2),size(sequence,2));
x=x(:);t_time=Ts*[0:length(x)];
figure(1);
subplot(235); plot(t_time,[x;0]);
axis([0 0.041 -1.2 1.2]);
xlabel('t [s]'); ylabel('A [V]')
title('Polar RZ code');
[y1,f]=psd(x,64,10000,64);
figure(2);
subplot(235); plot(f,y1);
title('Polar RZ code');
xlabel('frequency [Hz]');ylabel('PSD [W]')
pulse=ones(1,fs/Rb);
for ii = 1:no
    bipol(ii,1) =sequence(ii)*(-1)^(sum(sequence(1:ii)) - 1);
end
x=(bipol*pulse)';
x=x(:);t_time=Ts*[0:length(x)];
figure(1);
subplot(233);
plot(t_time,[x;0]);
axis([0 0.021 -1.2 1.2]);
xlabel('t [s]');ylabel('A [V]')
title('Bipolar WRZ code');
[y1,f]=psd(x,64,10000,64);
figure(2);
subplot(233) ;plot(f,y1);
xlabel('frequency [Hz]');
ylabel('PSD [W]')
title('Bipolar WRZ code');
pulse = ones(1,no);
pulse((n_middle + 1):(no)) = zeros(1,(no-n_middle));
for ii = 1:no
    bipol(ii,1) = sequence(ii)*(-1)^(sum(sequence(1:ii)) - 1);
end
x=(bipol*pulse)'; x=x(:);
t_time=Ts*[0:length(x)];
figure(1);
subplot(236); plot(t_time,[x;0]);
xlabel('t [s]'); ylabel('A [V]')
axis([0 0.041 -1.2 1.2]);
title('Bipolar RZ code');
[y1,f]=psd(x,64,10000,64);
figure(2);
subplot(236); plot(f,y1);
title('Bipolar RZ code');
xlabel('frequency [Hz]'); ylabel('PSD [W]')
```

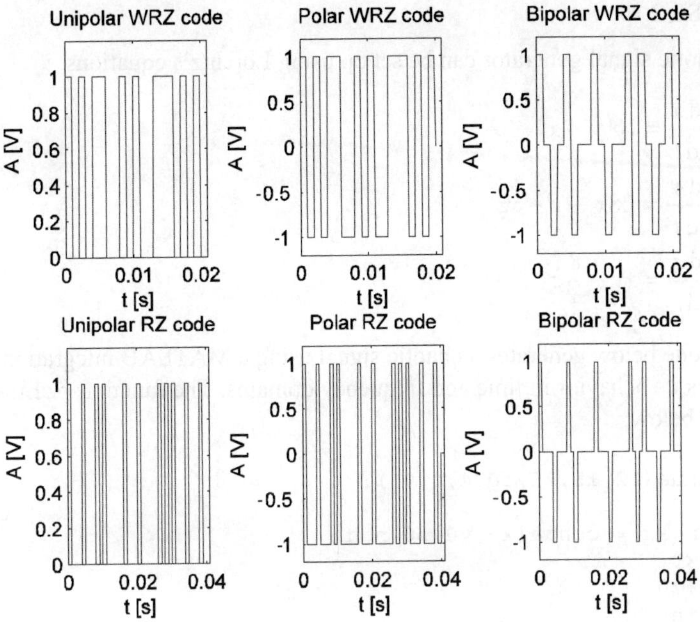

Figure 2.11. *Temporal representation of different codes*

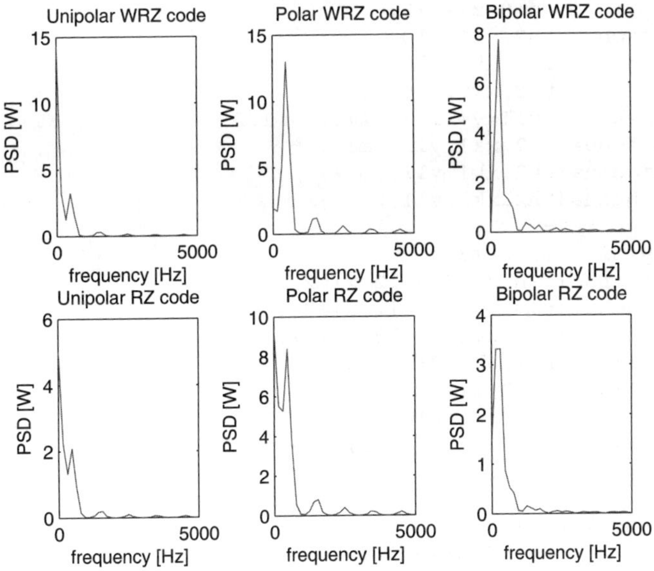

Figure 2.12. *Spectral representation of different codes*

EXERCISE 2.10.

A chaotic signal generator can be set up using Lorentz's equations:

$$\begin{cases} \dfrac{dx}{dt} = 10(y - x) \\ \dfrac{dy}{dt} = 28x - y - xz \\ \dfrac{dz}{dt} = xy - \dfrac{8}{3}z \end{cases}$$

The code below generates a chaotic signal using a MATLAB integration method, and shows its behavior in time and frequency domains. The function `cchua.m` is as indicated below:

```
xx = cchua(12,15,25,1024,0.05);

function xx = cchua(x0,y0,z0,n,h)
t=0; x=x0;
y=y0; z=z0;
for i=1:n
    k0=h*fchua(t,x,y,z);
    l0=h*gchua(t,x,y,z);
    m0=h*hchua(t,x,y,z);
    th2=t+h/2; xk0=x+k0/2;
    yl0=y+l0/2; zm0=z+m0/2;
    k1=h*fchua(th2,xk0,yl0,z+m0/2);
    l1=h*gchua(th2,xk0,yl0,zm0);
    m1=h*hchua(th2,xk0,yl0,zm0);
    xk1=x+k1/2; yl1=y+l1/2; zm1=z+m1/2;
    k2=h*fchua(th2,xk1,yl1,zm1);
    l2=h*gchua(th2,xk1,yl1,zm1);
    m2=h*hchua(th2,xk1,yl1,zm1);
    th=t+h; xk2=x+k2;
    yl2=y+l2; zm2=z+m2;
    k3=h*fchua(t+h,x+k2,y+l2,z+m2);
    l3=h*gchua(t+h,x+k2,y+l2,z+m2);
    m3=h*hchua(t+h,x+k2,y+l2,z+m2);
    x=x+(1/6)*(k0+2*k1+2*k2+k3);
    y=y+(1/6)*(l0+2*l1+2*l2+l3);
    z=z+(1/6)*(m0+2*m1+2*m2+m3);
    xx(i)=x; yy(i)=y; zz(i)=z;
    t=t+h; tt(i)=t;
end;
figure; subplot(211)
fs=1000000; tmp=0:1/fs:1023/fs;
plot(tmp,xx);
xlabel('time [s]'); ylabel('A');
```

```
title('Chaotic signal'); grid;
subplot(212)
fxx=abs(fftshift(fft(xx,1024)));
vf=((-511:512)/1024)*fs;
plot(vf,fxx,'-b');
xlabel('frequency [Hz]');
ylabel('A'); grid;
title('Signal spectrum');

%first equation
function fc = fchua(t,x,y,z)
fc=10*(y-x);

% second equation
function gc = gchua(t,x,y,z)
gc=28*x-y-x*z;
% third equation
function hc = hchua(t,x,y,z)
hc=x*y-(8/3)*z;
```

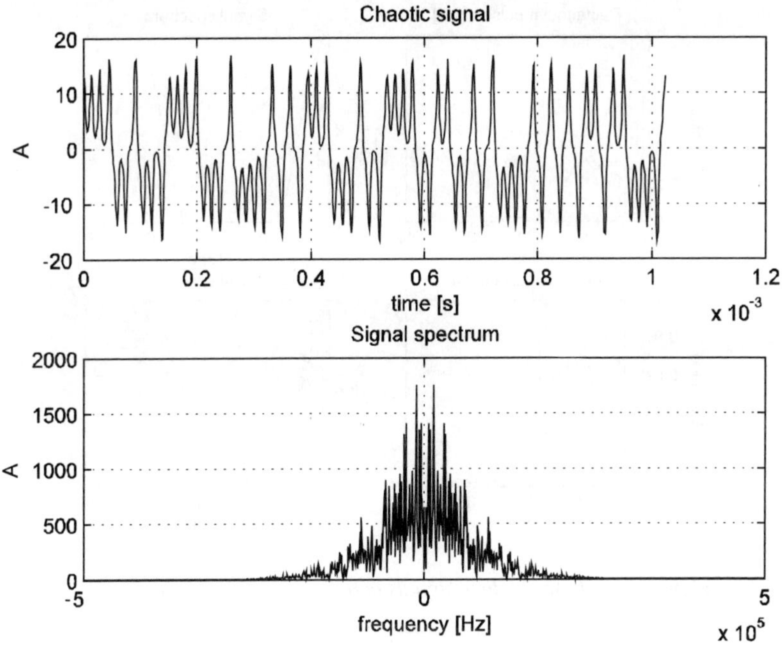

Figure 2.13. *Temporal and spectral representation of a chaotic signal*

EXERCISE 2.11.

The MATLAB code below is aimed at comparing the spectral representation of a 1D discrete-time signal to that obtained when it is periodized.

```
xd=[1,1,1,1,0,0,0,0,0,0,0,0,0,0,0,0];
xdf=abs(fft(xd,64)); f=0:63;
subplot(221);
stem([xd,zeros(1,16)]);
xlabel('Time'); ylabel('Amplitude');
title('Rectangular pulse signal');
subplot(222); plot(f,xdf); xlabel('Time');
ylabel('Amplitude'); title('Signal spectrum');
subplot(223);
stem([xd,xd]);
xlabel('Time'); ylabel('Amplitude');
title('Periodic pulse burst');
subplot(224); stem(xdf); xlabel('Time');
ylabel('Amplitude'); title('Signal spectrum');
```

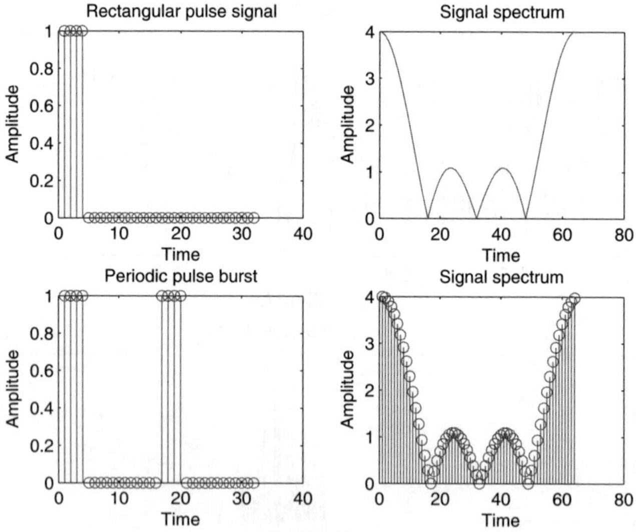

Figure 2.14. *Discrete-time signals and associated frequency spectra*

Notice that when a discrete-time signal is periodized, its spectrum is sampled.

Run the same code again to perform this comparative analysis for the following signals: triangular, sawtooth and exponential.

EXERCISE 2.12.

The DTFS of a periodical discrete-time signal is illustrated by the following MATLAB code:

```
x=[ones(1,32),zeros(1,32)]; N=64;
for k=0:N-1
    c(k+1)=0;
    for n=0:N-1
        c(k+1)=c(k+1)+x(n+1)*exp(-j*pi*2*k/N*n);
    end
    c(k+1)=c(k+1)/N;
end
stem(abs(c(1:N))); ylabel('Amplitude');
xlabel('coefficient index')
title('Absolute value of DTFS coefficients');
```

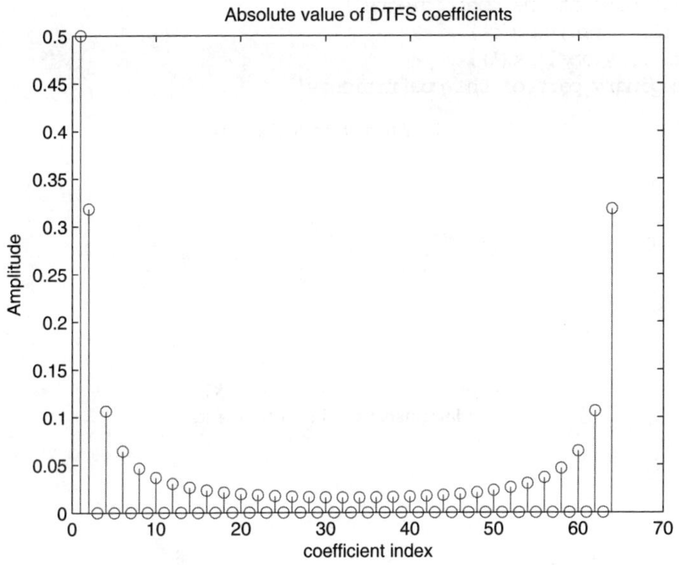

Figure 2.15. *Magnitude of the Fourier series coefficients*

Verify the following main properties of the DTFS coefficients corresponding to a periodical real discrete-time signal:

$$c(1)=\frac{1}{N}\sum_{n=0}^{N-1}x[n+1],\quad c(N/2)=\frac{1}{N}\sum_{n=0}^{N-1}x[n+1]\ (-1)^n,\quad c_{N-k}=c_k^*$$

Write a new code to calculate the DTFS coefficients of a 2D discrete-time signal.

```
x=zeros(16);x(1:4,1:4)=ones(4);
c=zeros(16); N=16;
for k1=1:N
    for k2=1:N
        for n1=1:N
            for n2=1:N
            expo=exp(-j*2*pi*(n1-1)*(k1-1)/N-j*2*pi*(n2-1)*(k2-1)/N);
            c(k1,k2)=c(k1,k2)+x(n1,n2)*expo;
            end
        end
    end
end
figure;
subplot(211); mesh(real(c))
xlabel('k1'); ylabel('k2');
title('Real part of the coefficients')
subplot(212); mesh(imag(c))
xlabel('k1'); ylabel('k2');
title('Imaginary part of the coefficients')
```

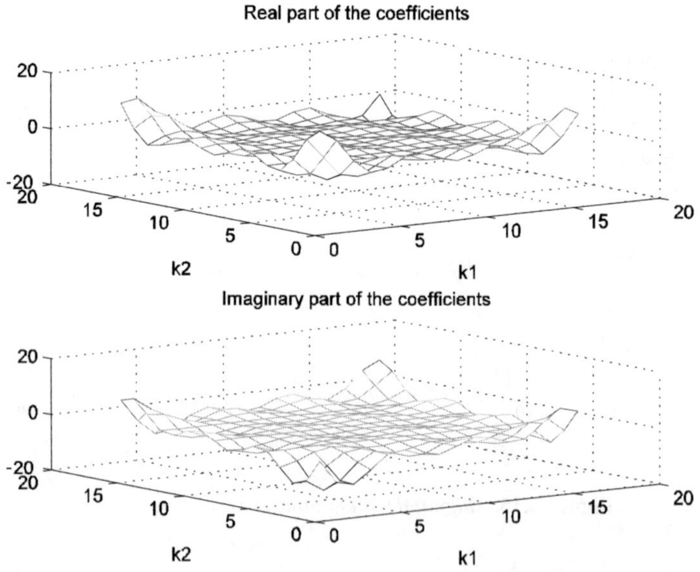

Figure 2.16. *Fourier series coefficients for a 2D discrete-time signal*

EXERCISE 2.13.

Write a MATLAB function for calculating the DTFT of a discrete-time finite signal $h[n]$ of length N, for N uniformly spaced frequencies on the unit circle.

```
function [H,W]=tftd(h,N)
W=(2*pi/N)*[0:N-1]';
mid=ceil(N/2)+1;
W(mid:N)=W(mid:N)-2*pi;
W=fftshift(W);H=fftshift(fft(h,N));
```

The following MATLAB code uses the new function `tftd.m` to calculate and plot the DTFT of the discrete-time signal $x[n] = 0.88^n$ in $N = 128$ points.

```
nn=0:40; xn=0.88.^nn;
[X,W]=tftd(xn,128);
subplot(211); plot(W/2/pi,abs(X))
xlabel('normalized frequency');
ylabel('amplitude [V]')
subplot(212);
plot(W/2/pi,180/pi*angle(X))
xlabel('normalized frequency')
ylabel('phase [deg.]')
```

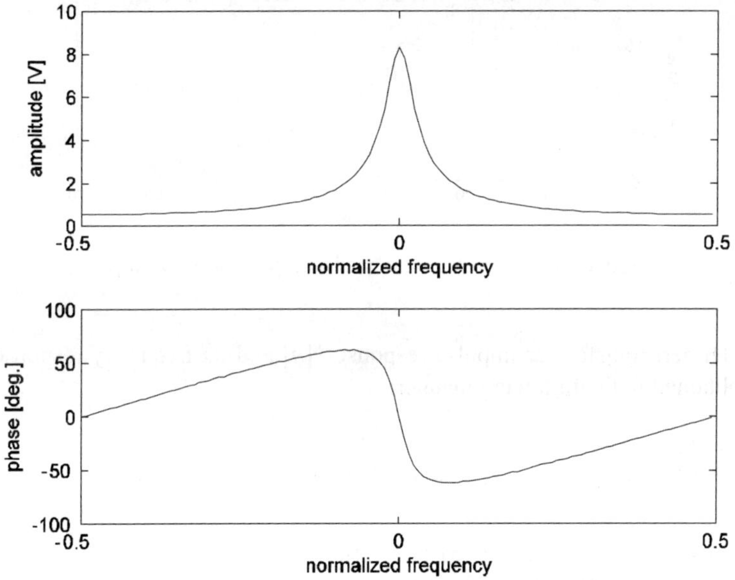

Figure 2.17. *Magnitude (top) and phase (bottom) spectrum*

EXERCISE 2.14.

Consider a discrete-time real signal defined by: $x[n] = \cos(2\pi n/N)$. Its Hilbert transform $y[n]$ is the imaginary part of the analytical signal: $x[n] + j{\cdot}y[n]$. The MATLAB code below allows the calculation of the analytical signal components.

```
N=64;n=0:N-1;
x=cos(2*pi/N*n);stem(x)
y=hilbert(x);
subplot(211);stem(real(y));
title('Real signal');
xlabel('n');ylabel('Amplitude');
subplot(212);stem(imag(y));
title('Signal Hilbert transform');
xlabel('n'); ylabel('Amplitude');
```

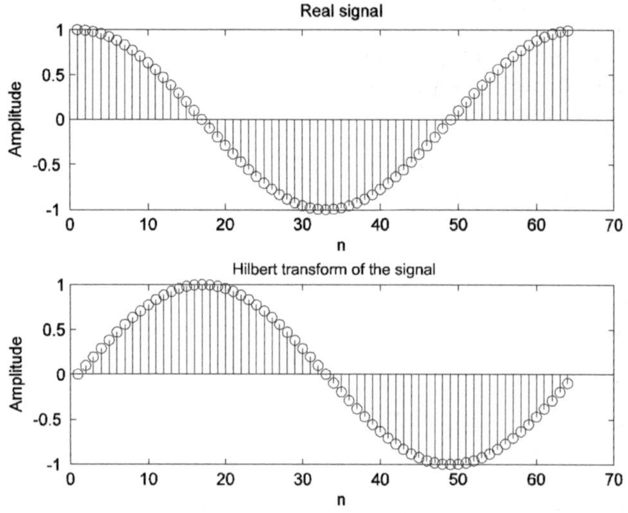

Figure 2.18. *Discrete-time signal and its Hilbert transform*

The Hilbert transformer impulse response $h[n]$ and its frequency response $H[\upsilon]$ can be obtained in the following manner:

```
for nn=-31:1:32;
    h(nn+32)=2*pi./nn.*((sin(pi*nn/2)).^2);
end
h(32)=0;
[H,f]=freqz(h); amp=20*log10(abs(H));
phase=unwrap(angle(H))*180/pi;
subplot(311); stem(h);
title('Impulse response'); xlabel('n')
subplot(312); semilogy(f,amp);
xlabel('Normalized frequency');
ylabel('Amplitude (dB)'); grid
subplot(313); plot(f,phase);
xlabel('Normalized frequency');
ylabel('Phase (deg.)'); grid
```

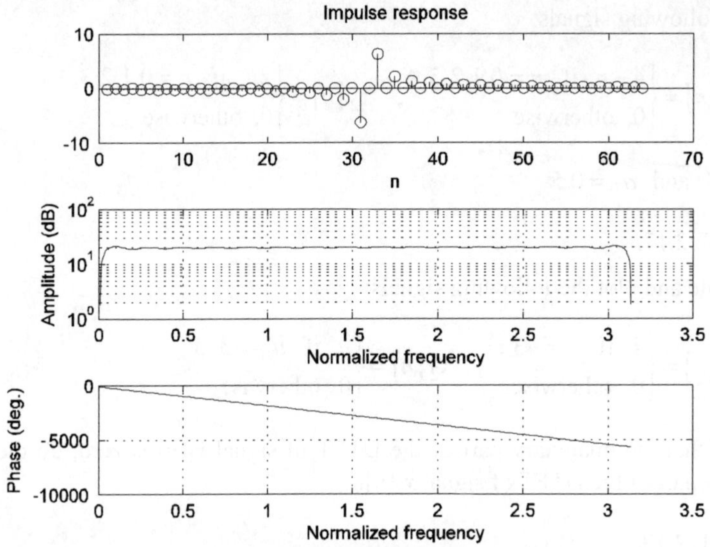

Figure 2.19. *Time (top) and frequency (bottom) representations of Hilbert operator*

The impulse response of the Hilbert transformer can be also obtained using the following MATLAB command:

```
h=remez(64,[0.1,0.9],[1,1],'Hilbert');
```

2.3. Exercises

EXERCISE 2.15.

Calculate and plot the spectrum of a periodic rectangular pulse of period $N = 16$, if:

$$x[n] = \begin{cases} 1 & \text{if} \quad n = 0..7 \\ 0 & \text{if} \quad n = 8..15 \end{cases}$$

Note the values of spectral components c_0 and $c_{N/2}$, and show that $c_{N-k} = c_k^*$ for $k = 0.15$.

EXERCISE 2.16.

Verify the linearity property of the $\text{DTFT}_{1\text{D}}$:

$$\text{DTFT}_{1\text{D}}\{a_1 x_1[n] + a_2 x_2[n]\} = a_1 \text{DTFT}_{1\text{D}}\{x_1[n]\} + a_2 \text{DTFT}_{1\text{D}}\{x_2(n)\}$$

using the following signals:

$$x_1[n] = \begin{cases} 4-n, & \text{if } n = 0,1,2,3,4 \\ 0, & \text{otherwise} \end{cases}, \quad x_2[n] = \begin{cases} 2n, & \text{if } n = 0,1,2,3,4 \\ 0, & \text{otherwise} \end{cases}$$

with $a_1 = 1$ and $a_2 = 0.5$.

EXERCISE 2.17.

Generate and plot the following signals:

$$x_1[n] = \begin{cases} 4, & \text{if } n = -3..3 \\ 0, & \text{otherwise} \end{cases}, \quad x_2[n] = \begin{cases} n, & \text{if } n = -3..3 \\ 0, & \text{otherwise} \end{cases}$$

Verify that the imaginary part of the DTFT of signal $x_1[n]$ is zero; do the same for the real part of the DTFT of signal $x_2[n]$.

EXERCISE 2.18.

Show that the DTFT_{1D} is the same for the two discrete-time signals defined below:

$$x_1[n] = a^n u[n]\ ;\ x_2[n] = -a^n u[-n-1]$$

EXERCISE 2.19.

Write a MATLAB code to verify Parseval's theorem:

$$\sum_{n=0}^{N-1} |x[n]|^2 = \frac{1}{N} \sum_{k=0}^{N-1} |X[k]|^2$$

for the signal:

$$x[n] = \begin{cases} n & \text{if } n = 0..7 \\ 0 & \text{otherwise} \end{cases}$$

EXERCISE 2.20.

Write a MATLAB code to verify that the following exponential signal family:

$$\varphi_k[n] = e^{jk(2\pi/N)n}, \quad k = 0..N-1$$

a) contains only N different signals $\varphi_k[n]$;

b) each signal $\varphi_k[n]$ is periodical, so that:

$$\varphi_k[n] = \varphi_k[n+N] \ , \forall n \in [0, N-1]$$

c) the N different signals $\varphi_k[n]$ form an orthogonal basis, so that:

$$\sum_{n=0}^{N-1} e^{jk(2\pi/N)n} = \begin{cases} N, & \text{if } k = 0, \pm N, \pm 2N, \ldots \\ 0, & \text{otherwise} \end{cases}$$

Chapter 3

Discrete-Time Random Signals

3.1. Theoretical background

3.1.1. *Introduction*

Random signals form a particularly important signal class because they are the only signals with the capability of transmitting information (this is a basic axiom of information theory).

The apparent division between signal and noise is artificial and depends on the criteria of the user. Some electromagnetic phenomena of galactic origin recorded by electrical antennae are considered as noise by telecommunication engineers, while they are very important signals for radioastronomers. The signal produced by a ship could be considered as a noise, but from a passive sonar point of view it is the information source that may allow localizing or even characterizing the ship. In fact, the useful or noisy nature of the captured signal is relative and is related to the observer's objectives.

In the framework of statistical theory the random variable concept is associated with a static study of the statistical phenomena. This is not often enough in practice because the probability distributions may vary in time or space. A more general concept is the stochastic process, which is defined as a system, or any variable set representing it, submitted to random influences.

In the case of a stochastic process, a random experience is associated with a function instead of a scalar or a vector, as in the case of a random variable. Although this function may depend on several parameters, we will consider only one in the following, which will be called time and will be denoted by t.

From a theoretical point of view a stochastic process can be represented as a two arguments function $(\omega, t) \to X(\omega, t)$, where ω is an element of a probability space (Ω, F, P) and t an element of a metric space.

The state of the system corresponding to given value t is a random variable. It is thus required that the applications:

$$X(t): w \to X(t, w) = X_t(w) \qquad [3.1]$$

be measurable for any value of t.

The random variable X_t is known as the t-section of the process. It represents the possible states of the system at the time t.

An "outcome" or a "trajectory" of the stochastic process is associated with each value of w: $t \to X_w(t) = X(w, t)$. It depends on the parameter t and has a random nature. That is why $t \to X_w(t)$ is often called a random function and is denoted by $X(t)$. A set of equal duration recordings form an outcome family associated with this random function.

3.1.2. *Real random variables*

A real random variable can be fully characterized by different means: cumulative distribution function (cdf), probability density function (pdf), characteristic function, moments, etc.

The cumulative distribution function of a real random variable X is denoted by $F_X(x)$ and is defined as indicated below:

$$F_X(x) = P(X \le x) \qquad [3.2]$$

This function has the following properties:

– F_X is a right-continuous, bounded, increasing function,

– $P(a \le X \le b) = F_X(b) - F_X(a)$,

– $\lim_{x \to +\infty} F_X(x) = 1, \quad \lim_{x \to -\infty} F_X(x) = 0$.

The probability density function $f_X(x)$ represents the probability distribution over the random variable outcomes and has the following properties:

– $F_X(x) = \int_{-\infty}^{x} f_X(u)du$

– $F_X(+\infty) = \int_{-\infty}^{+\infty} f_X(u)du = 1$

– $P[a \le X \le b] = \int_a^b f_X(u)du = F_X(b) - F_X(a)$

The nth order moment of the random variable X is defined as $E\left[X^n\right]$. $X - E[X]$ is the zero-mean variable corresponding to X. The nth order centered moment thus has the following expression: $E\left[\left(X - E[X]\right)^n\right]$.

The variance of a random variable is by definition its second order centered moment. In the case of a continuous random variable:

$$Var(X) = \int_{-\infty}^{+\infty} \left(x - E[X]\right)^2 f_X(x)dx \qquad [3.3]$$

The standard deviation $\sigma_X = \sqrt{Var[X]}$ is a probabilistic measure for the dispersion of a random variable X around its mean value, as is shown by the Bienaymé-Tchebycheff inequality:

$$P\left[\frac{|X - E[X]|}{\sigma_X} \ge t\right] < \frac{1}{t^2} \qquad [3.4]$$

The two characteristic functions of a random variable X are defined below:

$$\Phi_X(u) = E\left[e^{juX}\right], \ \Psi_X(u) = \ln\left(\Phi_X(u)\right) \qquad [3.5]$$

They are very useful for calculating the moments and the cumulants of a random variable in a straightforward way. It can easily be seen that if X is a continuous random variable, the first characteristic function $\Phi_X(u)$ is directly related to the Fourier transform of its pdf:

$$\Phi_X(u) = \int_{-\infty}^{+\infty} f_X(x)e^{jux}dx \qquad [3.6]$$

Random vectors

An n-dimensional random vector corresponds to n random variables, called marginals. The knowledge of the vector pdf involves the knowledge of the marginal pdfs, but the reciprocal is not generally true. A notable exception is represented by the case of independent marginals, when the multidimensional vector pdf is obtained as the product of the n marginal pdfs.

Let us consider in the following the simplified case of a two-dimensional continuous random vector denoted by $\{X, Y\}$. Its pdf is defined by:

$$f_{XY}(x,y) = \frac{\partial^2 F_{XY}(x,y)}{\partial x \partial y} \qquad [3.7]$$

$$P[x \le X \le x+dx \ \text{ and } \ y \le Y \le y+dy] = f_{XY}(x,y)dxdy \qquad [3.8]$$

Its statistical moments m_{rs} have the following form:

$$m_{rs} = E\left[X^r Y^s\right] = \int_{-\infty}^{+\infty}\int_{-\infty}^{+\infty} x^r y^s f_{XY}(x,y)dxdy \qquad [3.9]$$

The 1st and 2nd order moments, which are the most important, are defined here below:

$$m_{10} = E\left[X\right] = \int_{-\infty}^{+\infty}\int_{-\infty}^{+\infty} x f_{XY}(x,y)dxdy = \int_{-\infty}^{+\infty} x f_X(x)dx \qquad [3.10]$$

$$m_{01} = E\left[Y\right] = \int_{-\infty}^{+\infty} y f_Y(y)dy \qquad [3.11]$$

$$m_{20} = E\left[X^2\right] = \int_{-\infty}^{+\infty}\int_{-\infty}^{+\infty} x^2 f_{XY}(x,y)dxdy = \int_{-\infty}^{+\infty} x^2 f_X(x)dx \qquad [3.12]$$

$$m_{02} = E\left[Y^2\right] = \int_{-\infty}^{+\infty} y^2 f_Y(y)dy \qquad [3.13]$$

$$m_{11} = E\left\{XY\right\} = \int_{-\infty}^{+\infty}\int_{-\infty}^{+\infty} xy f_{XY}(x,y)dxdy \qquad [3.14]$$

The correlation and covariance of the random variable couple $\{X, Y\}$ can thus be calculated as:

$$\begin{aligned} &\operatorname{cor}\left[X,Y\right] = E\left[XY\right] = m_{11} \\ &\operatorname{cov}\left[X,Y\right] = E\left[(X-m_{10})(Y-m_{01})\right] = E\left[XY\right] - E\left[X\right]E\left[Y\right] = m_{11} - m_{10}m_{01} \end{aligned} \qquad [3.15]$$

According to the Schwarz inequality:

$$E\left[X^2\right]E\left[Y^2\right] \geq E^2[XY] \Leftrightarrow m_{02}m_{20} \geq m_{11}^2 = \left(\text{cor}[XY]\right)^2$$

$$E\left[(X-m_{10})^2\right]E\left[(Y-m_{01})^2\right] \geq E^2[(X-m_{10})(Y-m_{01})] \Leftrightarrow \sigma_X^2\sigma_Y^2 \geq \left(\text{cov}[XY]\right)^2$$

The correlation and covariance coefficients measure the underlying behavior similarity of the two variables and can be expressed in the following form:

$$r(X,Y) = \frac{\text{cor}(X,Y)}{\sqrt{E\left[X^2\right]E\left[Y^2\right]}}, \quad \rho(X,Y) = \frac{\text{cov}(X,Y)}{\sigma_X\sigma_Y} \qquad [3.16]$$

It can easily be seen from the inequalities above that their values are comprised between -1 and 1.

If Y is an affine function of X, i.e. $Y = aX + b$, it can be shown that:

$$\begin{cases} \text{if} \quad a > 0 \quad r(X,Y) = \rho(X,Y) = 1 \\ \text{if} \quad a < 0 \quad r(X,Y) = \rho(X,Y) = -1 \end{cases} \qquad [3.17]$$

If $\rho(X,Y) = 0$, X and Y are called uncorrelated random variables. The statistical independence is a sufficient, but not a necessary condition. Indeed, we may obtain $E[XY] = E[X]E[Y]$ even if X and Y are not independent.

Pdf examples

A continuous random variable is uniformly distributed if its pdf is constant over the interval $[a, b]$:

$$f_X(x) = \begin{cases} \dfrac{1}{b-a} & a \leq x \leq b \\ 0 & \text{otherwise} \end{cases} \qquad [3.18]$$

This results in:

$$E[X] = \frac{a+b}{2} \text{ (interval center); } \text{var}[X] = \frac{(b-a)^2}{12}; \qquad [3.19]$$

$$\sigma_X = \frac{b-a}{2\sqrt{3}}; \ \varphi_X(u) = \exp\left(ju\left(\tfrac{a+b}{2}\right)\right)\text{sinc}\left(\frac{(b-a)}{2}u\right) \qquad [3.20]$$

The random variable X is Gaussian or normally distributed if its pdf is bell shaped:

$$f_X(x) = \frac{1}{\sqrt{2\pi}\sigma} \exp\left(-\frac{(x-m)^2}{2\sigma^2} \right) \qquad [3.21]$$

Its mean, variance and characteristic function are given by:

$$\begin{aligned} E[X] &= m \\ \operatorname{var}[X] &= \sigma^2 \end{aligned} \qquad [3.22]$$

$$\varphi_X(u) = \exp(jum)\exp\left(\frac{-\sigma^2 u^2}{2} \right) \qquad [3.23]$$

This random variable is particularly interesting because it is fully characterized by its two first moments. Its cdf is tabulated.

Central limit theorem

Consider n iid (independent and identically distributed) random variables $\{X_k\}_{k=1..n}$ having the same mean and standard deviation, m and σ. Thus, the characteristic function of the random variable defined below:

$$X(n) = \frac{1}{\sqrt{n}} \sum_{k=1}^{n} \frac{X_k - m}{\sigma} \qquad [3.24]$$

tends towards the characteristic function of a zero-mean random Gaussian variable with unit variance.

This theorem provides an explanation for the fact that most physical random processes are Gaussian. Indeed, an observed random process can be considered as the sum of a large number of weak identically distributed random variables.

3.1.3. *Random processes*

The two most important properties of a random process are the stationarity and ergodicity. The strict sense stationarity (SSS) is related to the time-invariance of the random process statistics and can be expressed by:

$$\begin{aligned} & f_X(x_1, x_2, \cdots, x_n, t_1, \cdots, t_n) = f_X(x_1, \cdots, x_n, t_1 + t_0, \cdots, t_n + t_0) \\ \Leftrightarrow\ & f_X(x_1, x_2, \cdots, x_n, t_1, \cdots, t_n) = f_X(x_1, \cdots x_n, 0, t_2 - t_1, \cdots, t_n - t_1) \end{aligned} \qquad [3.25]$$

The specific 1D and 2D cases become:

$$f_X(x,\ t) = f_X(x,0) = f_X(x) \text{ time-independent} \qquad [3.26]$$

$$f_X\left(x_1,x_2,t_1,t_2\right) = f_X\left(x_1,x_2,0,t_2-t_1\right) = f_X\left(x_1,x_2,\tau\right) \text{ with } \tau = t_2 - t_1 \qquad [3.27]$$

It can be seen that the joint pdf corresponding to the 2D case depends only on the difference between the moments corresponding to the two observations.

The 2nd order stationarity, also known as wide sense stationarity (WSS), involves the time-invariance only for the 1st and 2nd order moments:

$$E\left[X(t)\right] = m_1(t) = m_1(0) = m_1,\ \forall\, t \qquad [3.28]$$

and $$E\left[X(t_1)X(t_2)\right] = m_{11}(\tau),\ \tau = t_1 - t_2 \qquad [3.29]$$

The wide sense stationarity does not involve the strict sense stationarity. However, if the random process is completely described by the 1st and 2nd order statistics, as in the Gaussian case, it is also a strict sense stationary if it is a wide sense stationary.

A random process is ergodic if its statistical means of any order are equal to the corresponding temporal means. The ergodicity is a very useful property because it allows the calculation of any statistical moment over all the process outcomes to be replaced by the calculation of a temporal average, over only one of them. Any particular outcome of the random process becomes in this case completely representative of it. From a practical point of view this means a lot of time and memory space can be saved because it is enough to record only one outcome in order to analyze a random process.

Two types of ergodicity are particularly interesting for a WSS random process: mean ergodicity and covariance ergodicity:

$$m_1 = E\left[X(t)\right] = \int_{-\infty}^{+\infty} x f_X(x,t)dx = \lim_{T\to\infty}\frac{1}{T}\int_{-T/2}^{T/2} X(t)dt \qquad [3.30]$$

$$\begin{aligned} m_{11}(t,t-\tau) &= E\left[X(t)X(t-\tau)\right] = \int_{-\infty}^{+\infty}\int_{-\infty}^{+\infty} x_1x_2 f_X(x_1,x_2,t,t-\tau)dx_1dx_2 \\ &= \lim_{T\to\infty}\frac{1}{T}\int_{-T/2}^{T/2} X(t)X^*(t-\tau)dt = \Gamma_X(\tau) \end{aligned} \qquad [3.31]$$

$\Gamma_X(\tau)$ is the autocorrelation function of a random process $X(t)$. It is a major tool for the analysis of the ergodic WSS random processes. In order to simplify their analysis they are generally zero-mean (this is easy to do as its mean is constant). The main properties of this type of random process are listed below:

a) the autocorrelation function is maximum at origin: $\Gamma_X(\tau) \leq \Gamma_X(0) \ \forall \tau$;

b) the autocorrelation function is Hermitian: $\Gamma_X^*(-\tau) = \Gamma_X(\tau)$. If $X(t)$ is real, $\Gamma_X(\tau)$ is real and even;

c) generally the following equation holds: $\lim_{\tau \to \infty} \Gamma_x(\tau) = |m|^2 = |E[X]|^2$. This means that for a large enough τ, the random variables $X(t)$ and $X^*(t - \tau)$ tend to be uncorrelated;

d) the power spectral density (PSD) is the Fourier transform of the autocorrelation function:

$$\gamma_X(\nu) = \int_{-\infty}^{+\infty} \Gamma_X(\tau) e^{-j2\pi\nu\tau} d\tau \qquad [3.32]$$

e) the PSD is a real, non-zero, positive function;

f) if $X(t)$ is real, $\gamma_X(\nu)$ is even. Since $\gamma_X(\nu)$ is also real, it results that $\Gamma_X(\tau)$ is real and even and thus:

$$\gamma_X(\nu) = \int_{-\infty}^{+\infty} \Gamma_X(\tau) \cos(2\pi\nu\tau) d\tau \qquad [3.33]$$

A random process is Gaussian if for any n and any moment series $t_1, \cdots, t_n$, the n-dimensional variable $X = [X(t_1), \cdots, X(t_n)] = [x_1, \cdots, x_n]$ is Gaussian. Its pdf is given by:

$$p(X) = \frac{1}{(2\pi)^{n/2} |\det \Sigma|^{1/2}} \exp\left[-\frac{1}{2}(X - \boldsymbol{\mu})^t \Sigma^{-1} (X - \boldsymbol{\mu})\right] \qquad [3.34]$$

where:

– $\boldsymbol{\mu} = [\mu_1, \cdots, \mu_n]$ is the mean vector with:

$\mu_1 = E[X(t_1)], \ldots, \mu_n = E[X(t_n)]$

– $\Sigma = \{\sigma_{ij}\}_{i=1..N, j=1..n}$ is the covariance matrix with:

$\sigma_{ij} = E\left[X(t_i)\, X(t_j)\right] - \mu_i \mu_j$

Notice that all the vectors considered above are column vectors and that Σ is a Hermitian matrix: $\sigma_{ij} = \sigma_{ji}^*$.

A natural example of a Gaussian random process is the thermal noise, which is related to the thermal chaotic motion of elementary particles in any physical system (such as the voltage fluctuations measured on an unpolarized resistor).

An important point is that the mathematical derivations are simplified in the case of random Gaussian processes. Their most important properties are recalled in the following:

a) a Gaussian random process is completely defined by its 1st and 2nd order statistics. In fact, the mean vector $\boldsymbol{\mu}$ as well as the covariance matrix Σ can be calculated for any series $t_1, \cdots, t_n$ from functions $\mu_X(t)$ and $\Gamma_X(t_1,t_2)$:

$$\boldsymbol{\mu} = \left[\mu_X(t_1), \cdots, \mu_X(t_n)\right] \qquad [3.35]$$

$$\sigma_{ij} = \Gamma_X(t_i,t_j) - \mu_X(t_i)\mu_X(t_j) \qquad [3.36]$$

b) a wide sense stationary Gaussian process is also a strict sense stationary because the temporal law is completely characterized by its two first moments. All the components of its mean vector are equal in this case and its covariance matrix has a Toeplitz structure;

c) the decorrelation involves the independence;

d) any linear combination of independent Gaussian variables is Gaussian. The Gaussian character is thus preserved by any linear transformation or linear filtering;

e) the Gaussian character is often associated with the white random processes defined below, but there is no implication between these two properties.

A random process $b(t)$ whose power spectral density $\gamma_b(\nu) = N_0/2$ is constant over all frequencies is called white noise. Its autocorrelation function can thus be obtained in the form:

$$\Gamma_{bb}(\tau) = TF^{-1}\{\gamma_b(\nu)\} = \frac{N_0}{2}\delta(\tau) \qquad [3.37]$$

In many applications, a signal $r(t)$ can be expressed as the sum of a useful component $s(t)$ and a noise component $b(t)$:

$$r(t) = s(t) + b(t) \qquad [3.38]$$

The signal-to-noise ratio (SNR) provides the information about the relative power of noise with respect to the useful signal component power. It can thus be defined as the ratio between the mean power levels of the useful signal and noise components:

$$RSB = \frac{P_s}{P_b} \qquad [3.39]$$

Notice that the noise mean power should be estimated within the useful signal frequency band.

3.2. Solved exercises

EXERCISE 3.1.

The aim of this exercise is to become more familiar with the Gaussian distribution, currently used in signal processing. Generate 1,000 zero-mean normally distributed random samples with the variance 1 using the function `randn`. Plot on the same figure the histogram and the theoretical pdf. Repeat the same experiment for a non-zero mean. Then consider a different value for the signal variance. Finally, decrease the number of samples from 1,000 to 20. Comment on the obtained results.

The Gaussian distribution occurs very frequently in the real world. For example, a mechanical factory may specify a Gaussian distribution of its manufactured wheel diameter around the nominal value of 1 m, with a standard deviation of 0.04 m. Let us consider that customers only accept wheels whose diameter is between 0.95 and 1.05 m. Give an approximation of the merchandise percentage which will not be sold by the manufacturer.

```
Ntotal=1000;
mu=0; sigma=1; %Mean and standard deviation
ech = randn(1,Ntotal) .* sigma + mu;
Nclas=8; dx=.01;
[N,X]=hist(ech,Nclas);
x=mu-4*sigma:dx:mu+4*sigma;
y=exp(-0.5*((x-mu)/sigma).^2)/(sqrt(2*pi)* sigma);
figure;clf;zoom on;
subplot(221); hold on;
bar(X,N/Ntotal); plot(x,y,'r');
title('N(0,1), 1000 samples ');
subplot(212);
hold on;plot(x,y,'r')

%% Change the mean
mu=2;sigma=1;x=mu-4*sigma:dx:mu+4*sigma;
```

```
y=exp(-0.5*((x-mu)/sigma).^2)/(sqrt(2*pi)* sigma);
subplot(212);
hold on;plot(x,y,'b--')

%% Change the variance (or the standard deviation)
mu=2;sigma=5;
x=mu-4*sigma:dx:mu+4*sigma;
y=exp(-0.5*((x-mu)/sigma).^2)/(sqrt(2*pi)*sigma);subplot(212);
hold on;plot(x(1:50:end),y(1:50:end),'g+');grid;
legend('N(0,1)','N(2,1)','N(2,25)')

%%Change the number of samples
Ntotal=20;mu=0;sigma=1;
ech=randn(1,Ntotal)*sigma+mu;
Nclas=8; [N,X]=hist(ech,Nclas);
x=mu-4*sigma:dx:mu+4*sigma;
y=exp(-0.5*((x-mu)/sigma).^2)/(sqrt(2*pi)* sigma);
subplot(222); hold on;
bar(X,N/Ntotal);
plot(x,y,'r');
title('N(0,1), 20 samples')
```

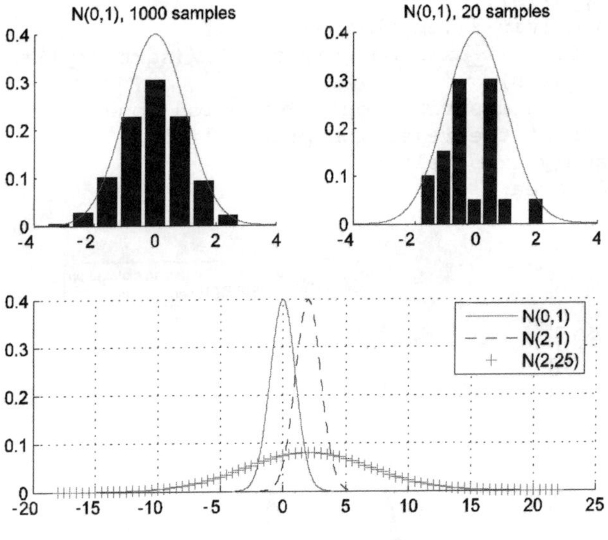

Figure 3.1. *Gaussian pdf examples*

It can be seen that the obtained histogram using 1,000 samples is bell shaped and close to the theoretical pdf, unlike in the case when only 20 samples are used. The quality of the statistical properties estimation for a random process is therefore directly related to the number of its recorded samples.

The third figure illustrates the concepts of mean and variance. Thus, when the mean is changed from 0 to 2, the initial curve (solid line) is shifted to the right (dashed line) and its shape remains the same, while when the standard deviation is modified the distribution becomes flatter. Notice that the surface under the pdf curve still equals 1.

```
% Wheel manufacturer
mu=1;sigma=0.04; dx=.01; x=mu-4*sigma:dx:mu+4*sigma;
y=exp(-0.5*((x-mu)/sigma).^2)/(sqrt(2*pi)* sigma);
figure;clf;zoom on;hold on; plot(x,y)
```

The probability of having an incorrect wheel (which will not be sold) can be calculated as the filled surface on the second figure, i.e. as the integral of the pdf on $(-\infty, 0.95]$ and $[1.05, +\infty)$. The value of this integral is estimated using the rectangle method.

```
%% Surface filling
deltax=0.001;
x_axis=x(1):deltax:0.95;
y_axis=exp(-0.5*((x_axis-mu)/sigma).^2)/(sqrt(2*pi)* sigma);
bar(x_axis,y_axis);
x_axis=1.05:deltax:x(length(x));
y_axis=exp(-0.5*((x_axis-mu)/sigma).^2)/(sqrt(2*pi)* sigma);
bar(x_axis,y_axis);
legend('Gaussian distribution','Rejection zone')
%% Estimation of the rejection probability
half_area=sum(y_axis*deltax);
proba=2*half_area
```

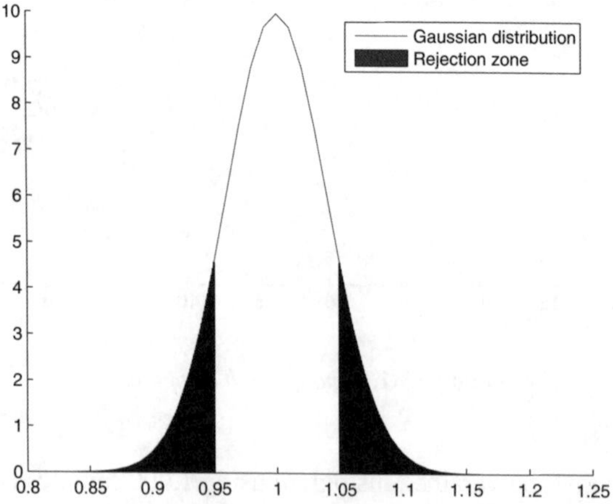

Figure 3.2. *Rejection test*

The obtained value for the rejection probability is 21.58%. The use of a table would give 21.12%. Thus, about 21% of the manufactured wheels will not be appropriate to be sold. Since this proportion is significant the manufacturer will have to improve its equipment or to increase the product price in order to balance his budget.

EXERCISE 3.2.

This exercise deals with the distribution of two independent random variables. Let us consider X and Y, two i.i.d. random variables, whose pdf is an exponential function with the parameter lambda = 2.

Determine the distribution of $Z = X+Y$. Compare the obtained distribution to the analytical one. Then calculate the convolution of the pdfs corresponding to Z and $-X$. How do you explain the difference with respect to the Y distribution knowing that $Z+(-X) = Y$?

```
%Pdf of X or Y on [0,6]
dx = .01 ; x = (0:dx:6);
npt = length(x);
lambda = 2;
Px = lambda*exp(-lambda*x);
```

Remember that the pdf of the sum of two i.i.d. random variables is the convolution of the corresponding pdf.

```
% Distribution of X+Y
Pz = conv(Px,Px)*dx;
figure;
clf;
plot(x,Px);
hold on;
plot(x(1:5:end),Pz(1:5:npt),'r+')

% Comparison with the theoretical result
Pzanalytic=lambda^2*exp(-lambda*x).*x;
plot(x(1:5:end),Pzanalytic(1:5:npt),'g*')

% Convolution of the pdfs corresponding to Z and -X
Czx = conv(Pz,Px(length(Px):-1:1))*dx;
plot([-x(length(x):-1:2),x],Czx(1:2*npt-1),'k--')
legend('Pdf of X (or Y)', 'Pdf of Z=X+Y', 'Analytical pdf of
Z','pdf(Z)*pdf(-X)')
```

Figure 3.3 shows that the distribution obtained from the convolution of the pdfs corresponding to X and Y and the pdf of Z are identical.

On the other hand, it can be seen that the convolution of pdfs corresponding to Z and $-X$ is not equal to the pdf of Y, although $Y = Z + (-X)$, because $-X$ and Z are not independent ($Z = X + Y$). The pdf of the sum can no longer be obtained by the convolution of the two pdfs.

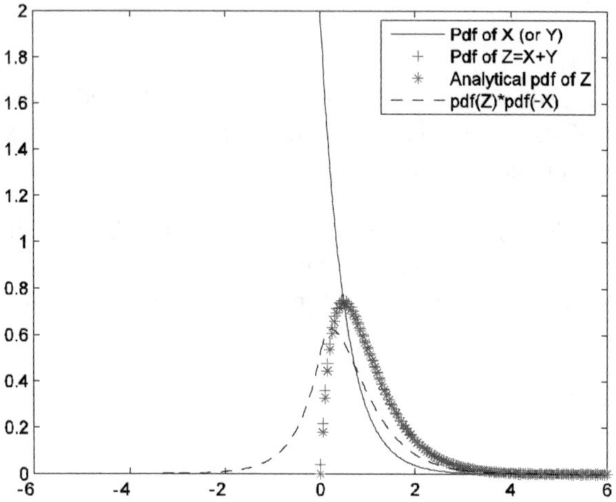

Figure 3.3. *Illustration of the pdf of the sum of two random variables*

EXERCISE 3.3.

The goal of this exercise is to use the central limit theorem. Let us consider a set of independent Bernoulli-distributed random variables X_i. If the parameter of the Bernoulli pdf is p, then $E[X_i] = p$ and the $var[X_i] = p - p^2$.

Write a MATLAB function for calculating recursively the pdf of $X_1 + X_2 + ... + X_i = Z_i$ as the result of the convolution of the pdfs corresponding to the random variables Z_i -1 and X_i. Compare the pdf of Z_i to the normal pdf $N(m, s)$ with $m = i \cdot E[X_i]$ and $s^2 = i \cdot var[X_i]$, on the interval $[m-4s, m+4s]$.

Use this function for $p = 0.6$, then $p = 0.9$, $p = 0.99$ and comment on the results obtained.

```
function sumlaw(x,y,m,s,Nsum);
%   x: domain of the random variable x=[0,1]; Bernoulli pdf)
%   y: pdf(x) (y = [1-p p]; Bernoulli pdf)
%   m: mean value (m = p; Bernoulli pdf)
%   s: standard deviation (s = sqrt(p-m^2); Bernoulli pdf)
%   Nsum: number of added variables
%
%%% Normal pdf set up
```

```
dx = .01; xn= (m-4*s):dx: (m+4*s);
yn = exp(-0.5 * ((xn-m)/s) .^2) ./ (sqrt(2*pi) .* s);
plot(xn,yn,'r');hold on;stem(x,y);
hold off;
%%% Recursive calculation of the sum pdf
yp = y; xp = x;
for ii=2:Nsum
  yp = conv(yp,y); xp =
(x(1)+xp(1):x(length(x))+xp(length(xp)));
% Pdf sum
  mn = ii*m;
  sn = sqrt(ii)*s;
  xn=(mn-4*sn):dx:(mn+4*sn);
  yn=exp(-0.5*((xn-mn)/sn).^2)/(sqrt(2*pi).*sn);%Normal pdf
% Visualization
  plot(xn,yn,'r'),
  hold on,
  set(gca,'xlim',[xn(1),xn(length(xn))])
  ind = find(xp>mn-4*sn&xp<mn+4*sn);
  stem(xp(ind),yp(ind))
  hold  off
  drawnow
end

% Use of the obtained pdf for different values of p
p = .6; x = [0:1]; y = [1-p p];
m = p; s = sqrt(p-m^2);
figure; clf; subplot(221);
sumlaw(x,y,m,s,20);
title('p=0.6, N=20')
p = .9;x = (0:1); y = [1-p p];
m = p; s = sqrt(p-m^2);
subplot(222);
sumlaw(x,y,m,s,20);
title('p=0.9,N=20')
p = .99;x = (0:1); y = [1-p p];
m = p; s = sqrt(p-m^2);
subplot(223);
sumlaw(x,y,m,s,20);title('p=0.99,N=20')
p = .9;x = (0:1); y = [1-p p];
m = p; s = sqrt(p-m^2);
subplot(224);
sumlaw (x,y,m,s,70);title('p=0.9,N=70');
legend('Normal pdf','Sum pdf')
```

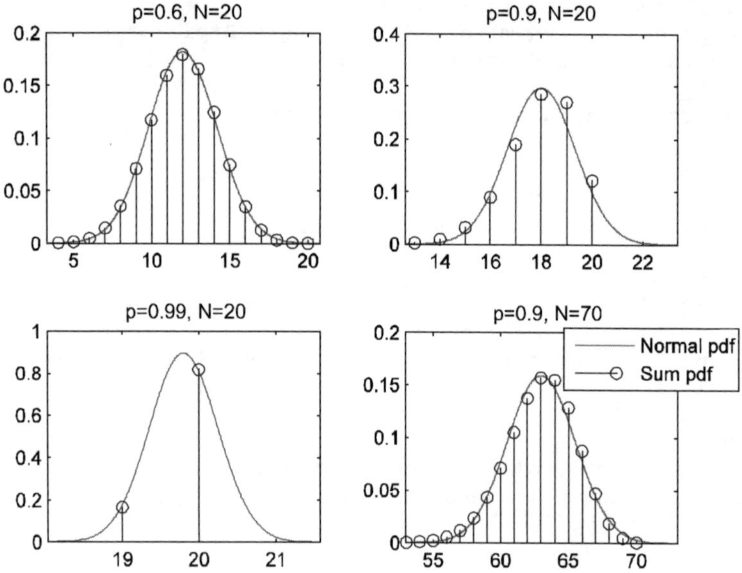

Figure 3.4. *Central limit theorem illustration*

According to the obtained results, the sum of N = 20 independent Bernoulli-distributed random variables asymptotically converges to N(m, s), with $m = N \cdot \mathrm{E}[X_i]$ and $s^2 = N \cdot \mathrm{var}[X_i]$, for $p = 0.6$.

However, the more p moves away from 0.5, the less good the results are. In fact, the binomial pdf (Bernoulli-distributed random variables sum) can be approximated by a Gaussian pdf only for large enough N, $N \cdot p > 5$ and $N \cdot (1-p) > 5$. Thus, $p = 0.9$ is not an appropriate value because $20 \cdot (1-p) = 2$.

In order to meet these conditions N has to be larger than 50. This statement is sustained by the result depicted on the figure above for $N = 70$, $p = 0.9$. In the same way, it would be necessary to have $N > 500$ for obtaining appropriate results with $p = 0.99$. The explanation comes from the fact that the more p is different from 0.5, the less symmetric the initial distribution is and the more random variables have to be added in order to obtain the convergence.

EXERCISE 3.4.

Generate and plot 4 outcomes of the random process $x(t)$, sampled at 100 kHz, during 0.01 s, $x(t, \varphi) = \cos(2\pi \cdot 1{,}000 \cdot t + \varphi)$, where φ takes the following values: $0,\ \pi/2,\ \pi,\ 3\pi/2$.

Calculate the mean value $E[x(t)]$ and the 2nd order statistical moment $E\left[x^2(t)\right]$.

```
phas=[0 pi/2 pi 3*pi/2]; leng=1000;
Fs=100000; t=[1:1000]/(Fs);
nb_real= length(phas);
for ii = 1:nb_real
   x(ii,:) = cos(2*pi*1000*t + phas(ii))
end
subplot(411);plot(t,x(1,:));title('phase at origin = 0')
subplot(412);plot(t,x(2,:));title('phase at origin = pi/2')
subplot(413);plot(t,x(3,:));title('phase at origin = pi')
subplot(414);plot(t,x(4,:));
title('phase at origin = 3*pi/2');xlabel('time [s]')
```

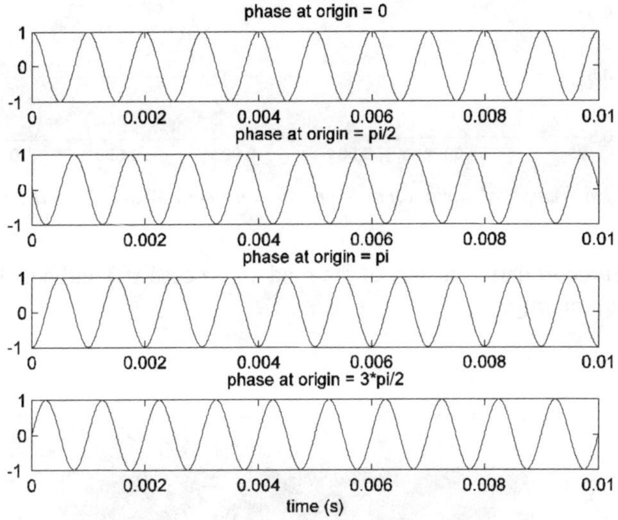

Figure 3.5. *Random phase sinusoids*

The 1st and 2nd order moments are time-invariant. This random process is therefore wide-sense stationary. Thus, it can also be demonstrated that its autocorrelation function depends only on the difference between the observation moments.

```
ave=mean(x); eqm=mean(x.^2);
figure
subplot(211); plot(t,ave);
axis([0 0.01 -0.50 .5]);
title('statistical mean variation')
subplot(212); plot(t,eqm);
```

```
title('2^n^d order moment variation')
```

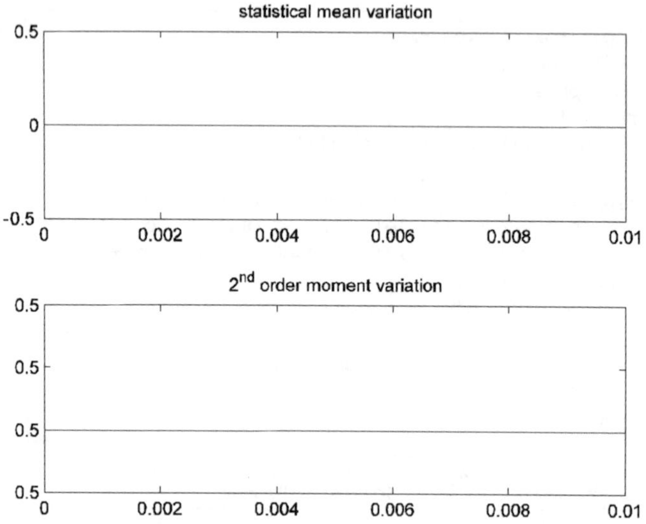

Figure 3.6. *Temporal variation of the random process 1st and 2nd order moments*

Plot the autocorrelation matrix of the random process $x(t)$ and conclude about its wide-sense stationarity.

```
corre=corrcoef(x(:,1:300));
figure
imagesc(t(1:300),t(1:300),corre)
```

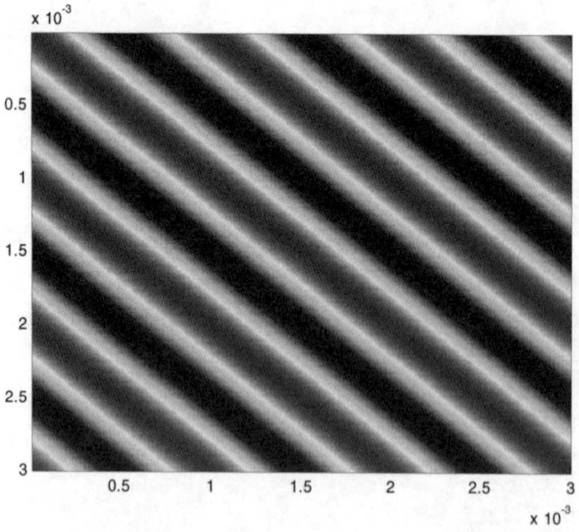

Figure 3.7. *Autocorrelation matrix representation*

As can be seen, the autocorrelation matrix $\Gamma_x(i,j) = E\left[x(t_i,\varphi)x(t_j,\varphi)\right]$ has a Toeplitz-like structure. Consequently, the autocorrelation matrix only depends on the difference among observation moments. Note that the curve corresponding to the 2nd order moment variation is obtained here as the main diagonal of the autocorrelation matrix.

Test the 1st and 2nd order ergodicity property for this random process.

```
ave_erg=mean(x')
ave_erg =
  1.0e-015 *
   -0.0893     0.0919     0.0560    -0.1225
var_erg=mean(x'.^2)
var_erg =
    0.5000     0.5000     0.5000     0.5000
```

It can be seen that the four temporal means are equal (up to the estimation error), to the 1,000 statistical means obtained from the four process outcomes, for both the 1st and 2nd order moments. Thus, the process is 2nd order ergodic.

EXERCISE 3.5.

Generate a white Gaussian random process with the mean 3 and the variance 3 on 1,092 points. Plot its autocorrelation function and its power spectral density.

Filter the previously generated random process after removing its mean. Use an IIR filter having a zero in 0 and a real simple pole in 0.95. Plot its autocorrelation function and its power spectral density. The two signals will be considered 2nd order ergodic.

Comment on the results obtained and conclude about the link between the signal predictability and the shape of its autocorrelation function.

```
rho=0.95;
var=3;
meanval=0;
b = [ 1];
a = [ 1 -rho ];
y = filter(b,a,(sqrt(var)*(randn(1,1092))))+meanval;
subplot(221);
plot(y);
axis([0 1091 -20 20])
xlabel('time [s]');
ylabel('amplitude (V)');
title('filtered signal');
corre=xcorr(y,'unbiased');
subplot(223);
```

```
plot([-400:400],corre(692:1492))
xlabel('time [s]');
ylabel('amplitude (V*V)');
title('autocorrelation function')
meanval=3;
y1 = sqrt(var)*(randn(1,1092))+meanval;
subplot(222);
plot(y1);
axis([0 1091 -5 10])
xlabel('time [s]');
ylabel('amplitude (V)');
title('signal before filtering');
correl=xcorr(y1,'unbiased');
subplot(224);
plot([-400:400],correl(692:1492))
xlabel('time [s]');
ylabel('amplitude (V*V)');
title('autocorrelation function')
```

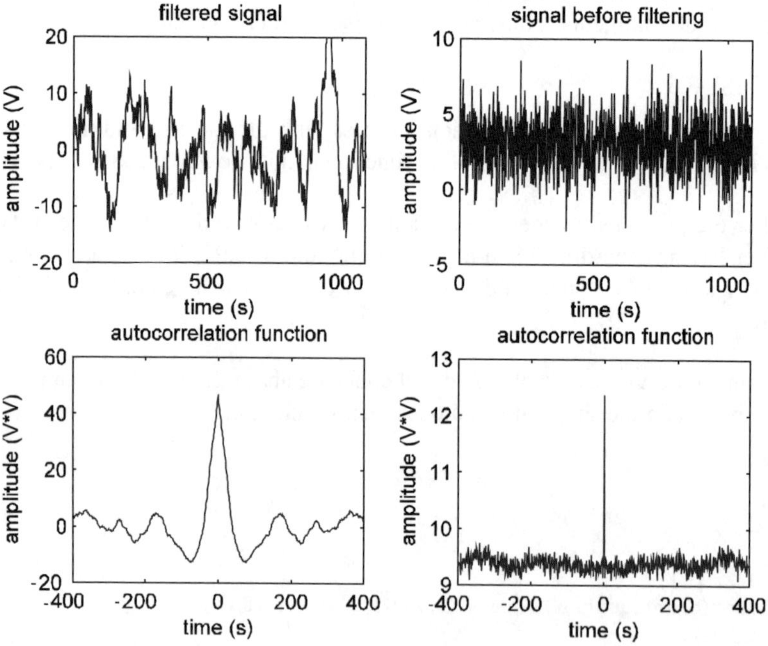

Figure 3.8. *Signals and associated autocorrelation functions*

The autocorrelation function of the white Gaussian noise is a Dirac pulse, as for large temporal shifting it tends towards the square of the mean value, i.e. 9.

In the case of the filtered noise, the autocorrelation function decreases very slowly. It also tends towards the square of the mean value, which is zero in this case. The lowpass filtering correlates the random process samples as it can be seen on the corresponding autocorrelation function depicted on the left side of the figure above.

The effect of the convolution introduced by the linear filter depends on its pulse response and it would be easy to verify that the lower its cutoff frequency, the larger the main lobe of the autocorrelation function is and thus, the stronger the correlation introduced by the lowpass filter. It is obvious then that the filtered signal can be partially predicted, while the white noise is completely unpredictable.

```
freq=[0:1/1024:511/1024];
y_f=abs(fft(y,1024));
figure;
subplot(211);
semilogy(freq,y_f(1:512));
xlabel('Normalized frequency');
ylabel('amplitude (W/Hz)');
title ('Filtered signal PSD')
subplot(212);
y1_f=abs(fft(y1,1024));
semilogy(freq,y1_f(1:512));
xlabel('Normalized frequency');
ylabel('amplitude (W/Hz)');
title ('White noise PSD')
```

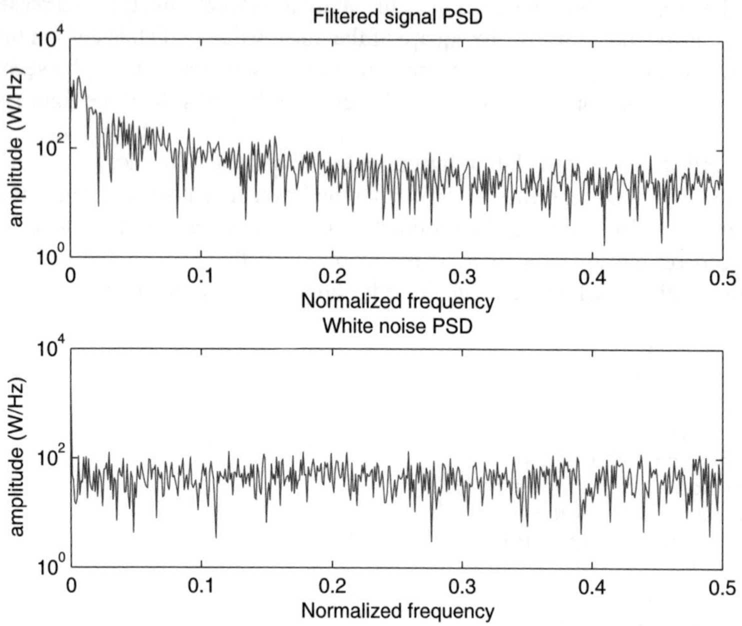

Figure 3.9. *PSDs of a filtered (top) and unfiltered (bottom) white noise*

The filtering effect can be easily identified on the PSDs plotted on Figure 3.9. Note the DC component in the spectrum of the white noise, which corresponds to its mean value.

The depicted PSDs are very fluctuant because their estimation is performed using a finite number of samples and no method is used to reduce the estimate variance (this topic will be discussed in Chapter 10).

EXERCISE 3.6.

1,000 people, from 20 to 50 years old, participate in a statistical survey. They are asked to provide their height and weight, which are compared with the standard values. The difference between the real and the standard values is a random Gaussian quantity, whose variance depends on the parameter considered.

Estimate the correlation coefficient for each pair of variables: (weight, height), (age, weight) and (age, height).

```
height = fix(randn(in,1)*12+168);
height(height<150) = 150 + randn(sum(height<150),1)*3;
```

```
weight = fix((height-168)*1.1 +64 + randn(in,1)*6+2);
weight(weight<45) = 45 + randn(sum(weight<45),1)*2;
weight(age>=35) = fix(weight(age>=35) + randn(sum(age>=35),1)*1+3);
weight(age>=40) = fix(weight(age>=40) + randn(sum(age>=40),1)*2+4);
weight(age>=45) = fix(weight(age>=45) + randn(sum(age>=45),1)*2+5);
x = [age height weight]; yx = x(:,3);yy = x(:,2);
subplot(311); plot(yx,yy,'*'); grid on;
xlabel('weight (Kg)'); ylabel('height (cm)')
title('height variation as a function of weight')
yx = x(:,1); yy = x(:,2);
subplot(312); plot(yx,yy,'*'); grid on;
xlabel('age'); ylabel('height (cm)')
title('height variation as a function of age')
yx = x(:,1); yy = x(:,3);
subplot(313);plot(yx,yy,'*'); grid on;
xlabel('age'); ylabel('weight (Kg)')
title('weight variation as a function of age')
```

Note that the height and the weight of any person are obviously correlated. Type the following code line in order to measure this correlation:

```
yx = x(:,3); yy = x(:,2); corrcoef(yx,yy)
```

This results in the correlation matrix of the two random variables:

```
ans =
    1.0000    0.8466
    0.8466    1.0000
```

Their correlation coefficient is therefore equal to 0.8466, which means that they are closely related. However, there is no statistical link between a person's age and his weight or his height, as can be shown from the results obtained below:

```
yx = x(:,1); yy = x(:,2); corrcoef(yx,yy)

ans =
    1.0000    0.0155
    0.0155    1.0000

yx = x(:,1); yy = x(:,3); corrcoef(yx,yy)

ans =
    1.0000    0.2655
    0.2655    1.0000
```

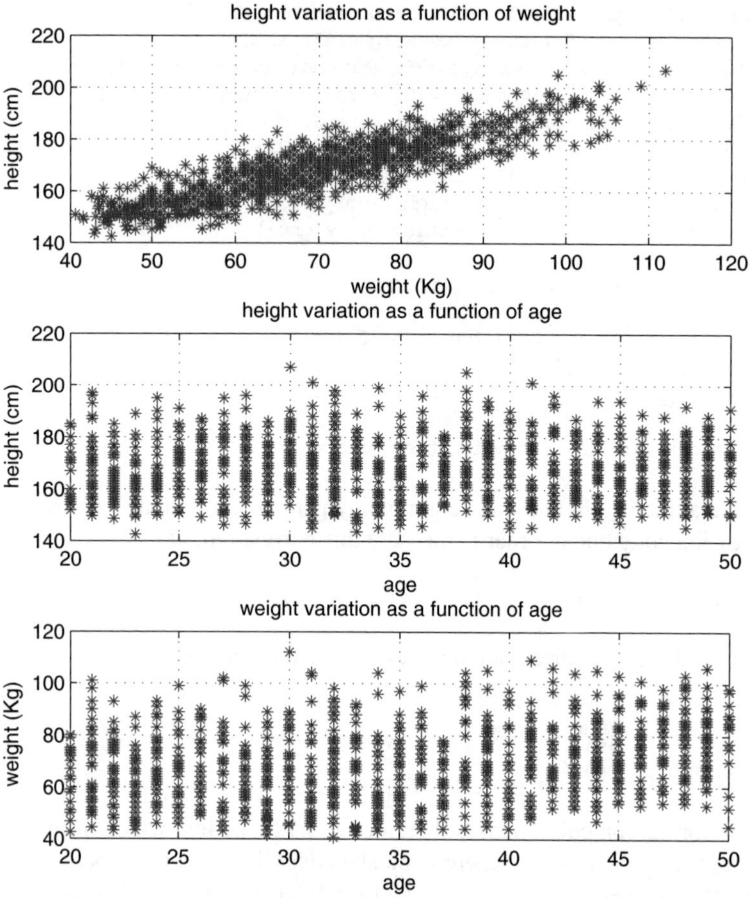

Figure 3.10. *Correlation illustration*

EXERCISE 3.7.

A linear system can be characterized using a white noise as input. The power spectral density of the output signal then provides an estimation of its transfer function.

Simulate a 2nd order digital bandpass filter, having a central frequency of 2 kHz and a frequency bandwidth of 1 kHz. Consider a sampling frequency of 10 kHz and a zero-mean white Gaussian noise, with the variance 1, as input signal.

Then define a bank of 6th order Butterworth filters, with a frequency bandwidth of 200 Hz, which will be useful to measure the filtered signal power in each frequency band. Finally, plot the power variation as a function of frequency.

```
Fs= 10000;
n = 1; fc = [1500 2500];
[b a] = butter(n,fc*2/Fs);
fs2 = Fs/2;
x0=randn(1024,1);
x = filter(b,a,x0);
rep_freq=freqz(b,a,25);
min_fract_bw = 200/5000;
min_fract_Hz = min_fract_bw * fs2;
normal= sqrt( length(x) );
f_deb = 0; f_fin = fs2;
bw = min_fract_Hz;
band = [f_deb, f_deb+bw];
freq_length = f_fin - f_deb;
no_bpf= fix(freq_length/bw);
for k = 1:no_bpf
    if ( min(band) == 0 )
      band_vr = max(band);
      f(k) = sum(band)/2;
      band_vr = band_vr/fs2;
      [b a] = butter(6,band_vr);
      y(k) = (var(filter(b,a,x))+ (mean(filter(b,a,x)).^2))/normal;
    elseif ( max(band) == fs2 )
      band_vr = min(band);
      f(k) = sum(band)/2;
      band_vr = band_vr/fs2;
      [b a] = butter(6,band_vr,'high');
      y(k) = (var(filter(b,a,x))+ (mean(filter(b,a,x)).^2))/normal;
    else
      f(k)=sum(band)/2; band_vr=band/fs2;
      [b a]=butter(3,band_vr);
      y(k) = (var(filter(b,a,x))+ (mean(filter(b,a,x)).^2))/normal;
    end;
    band = band + bw;
end
subplot(211);
semilogy(f/1000,abs(rep_freq).^2);
xlabel('Frequency [kHz]');
ylabel('Amplitude');grid
title('Theoretical transfer function of the system')
subplot(212);
semilogy(f/1000,y) ;
xlabel('Frequency [kHz]');
grid ; ylabel('Power (W)');
title('Estimated transfer function of the system')
```

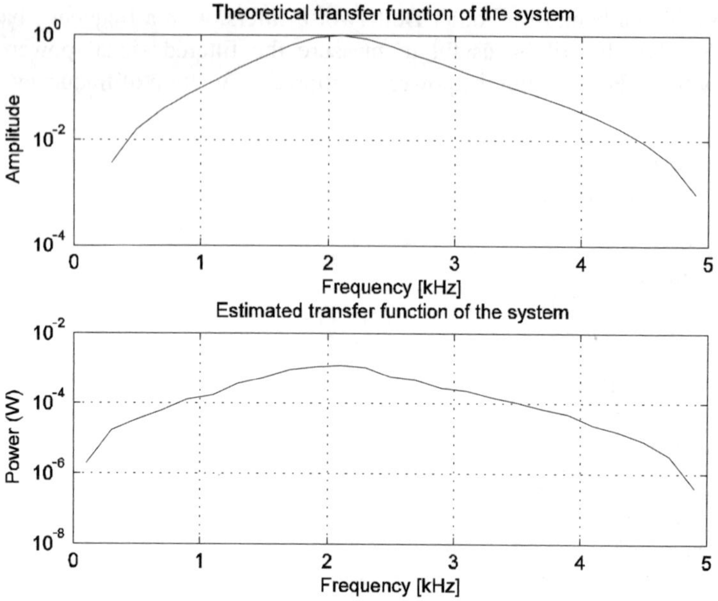

Figure 3.11. *Transfer functions*

The filter bank outputs illustrate properly the frequency system behavior. More accurate results could be obtained by the reduction of the filter frequency bandwidth, but this increases the processing time and results in a more complex filter synthesis (instability risk). The increased signal duration should also be considered to integrate the measured power properly (system identification problem: trade-off between the frequency bandwidths of the bandpass filters and the output integrator filter).

3.3. Exercises

EXERCISE 3.8.

Consider a zero-mean random Gaussian signal with the variance 3 at the input of the following systems:

– thresholder;

– rectifier;

– quadratic filter.

Determine the pdf of the output signals and estimate their 1st and 2nd order moments.

EXERCISE 3.9.

Go back to exercise 3.4 and consider a constant initial phase, but a uniformly distributed random amplitude, with the mean value 3 and standard deviation 0.5. Comment on the 2nd order stationarity of the new random process? Does the ergodicity have any sense in this case?

EXERCISE 3.10.

Write a MATLAB code for generating N samples of a random variable X having the following pdf:

$$f_X(x) = \begin{cases} 2-2x & 0 \leq x \leq 1 \\ 0 & \text{otherwise} \end{cases}$$

Find out the mean value and the variance of this random variable.

EXERCISE 3.11.

A linear system such as the one used in exercise 3.5 is driven by a zero-mean white uniform noise having a standard deviation of 5. Calculate and comment on the output signal pdf for ρ values between 0 and 0.999, such as $\rho = 0.001$, $\rho = 0.01$, $\rho = 0.1$, $\rho = 0.5$ and $\rho = 0.95$.

Can you explain the different obtained pdfs? Calculate the mean values and the variances of all variables. Let $\rho = -0.7$ and comment on the mean value and the variance.

EXERCISE 3.12.

Simulate the effect of decreasing the number of quantification bits for a 32 bit coded signal. Plot the quantification error and its pdf.

Calculate the SNR and demonstrate that it increases with 6 dB for each additional quantification bit.

EXERCISE 3.13.

Calculate the autocorrelation function of a RZ and then a NRZ binary coded random signal. Consider 1,024 points and 100 outcomes. Use this result for obtaining the signal power spectral density.

What condition do these codes have to meet in order to obtain a wide-sense stationary random signal?

Chapter 4

Statistical Tests and High Order Moments

4.1. Theoretical background

In everyday life, we are involved in different situations and we have to make decisions in order to act, react or interact with the real world. These decisions are generally subjective. Thus, for a simple task like crossing the street, the decision is always individual and depends on the culture (driving on the right side in France and driving on the left side in Great Britain), the age or on many other elements (street type, number of cars and speed, traffic light state, etc.).

Probability theory can help us to make these decisions more objective. It is suitable for the study of any partially or completely unpredictable process and forms the basis of different applications: quantum mechanics, reliability, meteorological predictions, radar, sonar, telecommunications, electronic warfare, noise filtering, games of chance, etc. Statistical methods are generally considered to be the most appropriate tools to exploit probabilistic concepts for real world applications. These methods can be found in an increasing number of modern world applications, such as information theory, industry, agriculture, science, politics, etc.

The distribution of an observed random variable is very often unknown in practical situations. Several probability density function estimates are proposed in the literature. However, most of them are not easy to implement and are time-consuming. On the other hand, many recent algorithms do not use variable distributions directly, but other equivalent quantities, which can be easily estimated, such as statistical moments and cumulants. They are calculated using the characteristic functions defined in Chapter 3 (see equation [3.5]).

4.1.1. *Moments*

Let us consider a continuous random variable X and its probability density function $f_X(x)$. Their k^{th} order and the kth order centered moments are thus defined as:

$$m_k = E\left[X^k\right] = \int_{\mathbf{R}} x^k f_X(x)dx = (-j)^k \left.\frac{\partial^k \Phi_X(u)}{\partial u^k}\right|_{u=0} \qquad [4.1]$$

$$\mu_k = E\left[\left(X - E[X]\right)^k\right] = \int_{\mathbf{R}} \left(x - m_1\right)^k f_X(x)\,dx \qquad [4.2]$$

The kth order absolute moments of X have the following form:

$$E\left[|X|^k\right] = \int_{\mathbf{R}} |x|^k f_X(x)dx \qquad [4.3]$$

Finally, the kth order generalized moments are expressed as:

$$\mu_k = E\left[\left(X - a\right)^k\right] = \int_{\mathbf{R}} \left(x - a\right)^k f_X(x)\,dx \qquad [4.4]$$

Note that if X has an even pdf then $m_{(2k+1)} = 0$.

4.1.2. *Cumulants*

In many recent applications, high order statistics are used in the form of cumulants instead of moments. Their algebraic properties are very interesting and allow the calculations to be simplified and algorithms' performance to be improved.

For a given random variable X, the rth order cumulant can be calculated using the moments up to the rth order. According to the Leonov-Shiryayev relationship:

$$\begin{aligned}\mathrm{Cum}_r(X) &= \mathrm{Cum}(X,\cdots,X) = (-j)^r \left.\frac{\partial^r \Psi_X(u)}{\partial u^r}\right|_{u=0} \\ &= \sum(-1)^{k-1}(k-1)!\,E\left[X^{\vartheta_1}\right]\cdots E\left[X^{\vartheta_p}\right]\end{aligned} \qquad [4.5]$$

where the sum is taken on all the permutations of ϑ_i, $1 = i = p = r$, ϑ_i form a partition of the set $\{1, ..., r\}$ and K is the number of partition elements.

High order statistics with orders higher than 4 are seldom used. Using equation [4.7] it can be shown that the 3rd and 4th order cumulants are expressed as follows:

$$\mathrm{Cum}_3(X) = E\left[X^3\right] - 3E[X]E\left[X^2\right] + 2E^3[X] \qquad [4.6]$$

$$\begin{aligned}\mathrm{Cum}_4(X) = {} & E\left[X^4\right] - 4E[X]E\left[X^3\right] - 3E^2\left[X^2\right] \\ & + 12E^2[X]E\left[X^2\right] - 6E^4[X]\end{aligned} \qquad [4.7]$$

For a zero-mean random variable X:

$$\mathrm{Cum}_4(X) = E\left[X^4\right] - 3E^2\left[X^2\right] \qquad [4.8]$$

4.1.3. *Cumulant properties*

For any random variable X, the following statements are valid:

1. The 2nd order cumulant is equal to the variance and measures the power of X.

2. The 3rd order cumulant is related to the symmetry of the distribution of X. The skewness is a measure of the asymmetry of the data around the sample mean:

$$S_k = \frac{\mu_3}{\sigma^3} \qquad [4.9]$$

3. The 4th order cumulant is related to the flattening of the distribution of X with respect to a Gaussian pdf. The kurtosis is a measure of how outlier-prone a distribution is:

$$\kappa(X) = \frac{\mu_4}{\sigma^4} - 3 \qquad [4.10]$$

4. The kurtosis and the 4th order cumulant are invariant to any linear transform. Thus, $\forall$ a and $b \in \mathbf{R}$, it can be shown that:

$$Cum_4(aX + b) = a^4 Cum_4(X) \qquad [4.11]$$

$$\kappa(aX + b) = \kappa(X) \qquad [4.12]$$

5. It can be shown that the cumulants whose order exceeds 2 are zero for any Gaussian random variable.

4.1.4. *Chi-square (Chi2) tests*

The Chi2 or χ^2 test was introduced by Karl Pearson in 1900. It is based on an asymptotical property of the multinomial distribution, which yielded the χ^2 distance. χ^2 tests are non-parametric, i.e. hypothesis H_i is not linked to parameters of the pdfs, but to the distributions themselves.

Let us consider the following H_0 hypothesis: "X is distributed according L_0". The non-parametric hypothesis H_0 is called *simple* if it fully specifies the L_0 pdf, otherwise it is called *composed*. For instance, if L_0 is a Gaussian pdf $N(m,\sigma)$, with known m and σ, the corresponding H_0 hypothesis is simple.

In the case of a composed hypothesis, the missing parameters have to be estimated to reduce it to a simple hypothesis. The observations are divided into several classes or bins C_1, C_2, … Let us denote by n_i the number of observations belonging to C_i and by p_i the probability that an observation belongs to C_i, under hypothesis H_0. The theoretical number of observations belonging to C_i is then given by np_i, where n is the total number of observations.

Usually, we only consider classes with $np_i = 5$. Some bins may be merged in order to meet this condition. Let us suppose r remaining classes and k parameters to be estimated under the hypothesis H_0. The quantity defined below thus measures the "distance" between the observed and the theoretical distribution:

$$\chi_0^2 = \sum_{i=1}^{r} \frac{\left(n_i - np_i\right)^2}{np_i} \qquad [4.13]$$

It can be shown that for a large number of samples, χ_0^2 is a χ^2 random variable with r-k-1 freedom degrees. Using a χ^2 table, the observed value can be considered acceptable or not, i.e. hypothesis H_0 is satisfied or not.

4.1.5. *Normality test using the Henry line*

The main idea of this test is given in Figure 4.1. In fact, if X is a Gaussian random variable with a mean value m and a standard deviation σ then its reduced and centered correspondent T obeys N(0,1). The linear relationship between T and X is known as the Henry straight line:

$$T = \frac{1}{\sigma} X - \frac{m}{\sigma} \qquad [4.14]$$

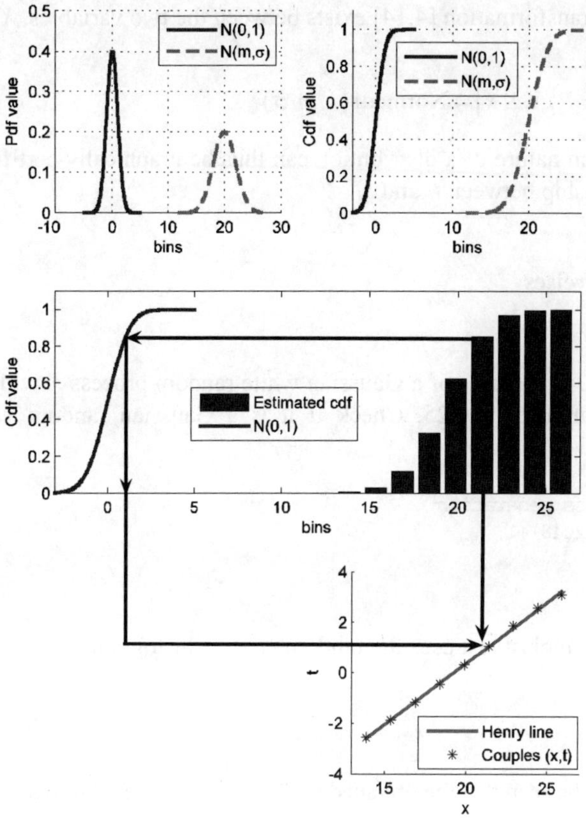

Figure 4.1. *Principle of the normality test using the Henry line*

The cumulative density function of T can be expressed as follows using the MATLAB notations:

$$F_T(t) = P[T \leq t] = \text{Normcdf}(t,0,1) \tag{4.15}$$

Let us now consider bins x_k and values a_k of the cumulative density function of X, estimated from the observed data:

$$a_k = \hat{F}_X(x_k) \tag{4.16}$$

These values are then used as values of a normalized Gaussian cdf.

$$a_k = F_T(t_k) \Rightarrow t_k = F_T^{-1}(a_k) \tag{4.17}$$

The affine transformation [4.14] exists between the two variables, X and T, if and only if:

$$F_X(x) = P[X \leq x] = \text{Normcdf}(x, m, \sigma) \qquad [4.18]$$

The Gaussian nature of a distribution can thus be graphically certified if there is a linear relationship between t_k and x_k.

4.2. Solved exercises

EXERCISE 4.1.

Consider 2,048 samples of a Gaussian white random process P with a unit mean value and a variance of 0.25. Check if it is a Gaussian random variable using different tests.

```
% Random process generation
P=0.5*randn(1,2048)+1;
mp = mean(P)
vp = var(P)
```

1st test: the simplest test uses the random process histogram.

```
bins = 20;
hist(P,bins);
```

It can easily be seen that the obtained histogram is very similar to a Gaussian pdf.

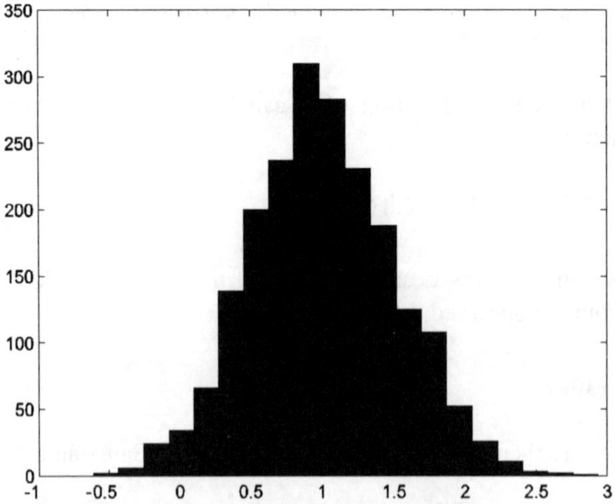

Figure 4.2. *Histogram of a Gaussian random process*

2nd test: Let us calculate 3rd and 4th order cumulants of the random process P using equations [4.6] and [4.7].

```
Cum3 = mean(P.^3) - 3*mean(P)*mean(P.^2)+2* mean(P)^3
Cum4 = mean(P.^4) - 4*mean(P)*mean(P.^3) - 3*mean(P.^2)^2 +
12*mean(P)^2*mean(P.^2)- 6*mean(P)^4

Cum3 =

    0.0234

Cum4 =

    0.0080
```

The obtained values for the two cumulants are close to zero, so the Gaussianity hypothesis on P is reinforced.

3rd test: Henry line.

```
nb_bins = 20;
[hv,x]=hist(P,nb_bins);
F=cumsum(hv)/sum(hv); F(end)=0.999 ;% empirical cdf
y=norminv(F,0,1); % percentile of a N(0,1) distributed variable
% Graphical matching
p=polyfit(x,y,1); % linear matching of x and y
s=1/p(1) % standard deviation estimation = inverse of the slope
z=polyval(p,x); % values on the Henry line
figure;
hold on;
plot(x,z,'-r');
plot(x,y,'*');
legend('Henry line','Couples (x,t)');
xlabel('x');
ylabel('t');
```

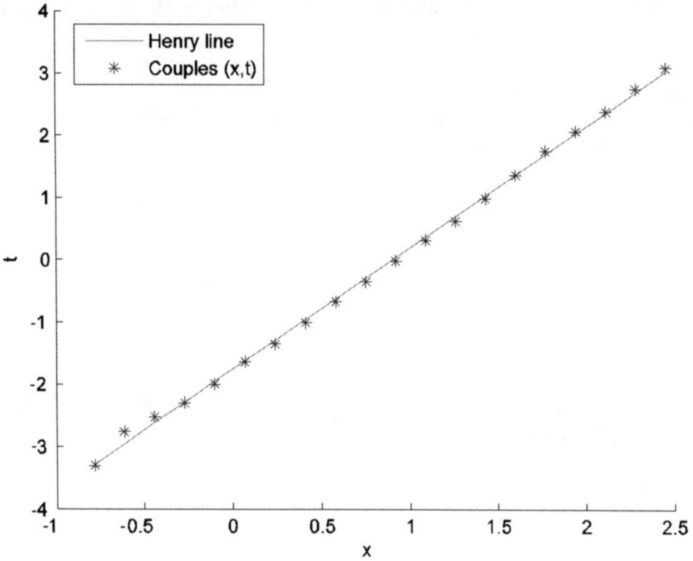

Figure 4.3. *Henry line test*

The linear relationship between x and t is obvious and thus the Gaussianity of the random process is certified.

4th test: Chi2 test.

```
nb_bins = 20;
[hv,x]=hist(P,nb_bins);
Lc = x(2)-x(1);
My = sum(x .* hv) / sum(hv)
My2 = sum(x.^2 .* hv) / sum(hv);
vare =My2-My^2 % Variance estimate
sde =sqrt(vare) % Standard deviation estimate
```

The estimation of the theoretical number of samples in each histogram bin is the first step of the Chi2 test. For any Gaussian random variable X with the mean value m and standard deviation σ, the probability of $X \in [a, b]$ is given by:

$$P(a \leq X \leq b) = \text{Normcdf}\left(\frac{b-m}{\sigma}\right) - \text{Normcdf}\left(\frac{a-m}{\sigma}\right)$$

The theoretical number of samples in each histogram bin can thus be calculated in this case using the following code lines:

```
eft = (normcdf((x+0.5*Lc-My)/sde)-normcdf((x-0.5*Lc-My)/sde));
eft = eft*sum(hv);
```

Note that this last calculation depends on the distribution considered.

```
while sum(eft < 10) > 0 % Verify if for any bin the number of
                              % samples >=5
                           % otherwise, the bins are merged
    [minef,indminef] = min(eft);Leff = length(eft);
    if indminef ==1 % first bin problem
        eft(2) = eft(2) +eft(1); % bins fusion
        hv(2) = hv(2) + hv(1);
        eft(1) = [];hv(1) = [];
    elseif indminef ==  Leff % last bin problem
        eft(Leff-1) = eft(Leff-1) +eft(Leff); % bins fusion
        hv(Leff-1) = hv(Leff-1) + hv(Leff);
        eft(Leff) = []; hv(Leff) = [];
    else
        Minval = eft(indminef+1) < eft(indminef-1);
        eft(indminef-1+2*Minval)=eft(indminef-1+2*Minval)+eft(indminef);
        hv(indminef-1+2*Minval)=hv(indminef-1+2*Minval)+hv(indminef);
        eft(indminef) = [];hv(indminef) = [];
    end
end
Leff = length(eft)  % Number of remaining bins
% Calculation of the quantity used in the Chi2 test

chi2 = sum(((eft-hv).^2)./eft)

% Chi2 test
k=2; % two estimated parameters(mean value and standard deviation)
deglib = Leff-k-1 % freedom degrees
NivConf = 0.95; % confidence level
if  chi2 < chi2inv(NivConf,deglib)
    disp('The distribution is Gaussian')
else
    disp('The distribution is not Gaussian')
end
```

EXERCISE 4.2.

A factory supplies series of 1,000 pieces of a given type of product. Some randomly chosen series, are checked. The table indicated below provides the number of checked series n_k having k faulty pieces:

k	0	1	2	3	4	5	6	7	8	9	10	11	12
n_k	7	12	23	27	41	36	25	20	9	5	2	1	2

Use the Chi2 test for validating or rejecting the hypothesis that the number of faulty pieces per series is Poisson distributed.

```
k = 0:12;
hv = [7  12  23  27  41  36  25  20  9  5  2  1  2]
L =length(hv);
My = sum(k .* hv)/sum(hv)
My2 = sum(k.^2 .*hv)/sum(hv);
vare =My2-My^2;
```

For any Poisson distributed random variable X with the parameter λ:

$$E[X]=\lambda$$

```
Lam = My;
eft =poisspdf(k,Lam)* sum(hv);
while sum(eft < 5) > 0
    [minef,indminef] = min(eft);
    Leff = length(eft);
    if indminef ==1
        eft(2) = eft(2)+eft(1);
        hv(2) = hv(2) + hv(1);
        eft(1) = [];
        hv(1) = [];
    elseif indminef ==  Leff
        eft(Leff-1) = eft(Leff-1) +eft(Leff);
        hv(Leff-1) = hv(Leff-1) + hv(Leff);
        eft(Leff) = []; hv(Leff) = [];
    else
        Minval = eft(indminef+1) < eft(indminef-1);
        eft(indminef-1+2*Minval)=eft(indminef-1+2*Minval)+eft(indminef);
          hv(indminef-1+2*Minval)=hv(indminef-1+2*Minval)+hv(indminef);
        eft(indminef) = [];
        hv(indminef) = [];
    end
end
Leff = length(eft)
chi2 = sum (((eft - hv).^2)./eft)
k=1; % only one estimated parameter (the mean value)
deglib = Leff - k - 1 % freedom degrees
NivConf = 0.95; % confidence level
if chi2 < chi2inv(NivConf,deglib)
    disp('The random variable is Poisson distributed')
else
    disp('The random variable isn't Poisson distributed')
end
```

EXERCISE 4.3.

The source separation is a recent problem in signal processing. It consists of retrieving p mutually independent unknown sources of which q mixed signals are used.

The p sources are denoted by a vector and are grouped in a vector $S(t) = \{s_i(t)\}$, with $1 \leq i \leq p$. Let $Y(t)$ be the observation vector with the components $y_i(t)$, $1 \leq i \leq q$, which are unknown functions of the p sources.

This exercise is aimed to stress the importance of high order statistics for performing a blind or semi-blind source separation.

Let us consider a vector of unknown sources S and an observation vector[1] Y (mixtures of source signals).

```
% Blind sources separation using high order statistics
% Simplified model with p = q = 3 and echoless transmission channel

Vect = wavread('melange1.wav');
Y(:,1) = Vect;
Ns = length(Vect);
Vect = wavread('melange2.wav');
Ns = min(length(Vect),Ns);
Y(1:Ns,2) = Vect(1:Ns);

Vect = wavread('melange3.wav');
Ns = min(length(Vect),Ns);
Y(1:Ns,3) = Vect(1:Ns);
Ys = Y(1:Ns,:);clear Y;clear vect;
Ys = Ys';

figure
hist(Ys(1,:)) % Histogram of a mixture of signals
```

Note that it is not possible to separate the signals in the mixture using its histogram.

1 The signals used in this exercise can be downloaded from the website: http://ali.mansour.free.fr/EXP.

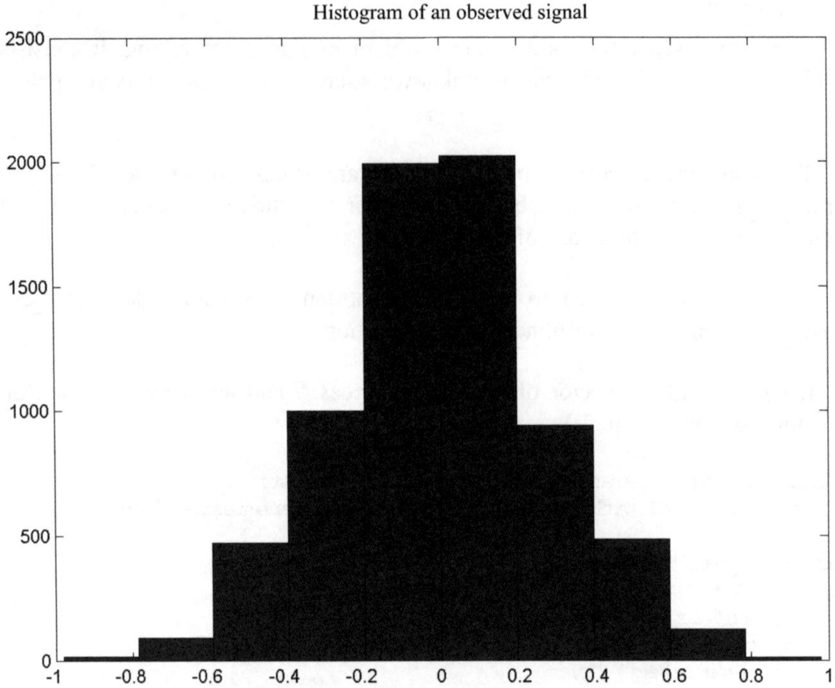

Figure 4.4. *Histogram of a mixture of signals*

```
figure;
plot(Ys(1,:)); % temporal variation of a mixture of signals
```

The figure below shows that the observed signal cannot be decomposed into its components using a temporal approach.

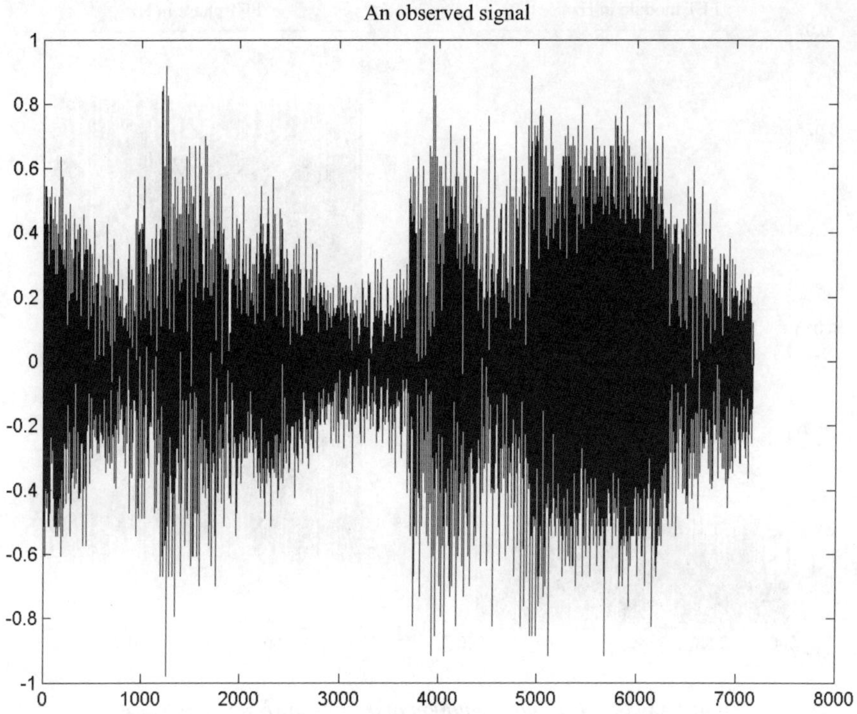

Figure 4.5. *Unknown mixture of temporal signals*

```
% Plotting the frequency signal representation
S = Ys(1,1:4096);
l=length(S); t=0:l-1; ts=t/fe; FFT = fft(S);
FFTshift = fftshift(FFT)/l;
faxe=t/l-0.5; %recenter the curve
figure ; subplot(121),plot(fe*faxe,abs(FFTshift));
title('Amplitude spectrum [Hz]');
subplot(122); plot(fe*faxe,angle(FFTshift));
title('Phase spectrum [Hz]'); clear FFT FFTshift faxe t ts;
```

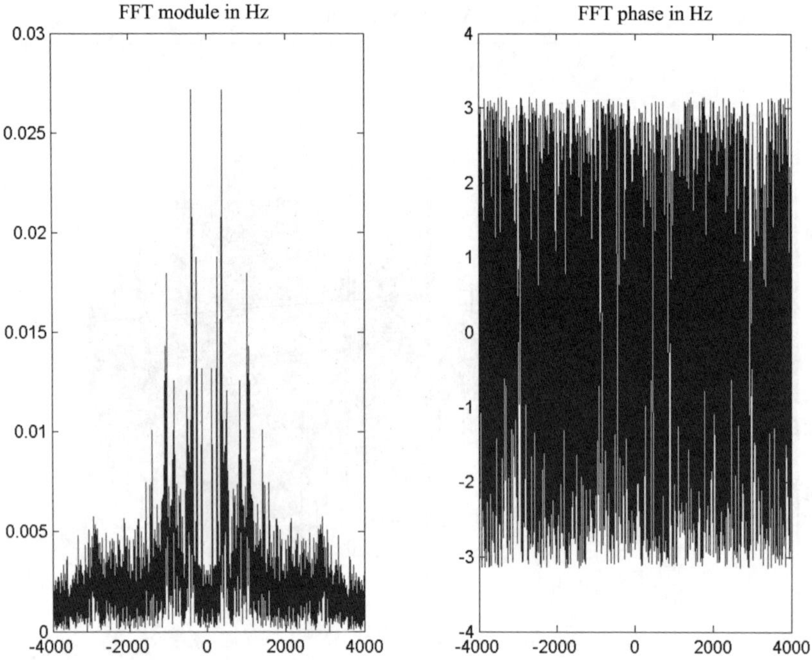

Figure 4.6. *Spectral representation of the unknown signal mixture*

It is obvious that a standard filter is not appropriate for our problem. The source separation thus cannot be performed by simply filtering the signal mixture. More powerful tools such as the time-frequency representations[2] can also be considered.

```
tfrpwv(S'); % Pseudo Wigner-Ville representation for
            % non-stationary signals
clear S;
```

However, Figure 4.7 shows that even the time-frequency representation is not able to separate the sources properly.

Let us now try high order cumulant based approaches. In fact, since the sources are independent, high order cross-cumulants are equal to zero. Using this property, Cardoso and Soulamiac proposed a source separation algorithm based on the simultaneous diagonalization of several eigenmatrices of the 4th order cumulant tensor.

---

2 The MATLAB codes corresponding to some time-frequency representations can be downloaded from the website http://tftb.nongnu.org/.

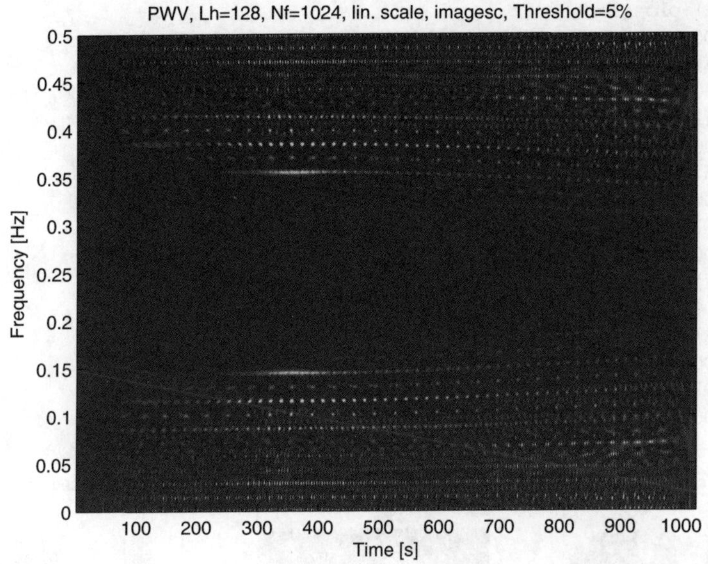

Figure 4.7. *Time-frequency representation of the unknown signal mixture*

Diagonalization is performed using the algorithm JADE[3] "*Joint Approximate Diagonalization of Eigenmatrices*". The new algorithm is able to separate several signals of different natures (acoustical or telecommunication signals).

```
% Source separation using the JADE algorithm
[A,Xs]=jade(Ys,3);
[Vect,fs] = wavread('source1.wav'); % fs is the sampling frequency
S(:,1) = Vect; Ns = length(Vect);
Vect = wavread('source2.wav');
Ns = min(length(Vect),Ns);
S(1:Ns,2) = Vect(1:Ns);
Vect = wavread('source3.wav');
S(1:Ns,3) = Vect(1:Ns);
Ns = min(length(Vect),Ns);
Ss = S(1:Ns,:); clear S;
Ss = Ss'; % the rows of the Ss matrix are the separated signals
```

The separation quality here is evaluated graphically or acoustically, although much more complex and objective criteria can also be used.

```
figure; hold on; subplot(331); plot(Ss(1,:));
```

3 Jade can be downloaded on the website http://www.tsi.enst.fr/~cardoso/guidesepsou.html.

```
axis([0 7100 -1.5 1.5]); title('1st Source');
subplot(332);plot(Ss(2,:));
axis([0 7100 -1.5 1.5]); title('2nd Source');
subplot(333);plot(Ss(3,:));
axis([0 7100 -1.5 1.5]); title('3rd Source');
subplot(334);plot(Ys(1,:));
axis([0 7100 -1.5 1.5]); title('1st observed signal');
subplot(335);plot(Ys(2,:));
axis([0 7100 -1.5 1.5]); title('2nd observed signal');
subplot(336);plot(Ys(3,:));
axis([0 7100 -1.5 1.5]); title('3rd observed signal');
subplot(337);plot(Xs(1,:));
axis([0 7100 -4 4]); title('1st estimated signal');
subplot(338);plot(Xs(2,:));
axis([0 7100 -4 4]); title('2nd estimated signal');
subplot(339);plot(Xs(3,:));
axis([0 7100 -4 4]); title('3rd estimated signal');
```

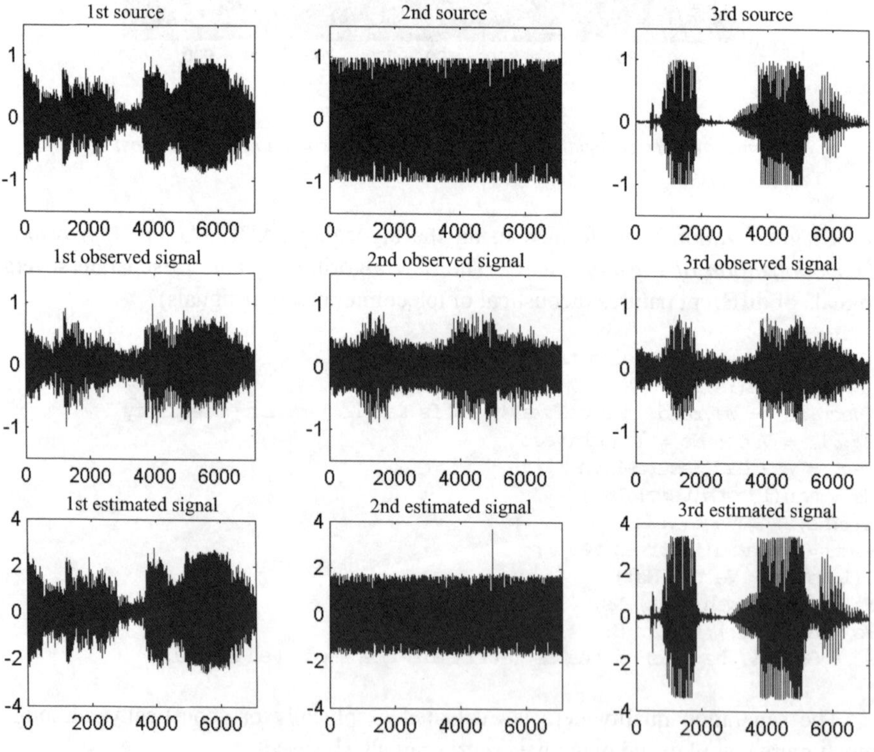

Figure 4.8. *Extracted signals after deconvolution*

The signal separation can also be tested acoustically.

```
disp('1st Source'); pause; sound(Ss(1,:),fe);
disp('2nd Source'); pause; sound(Ss(2,:),fe);
disp('3rd Source'); pause; sound(Ss(3,:),fe);

disp('1st Mixed signal'); pause; sound(Ys(1,:),fe);
disp('2nd Mixed signal'); pause; sound(Ys(2,:),fe);
disp('3rd Mixed signal'); pause; sound(Ys(3,:),fe);

disp('1st Recovered signal'); pause; sound(Xs(1,:),fe);
disp('2nd Recovered signal'); pause; sound(Xs(2,:),fe);
disp('3rd Recovered signal'); pause; sound(Xs(3,:),fe);
```

4.3. Exercises

EXERCISE 4.4.

Suppose that the probability of measuring k points of a random process X within an interval T, is $P_k = c\, a^k$, for $k \geq 0$.

1. Determine the value of the constant c as a function of parameter a, which belongs to an interval to be specified.

2. Demonstrate that $E[K] = a/(1-a)$.

3. Calculate the probability of having at least k points within the interval T.

4. The following results are issued from process observations:

k	0	1	2	3	4	5	6	7	8	≥ 9
Number of samples	29	22	15	12	9	7	6	3	2	4

a. Give an estimation of a.

b. Verify if the measured values obey the distribution defined above.

c. Explain why these values are trustworthy or not.

EXERCISE 4.5.

In order to control the homogenity of a factory production, 1,000 samples of a manufactured mechanical part are randomly selected. The size X measured for each of them is recorded in the table below:

X [cm]	3.3	3.4	3.5	3.6	3.7	3.8	3.9	4	4.1
Sample number	2	1	2	4	10	22	60	148	235

X [cm]	4.2	4.3	4.4	4.5	4.6	4.7	4.8
Sample number	207	160	92	41	11	2	3

The problem is to know if the distribution of X can be considered as Gaussian.

1. Calculate the empirical mean value, variance and standard deviation.
2. Plot the histogram and conclude.
3. For a Gaussian variable, what is the probability of having a value up to 2σ from its mean value?
4. Give a probability for $4 \leq X \leq 4.2$ and conclude.
5. Validate your conclusion using one or two statistical tests.

EXERCISE 4.6.

The following values are recorded with a receiver:

5.97	5.5	5.45	5.6	4.8	3.4	6.3	6.73	3.71	4.91
7.54	2.05	5.48	5.39	1.01	4.47	5.87	6.31	9.37	9.11
2.98	3.93	3.66	8.98	6.73	8.75	3.6	2.38	4.03	4.96

1. Sort these values in increasing order.
2. Group these values using bins of length 1: from 1 to 2, from 2 to 3, etc.
3. Determine the number of samples corresponding to each bin and calculate the cumulated number of samples, the frequency[4] of occurrence and the cumulated frequency of occurrence.
4. Plot the histogram and find out the median value[5].
5. Find out the mean value and standard deviation using a table of calculations.
6. Consider the following hypothesis H_0: "the received signal $s(t)=m+n(t)$, where m is a constant value and $n(t)$ is a Gaussian zero-mean noise".
    a. Estimate the value of m.
    b. Test the hypothesis H_0 using two different methods.

EXERCISE 4.7.

According to the central limit theorem the output signal of a MA FIR filter tends to be Gaussian.

Generate a white random signal, uniformly distributed over [-2, 2], using the function rand. Let Y be the signal obtained at the output of an nth order FIR filter with the transfer function $H(z)$. For different values of n:

---

4 The frequency of occurrence associated with a bin is the ratio between its sample number and the total number of samples.

5 The median value divides the recorded samples, previously sorted in increasing order, into two equal parts.

1. Compare signals X and Y.
2. Plot the histograms for the two signals.
3. Plot, on the same figure, the Henry line for X and Y.
4. Calculate the 3rd and 4th order cumulants of X and Y.

Repeat the same exercise for a RII filter and conclude.

EXERCISE 4.8.

The cumulants defined in this chapter are often also called auto-cumulants. It has already been shown (see exercise 4.1) that they are useful to certify the Gaussian character of a random process.

The cross-cumulants may be used for evaluating the statistical independence of a signal set. For two zero-mean real signals X and Y, three 4th order cross-cumulants can be defined:

$$\mathrm{Cum}_{1,3}(X,Y) = E\left[XY^3\right] - 3E\left[X^2\right]E[XY]$$

$$\mathrm{Cum}_{3,1}(X,Y) = E\left[X^3Y\right] - 3E\left[Y^2\right]E[XY]$$

$$\mathrm{Cum}_{2,2}(X,Y) = E\left[X^2Y^2\right] - E\left[X^2\right]E\left[Y^2\right] - 2E^2[XY]$$

If X and Y are independent then their cross-cumulants are equal to zero.

The objective of this exercise is to exploit the cross-cumulant properties in order to conceive a simple blind source separation algorithm.

Let $S(t)$ be the vector of the two source signals[6] $s_1(t)$ and $s_2(t)$ and $Y(t)$ the mixture vector obtained from $S(t)$. If $H(t)$ denotes the mixture matrix[7] then:

$$Y(t) = \begin{pmatrix} y_1(t) \\ y_2(t) \end{pmatrix} = H(t)S(t) = H(t)\begin{pmatrix} s_1(t) \\ s_2(t) \end{pmatrix}$$

6 Consider the same signals as those used in exercise 4.3.

7 The mixture matrix takes into account the effect of the transmission channel. In the case of an ideal transmission channel, without echo, delay or frequency fading, this effect can be represented by a simple matrix product.

where:

$$H(t) = H = \begin{pmatrix} 1 & a \\ 0 & 1 \end{pmatrix}$$

and a is a given real parameter.

1. For several values of $a \in [-1, 1]$ calculate the following cost function:

$$f(a) = \text{Cum}_{1,3}^2(y_1(t), y_2(t)) + \text{Cum}_{2,2}^2(y_1(t), y_2(t)) + \text{Cum}_{3,1}^2(y_1(t), y_2(t))$$

2. Find the value a_{min} that minimizes the cost function above. What do you notice?

3. Write a MATLAB algorithm for separating the two sources $s_1(t)$ and $s_2(t)$.

4. Repeat the same exercise for the following mixture matrix:

$$H(t) = H = \begin{pmatrix} 1-a & a \\ a & 1-a \end{pmatrix}$$

5. Consider the cases where the two mixed signals are generated using:
   a. the MATLAB function `rand`,
   b. the MATLAB function `randn`.

What do you notice and why?

Chapter 5

Discrete Fourier Transform of Discrete-Time Signals

5.1. Theoretical background

The discrete Fourier transform (DFT) is a basic tool for digital signal processing. From a mathematical point of view the DFT transforms a digital signal from the original time domain into another discrete series, in the transformed frequency domain. This enables the analysis of the discrete-time signal in both the original and (especially) the transformed domains.

The frequency analyses of a digital filter and of a discrete-time signal are very similar problems. It will be seen in the next chapter that, for a digital filter, it consists of evaluating the transfer function $H(z)$ on the unit circle $\left(z = e^{j2\pi\nu T}\right)$. This is the same for the digital signal analysis using the z-transform, which is closely related to the DFT.

As the calculations are performed on a digital computer there are 3 types of errors related to the transition from the Fourier transform of the continuous-time signal to the DFT of the discrete-time signal:

– errors due to time sampling (transition from $x(t)$ to $x_s(t)$),

– errors due to time truncation (transition from $x_s(t)$ to $x_s(t)\Pi_T(t)$),

– errors due to frequency sampling.

5.1.1. *Discrete Fourier transform of 1D digital signals*

The DFT of finite time 1D digital signals, denoted by DFT_{1D}, is defined by:

$$X[k] = TFD_{1D}\{x[n]\} = \sum_{n=0}^{N-1} x[n] W_N^{kn}$$
$$x[n] = TFD_{1D}^{-1}\{X[k]\} = \frac{1}{N}\sum_{k=0}^{N-1} X[k] W_N^{-kn} \quad [5.1]$$

where: $W_N = e^{-j2\pi/N}$ and $n, k = 0..N-1$.

In the above equations, the index of the vectors $x[n]$ and $X[k]$ should begin with 1 instead of 0, according to MATLAB notations, so that $n \rightarrow (n+1)$ and $k \rightarrow (k+1)$.

Just as in the case of the continuous Fourier transform the following properties hold for the DFT:

– linearity

– if $x[n] \leftrightarrow X(k)$ and $y[n] \leftrightarrow Y(k)$ then $ax[n] + by[n] \leftrightarrow aX(k) + bY(k)$

– time-delay

– if $x[n] \leftrightarrow X(k)$ then $x[n-m] \leftrightarrow X(k) W_N^{mn}$

– phase shifting

– if $x[n] \leftrightarrow X(k)$ then $x[n] W_N^{-mk} \leftrightarrow X(k-m)$

– time inversion (duality)

– if $x[n] \leftrightarrow X(k)$ then $(1/N) X[n] \leftrightarrow x(-k)$

– cyclic convolution of two N-periodical digital signals $x(n)$ and $y(n)$:

$$z[n] = \sum_{m=0}^{N-1} x[m]\, y[n-m]$$

– multiplication

– if $x[n] \leftrightarrow X(k)$ and $y[n] \leftrightarrow Y(k)$ then:

$$x[n]y[n] \leftrightarrow \frac{1}{N}\sum_{m=0}^{N-1} X(m)Y(k-m) = \frac{1}{N} X(k) * Y(k)$$

– complex conjugate

– if $x[n] \leftrightarrow X(k)$ then $x^*[n] \leftrightarrow X^*(k)$

– Parseval theorem

– if $x[n] \leftrightarrow X(k)$ then $\sum_{k=0}^{N-1} |x(n)|^2 = \frac{1}{N}\sum_{n=0}^{N-1} |X(k)|^2$

Discrete cosine transform for 1D digital signals

This transformation is defined by:

$$X[k] = \mathrm{DCT}_{1\mathrm{D}}\{x[n]\} = \sum_{n=0}^{N-1} 2x[n]\cos\left[\frac{\pi}{2N}k(2n+1)\right]$$
$$x[n] = \mathrm{DCT}_{1\mathrm{D}}^{-1}\{X[k]\} = \frac{1}{N}\sum_{k=0}^{N-1} w[k]X[k]\cos\left[\frac{\pi}{2N}k(2n+1)\right] \qquad [5.2]$$

where: $w[k] = \begin{cases} 1/2, & \text{if } k=0 \\ 1, & \text{if } 1 \le k \le N-1 \end{cases}$ and $n, k = 0..N-1$.

5.1.2. *DFT of 2D digital signals*

The $\mathrm{DFT}_{2\mathrm{D}}$ is used for the analysis of finite (rectangular) support 2D digital signals. If $W_{N_i} = e^{-j2\pi/N_i}, i \in \{1,2\}$ and $n_1 = 0..N_1 - 1$, $n_2 = 0..N_2 - 1$, then the direct and the inverse $\mathrm{DFT}_{2\mathrm{D}}$ are defined as follows:

$$X[k_1,k_2] = \sum_{n_1=0}^{N_1-1}\sum_{n_2=0}^{N_2-1} x[n_1,n_2] W_{N_1}^{n_1 k_1} \; W_{N_2}^{n_2 k_2}$$
$$x[n_1,n_2] = \frac{1}{N_1 N_2}\sum_{k_1=0}^{N_1-1}\sum_{k_2=0}^{N_2-1} X[k_1,k_2] W_{N_1}^{-n_1 k_1} \; W_{N_2}^{-n_2 k_2} \qquad [5.3]$$

Discrete cosine transform of 2D digital signals

This transformation is defined by:

$$X[k_1,k_2] = \sum_{n_1=0}^{N_1-1}\sum_{n_2=0}^{N_2-1} 4x[n_1,n_2]\cos\left[\frac{\pi}{2N_1}k_1(2n_1+1)\right]\cos\left[\frac{\pi}{2N_2}k_2(2n_2+1)\right]$$
$$x[n_1,n_2] = \frac{1}{N_1 N_2}\sum_{k_1=0}^{N_1-1}\sum_{k_2 0}^{N_2-1} w_1[k_1]w_2[k_2]X[k_1,k_2] \qquad [5.4]$$
$$\cos\left[\frac{\pi k_1}{2N_1}(2n_1+1)\right]\cos\left[\frac{\pi k_2}{2N_2}(2n_2+1)\right]$$

where: $w_i[k_i] = \begin{cases} 1/2, & \text{if } k_i = 0 \\ 1, & \text{if } k_i = 1..N_i - 1, \text{ with } i = 1,2 \end{cases}$

5.1.3. *Z-transform of 1D digital signals*

The direct and inverse unilateral z-transform (ZT) of 1D digital signals has the following expressions:

$$X(z) = TZ_{1D}\{x[n]\} = \sum_{n=0}^{N-1} x[n] z^{-n}$$
$$x[n] = \frac{1}{2\pi j} \int_C X(z) z^{n-1} dz \quad \text{where } C \in \text{RoC (region of convergence)} \qquad [5.5]$$

The ZT_{1D} allows the DFT_{1D} to be calculated if it is evaluated using N points on the unit circle. If it is evaluated on a spiral of equation $z = a[\exp(j2\pi/N)]^{-k}$, with $k = 0..N\text{-}1$ and $|a| < 1$, it defines the discrete *chirp* ZT.

5.1.4. *Z-transform of 2D digital signals*

The unilateral ZT of finite 2D digital signals have the following form:

$$X(z_1, z_2) = TZ_{2D}\{x[n_1, n_2]\} = \sum_{n_1=0}^{N_1-1} \sum_{n_2=0}^{N_2-1} x[n_1, n_2] z_1^{-n_1} z_2^{-n_2}$$
$$x[n_1, n_2] = \frac{1}{(2\pi j)^2} \oint_{C_1} \oint_{C_2} X(z_1, z_2) z_1^{n_1-1} z_2^{n_2-1} \, \mathrm{d}z_1 \, \mathrm{d}z_2 \qquad [5.6]$$

5.1.5. *Methods and algorithms for the DFT calculation*

Method of direct calculation of the DFT_{2D}

This method makes use of the fast DFT_{1D} algorithms and can be applied to a 2D digital signal if the DFT_{2D} can be expressed in one of the following two forms:

$$X_{k_1,k_2} = \sum_{n_1=0}^{N_1-1} W_{N_1}^{n_1 k_1} \left[\sum_{n_2=0}^{N_2-1} W_{N_2}^{n_2 k_2} x_{n_1,n_2} \right]$$
$$X_{k_1,k_2} = \sum_{n_2=0}^{N_2-1} W_{N_2}^{n_2 k_2} \left[\sum_{n_1=0}^{N_1-1} W_{N_1}^{n_1 k_1} x_{n_1,n_2} \right] \qquad [5.7]$$

It thus becomes possible to calculate the DFT_{2D} using the fast DFT_{1D}, firstly on each column (or row) and then on each row (or column).

Cooley-Tukey algorithm for DFT_{1D} calculation

This algorithm converts a vector of length $N = 2^m = N_1 N_2$ into a matrix of size $N_1 \times N_2$, using an index transformation. The most commonly used techniques are given below:

– decimation-in-time, for $N_1 = 2$ and $N_2 = 2^{m-1}$, so that:

$$\begin{aligned} n &= N_2 n_1 + n_2,\ n_{1,2} = 0..N_{1,2} - 1 \text{ and } n = 0..N-1 \\ k &= k_1 + N_1 k_2,\ k_{1,2} = 0..N_{1,2} - 1 \text{ and } k = 0..N-1 \end{aligned} \qquad [5.8]$$

– decimation-in-frequency, for $N_1 = 2^{m-1}$ and $N_2 = 2^m$, so that:

$$\begin{aligned} n &= n_1 + N_1 n_2,\ n_{1,2} = 0..N_{1,2} - 1 \text{ and } n = 0..N-1 \\ k &= N_2 k_1 + k_2,\ k_{1,2} = 0..N_{1,2} - 1 \text{ and } k = 0..N-1 \end{aligned} \qquad [5.9]$$

Consequently, for the calculation of $\text{DFT}_{1\text{D}}$ by decimation-in-frequency, the following equation is obtained:

$$X_{k_1,k_2} = \sum_{n_1=0}^{N_1-1} \beta^{n_1 k_1} \left[W_N^{n_1 k_2} \sum_{n_2=0}^{N_2-1} \gamma^{n_2 k_2} x_{n_1,n_2} \right] \qquad [5.10]$$

where $\beta = W_N^{N_2}$ and $\gamma = W_N^{N_1}$.

Equation [5.10] is equivalent to the following sequence of operations: $\text{DFT}_{1\text{D}}$ calculated in N_2 points on each column, multiplication of each element by $W_N^{n_1 k_2}$ and $\text{DFT}_{1\text{D}}$ calculated in N_1 points on each row.

Cooley-Tukey algorithm for the DFT_{2D} calculation

This method converts a 2D sequence of size $N \times N$ (with $N = 2^m = N'N''$), using the index transform:

$$x_{n_1,n_2} \to x_{n_1'+N'n_1'',n_2'+N'n_2''} = x_{n_1',n_1'',n_2',n_2''} \qquad [5.11]$$

into the 2D sequence:

$$X_{k_1,k_2} \to X_{N''k_1'+k_1'',N''k_2'+k_2''} = X_{k_1',k_1'',k_2',k_2''} \qquad [5.12]$$

according to the following equation:

$$X_{k_1',k_1'',k_2',k_2''} = \sum_{n_1'=0}^{N'-1}\sum_{n_2'=0}^{N'-1} \beta^{n_1'k_1'}\beta^{n_2'k_2'}\left[W^{n_1'k_1''}W^{n_2'k_2''}\sum_{n_1''=0}^{N''-1}\sum_{n_2''=0}^{N''-1}\gamma^{n_1''k_1''}\gamma^{n_2''k_2''}x_{n_1',n_1'',n_2',n_2''}\right] \quad [5.13]$$

The DFT_{2D} calculation is performed in 3 steps:

– calculate the DFT_{2D} in $N'' \times N''$ points for each n_1' and n_2',

– multiply each calculated element by $W^{n_1'k_1''}\,W^{n_2'k_2''}$,

– calculate the DFT_{2D} in $N' \times N'$ points for each index n_1'' and n_2''.

Rader's algorithm for DFT_{1D} calculation

The basic idea of this method is to transform the calculation of the DFT_{1D} for a digital signal having a prime number of samples N, into the calculation of a 1D convolution product using the variable changes below:

$$\begin{aligned} n &\to \rho^{-n} \bmod N = \left(\left(\rho^{-n}\right)\right)_N \\ k &\to \rho^{k} \bmod N = \left(\left(\rho^{k}\right)\right)_N \end{aligned} \quad [5.14]$$

where ρ is a primitive root of N, so that $\forall\, i \in [1, N-1]$ there exists a number $r(i) \in [1, N-1]$ satisfying $\rho^{r(i)}(\bmod N) = i$ and $\rho^{r(i)} \neq 0$.

With these variable changes the DFT_{1D} can be written in the form of the following convolution product:

$$X_{\left(\left(\rho^{k}\right)\right)_N} = X_0 + \sum_{n=0}^{N-2} W_N^{\left(\left(\rho^{k-n}\right)\right)_N} x_{\left(\left(\rho^{-n}\right)\right)_N}, \quad k = 0..N-2 \quad [5.15]$$

with $X_0 = \sum_{n=0}^{N-1} x_n$.

Winograd's algorithm for the DFT_{1D} calculation

This algorithm works in two steps:

– conversion of a DFT_{1D}, by data block of length N – prime number, into a convolution product using the Rader permutation,

– fast calculation of the obtained convolution product.

NOTE.– Although the same principle is used, the Winograd algorithm is different according as N is a prime number, a power of a prime number or a power of 2.

5.2. Solved exercises

EXERCISE 5.1.

Calculate and plot the spectrum of the 1D digital signal:

$$x[n]=\begin{cases}1, \text{ if } & n=0..7\\0, & \text{otherwise}\end{cases}$$

```
x=ones(1,8) ;
X=fft(x,128);
subplot(3,1,1),
stem(x),
xlabel('n'),ylabel('x[n]')
subplot(3,1,2),
stem(abs(X)),
xlabel('k'),ylabel('abs(X[k])')
subplot(3,1,3),
stem(angle(X)),
xlabel('k'),ylabel('angle(X)')
```

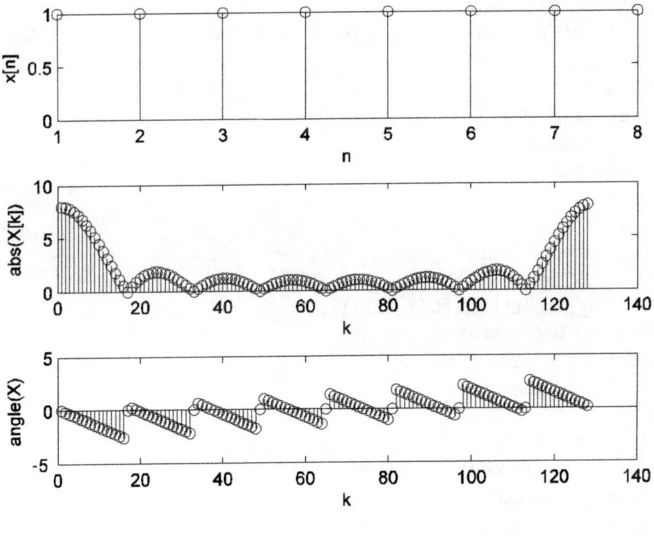

Figure 5.1. *Time and frequency representation of a digital pulse signal*

EXERCISE 5.2.

This exercise is aimed at stressing how important the phase spectrum is in the case of an image.

Write a MATLAB code to evaluate the two-dimensional Fourier transform of the two images "Clown" and "Gatlin2" from MATLAB. Plot the corresponding frequency representations and reconstruct the two images by inverse transformation, but exchanging their phase spectra.

```
clear; clf
%%%% IMAGE LOADING
% Loading the image x from the file clown.mat
load clown; x = X;
% Loading the image x from the file gatlin2.mat
load gatlin2; y = X;
% Resizing x and y to have the same size
l = min(size(x,1),size(y,1));
c = min(size(x,2),size(y,2));
x = x(1:l,1:c);y = y(1:l,1:c);
%%%%% RECONSTRUCTION OF THE NEW IMAGES
% Calculation of the FT of x and y
X=fft2(x); Y=fft2(y);
% Reconstruction of z1 using the magnitude of X and the phase of Y
z1 = real(ifft2(abs(X).*exp(i*angle(Y))));
% Reconstruction of z2 using the magnitude of Y and the phase of X
z2 = real(ifft2(abs(Y).*exp(i*angle(X))));
%%%%%% DISPLAYING THE IMAGES
% Displaying the FT of x and y: magnitude and phase
fh = ([1:l]-l/2)/l;
fv = ([1:c]-c/2)/c;
figure(1); subplot(3,2,1);
image(x); axis('image');
title('image x'); axis off
subplot(3,2,2);
image(y); axis('image');
title('image y'); axis off
subplot(3,2,3);
imagesc(fh,fv,log(abs(fftshift(X.^2))));
axis('image'); title('Magnitude of X');
xlabel('horizontal frequencies');
ylabel('vertical frequencies')
subplot(3,2,4);
imagesc(fh,fv,log(abs(fftshift(Y.^2))));
axis('image'); title('Magnitude of Y');
xlabel('horizontal frequencies');
ylabel('vertical frequencies')
subplot(3,2,5);
imagesc(fh,fv,angle(fftshift(X))); axis('image');
title('X phase');
xlabel('horizontal frequencies');
ylabel('vertical frequencies')
subplot(3,2,6);
imagesc(fh,fv,angle(fftshift(Y))); axis('image');
title('Y phase');
```

```
xlabel('horizontal frequencies');
ylabel('vertical frequencies')
% Displaying the reconstructed images z1 and z2
figure(2)
subplot(2,2,1); image(x); axis('image');
title('image x'); axis off
subplot(2,2,2); image(y); axis('image');
title('image y'); axis off
subplot(2,2,3); image(z1);
axis('image'); axis off
title('magnitude of x & phase of y')
subplot(2,2,4); image(z2);
axis('image'); axis off
title('magnitude of y & phase of x')
colormap('gray')
```

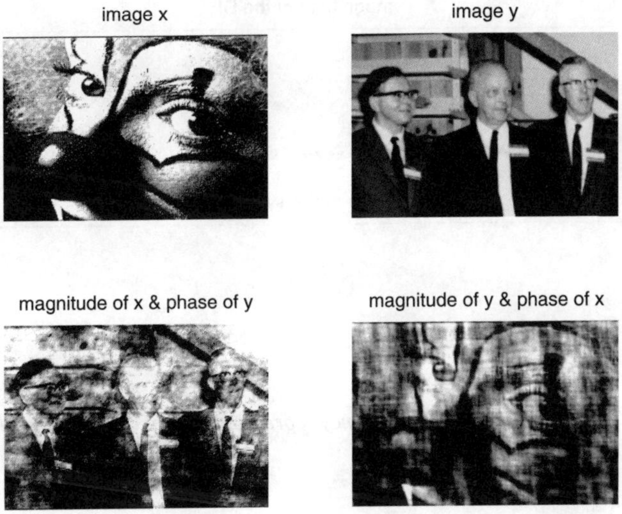

Figure 5.2. *Original and reconstructed images after exchanging the phase spectra*

EXERCISE 5.3.

Calculate the DFT of a 1D cosine defined on $N = 50$ points, and with the normalized frequency 2.25/50.

```
n=(0:49)';
x=cos(2*pi*2.25/50*n);
X=abs(fft(x));
Xsh=fftshift(X);
subplot(3,1,1);
stem(x);
```

```
title('discrete-time signal')
subplot(3,1,2);
stem(X);
title('magnitude of the DFT')
subplot(3,1,3),
stem(Xsh);
title('magnitude of the symmetric DFT')
```

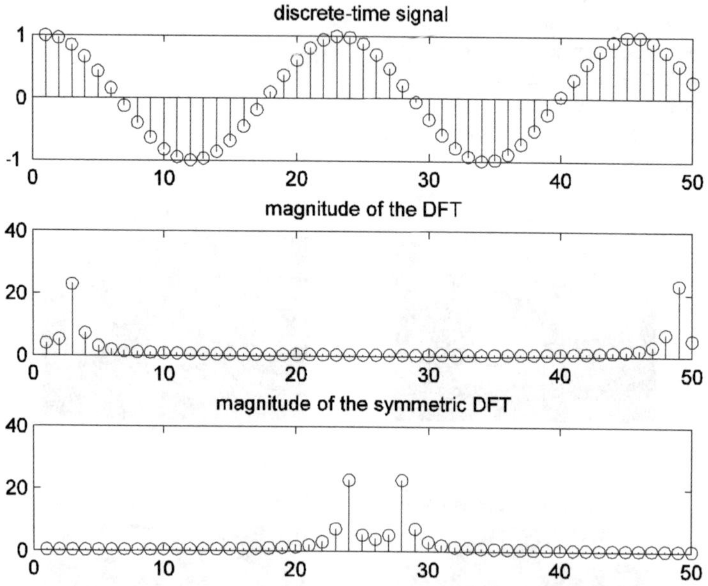

Figure 5.3. *Time and frequency representations of a cosine signal*

EXERCISE 5.4.

Write a MATLAB code to illustrate the following DFT properties: time-delay, frequency shifting, modulation and time inversion, using the digital signal:

$$x[n]=\begin{cases} n & \text{if} \quad n=1..8 \\ 0 & \text{if} \quad n=9..16 \end{cases}$$

```
% Time-delay property illustration
w = -pi:2*pi/255:pi;
wo = 0.4*pi; D = 10;
num = [1 2 3 4 5 6 7 8 9];
h1 = freqz(num, 1, w);
h2 = freqz([zeros(1,D) num], 1, w);
subplot(2,2,1)
```

```
plot(w/(2*pi),abs(h1));
grid
title('Original signal magnitude spectrum')
subplot(2,2,2)
plot(w/(2*pi),abs(h2));
grid
title('Time-delayed signal amplitude spectrum')
subplot(2,2,3)
plot(w/(2*pi),angle(h1));
grid
title('Original signal phase spectrum')
subplot(2,2,4)
plot(w/(2*pi),angle(h2));
grid
title('Time-delayed signal phase spectrum')
```

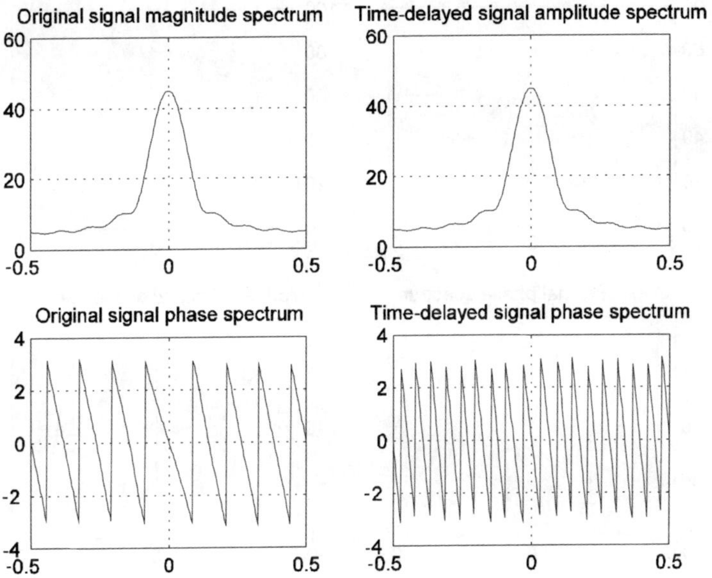

Figure 5.4. *Illustration of the time-delay property of the DFT*

```
% Frequency shifting property illustration
clf;
w = -pi:2*pi/255:pi;
wo = 0.4*pi;
num1 = [1 3 5 7 9 11 13 15 17];
L = length(num1);
h1 = freqz(num1, 1, w);
n = 0:L-1;
```

```
num2 = exp(wo*i*n).*num1;
h2 = freqz(num2, 1, w);
subplot(2,2,1)
plot(w/(2*pi),abs(h1));grid
title('Original signal magnitude spectrum')
subplot(2,2,2)
plot(w/(2*pi),abs(h2));grid
title('Modulated signal amplitude spectrum')
subplot(2,2,3)
plot(w/(2*pi),angle(h1));grid
title('Original signal phase spectrum')
subplot(2,2,4)
plot(w/(2*pi),angle(h2));grid
title('Modulated signal phase spectrum')
```

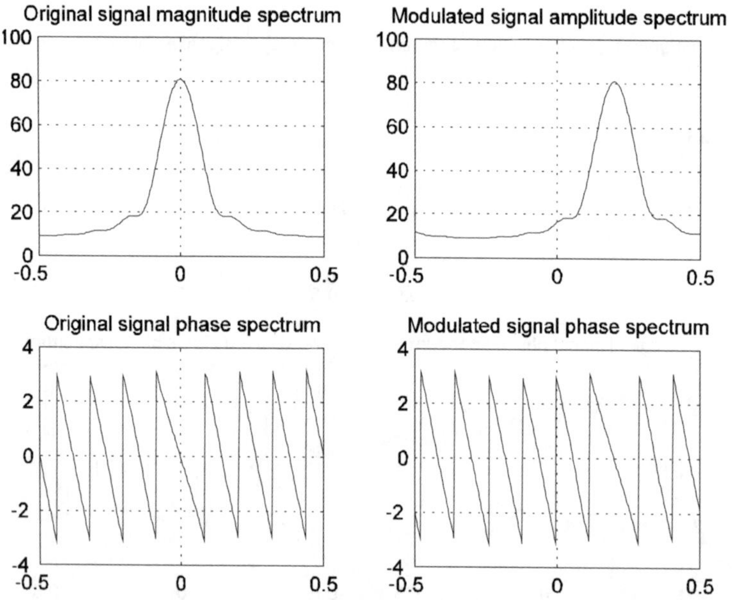

Figure 5.5. *Illustration of the frequency shifting property of the DFT*

```
% Modulation property illustration
clf;
w = -pi:2*pi/255:pi;
x1 = [1 3 5 7 9 11 13 15 17];
x2 = [1 -1 1 -1 1 -1 1 -1 1];
y = x1.*x2;
h1 = freqz(x1, 1, w);
h2 = freqz(x2, 1, w);
h3 = freqz(y,1,w);
```

```
subplot(3,1,1)
plot(w/(2*pi),abs(h1));
grid
title('1^s^t signal magnitude spectrum')
subplot(3,1,2)
plot(w/(2*pi),abs(h2));
grid
title('2^n^d signal magnitude spectrum')
subplot(3,1,3)
plot(w/(2*pi),abs(h3));
grid
title('Product signal magnitude spectrum')
```

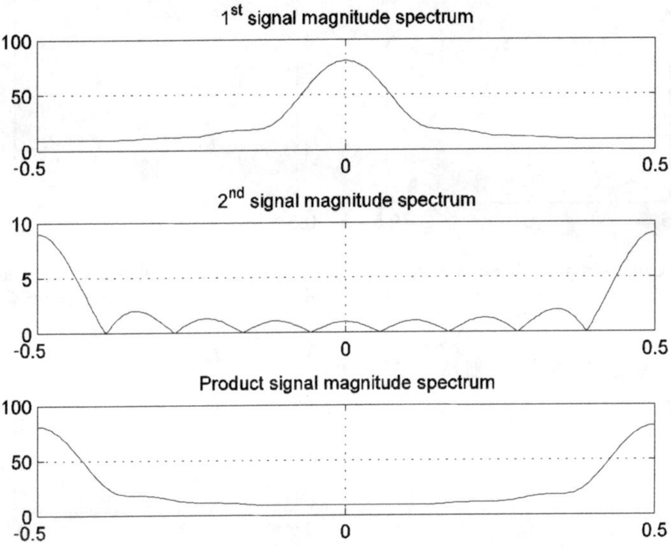

Figure 5.6. *Illustration of the modulation property of the DFT*

```
% Time inversion property illustration
clf;
w = -pi:2*pi/255:pi;
num = [1 2 3 4];
L = length(num)-1;
h1 = freqz(num, 1, w);
h2 = freqz(fliplr(num), 1, w);
h3 = exp(w*L*i).*h2;
subplot(2,2,1);
plot(w/(2*pi),abs(h1));grid
title('Original signal magnitude spectrum')
```

```
subplot(2,2,2);
plot(w/(2*pi),abs(h3));grid
title('Time inverted signal amplitude spectrum')
subplot(2,2,3);
plot(w/(2*pi),angle(h1));grid
title('Original signal phase spectrum')
subplot(2,2,4);
plot(w/(2*pi),angle(h3));grid
title('Time inverted signal phase spectrum')
```

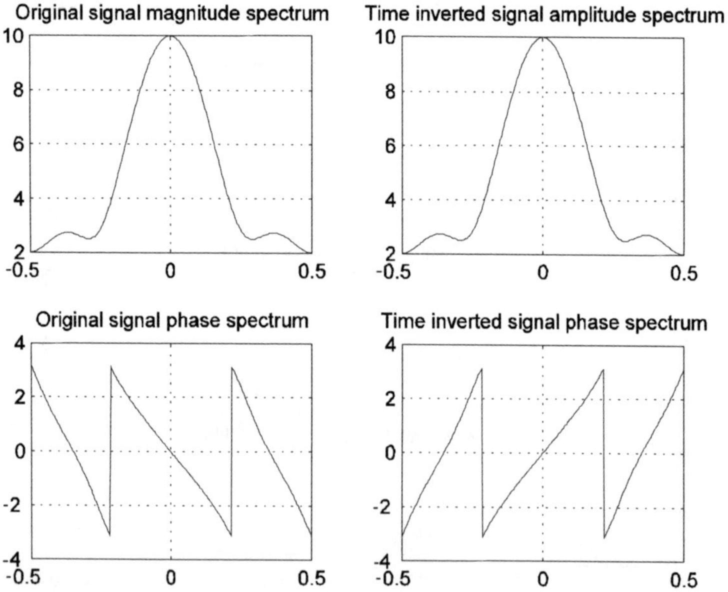

Figure 5.7. *Illustration of the time inversion property of the DFT*

EXERCISE 5.5.

Write a MATLAB code for illustrating the influence of the observation time of a signal on its spectrum. Consider a 10 Hz sine signal, sampled at 100 Hz, and deal with the following cases:

– the number of signal periods is complete,

– the signal is zero-padded before its spectrum calculation,

– the number of signal periods is not complete.

```
% Time support definition (50 samples = 5 periods)
t = 0:0.01:0.5-0.01;
```

```
% Signal definition
x = cos(20*pi*t);
% Spectrum calculation
N = length(x); X = fft(x,N);
% Amplitude spectrum plot
fp = (0:N-1)/N/0.01; fp = fp - 1/2/0.01;
stem(fp,fftshift(abs(X))); axis([ -1/(2*0.01) 1/(2*0.01) 0 25]);
xlabel('Frequency [Hz]'); ylabel('Amplitude spectrum')
axis([-50 50 0 30])
```

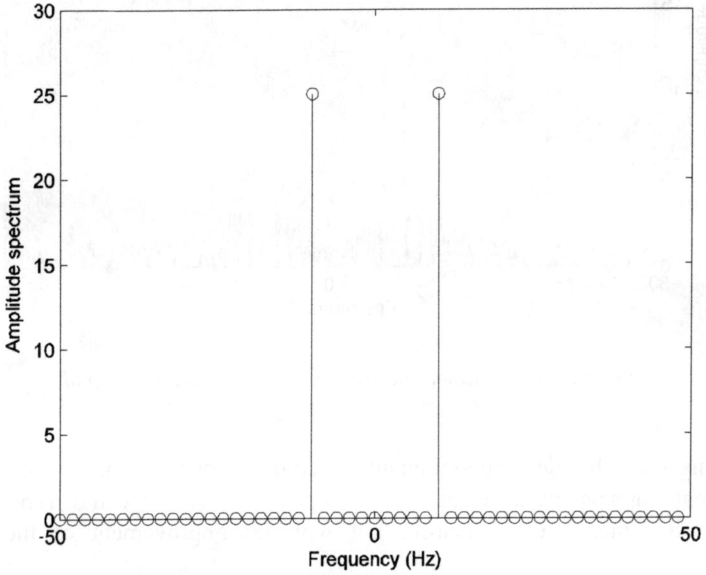

Figure 5.8. *Amplitude spectrum of a sine signal*

Figure 5.8 shows only one spectral component, at the frequency of the sine signal.

```
% Zero-padding on 100 periods
X = fft(x,20*N);
N = length(X);
fp = (0:N-1)/N/0.01;
fp = fp - 1/2/0.01;
plot(fp,abs(fftshift(X)));
axis([ -1/(2*0.01) 1/(2*0.01) 0 25]);
xlabel('Frequency [Hz]')
ylabel('Amplitude spectrum')
axis([-50 50 0 30])
```

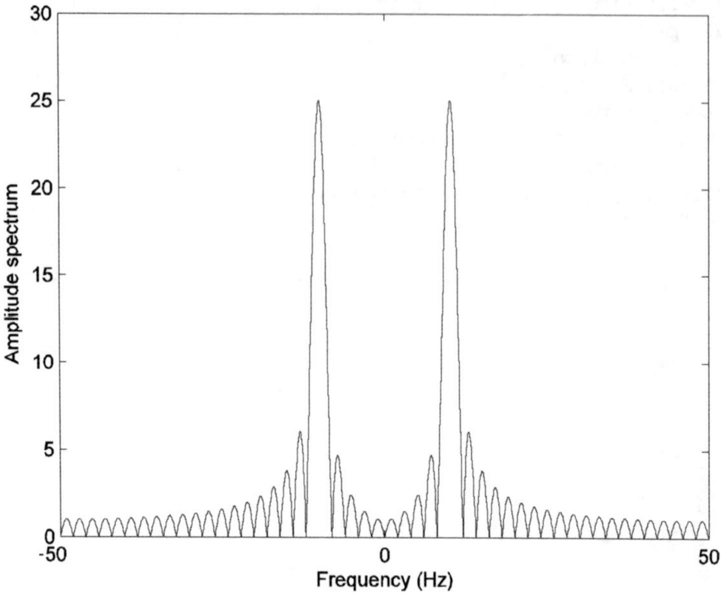

Figure 5.9. *Amplitude spectrum of a zero-padded sine signal*

In this case, besides the dominant spectral component, many other spectral components appear because of the zero-padding. The spectral representation accuracy has thus been enhanced, but with no improvement of the spectral resolution.

```
% Time support definition (45 samples = 4.5 signal periods)
t = 0:0.01:0.5-0.01-1/20;
x = cos(20*pi*t);
N = length(x); X = fft(x,N);
fp = (0:N-1)/N/0.01;
fp = fp - 1/2/0.01;
stem(fp,abs(fftshift(X)));
axis([ -1/(2*0.01) 1/(2*0.01) 0 25]);
xlabel('Frequency [Hz]')
ylabel('Amplitude spectrum')
axis([-50 50 0 20])
```

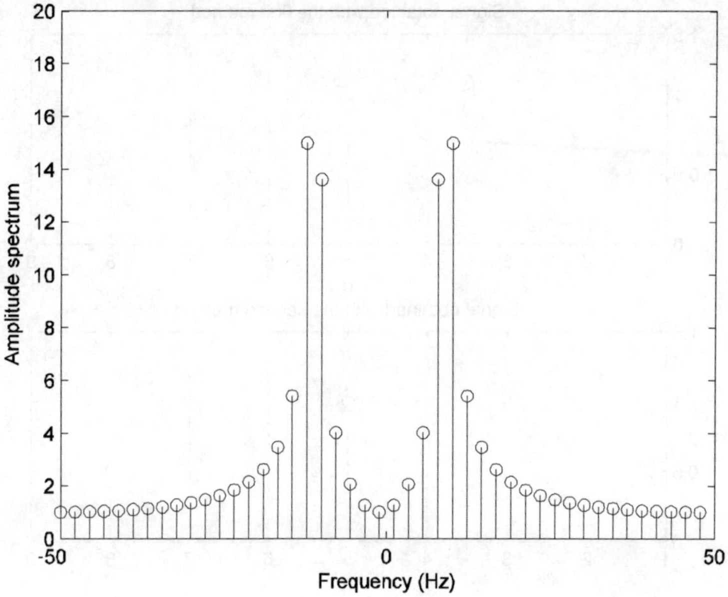

Figure 5.10. *Amplitude spectrum of a truncated sine signal*

As can be seen in Figure 5.10, the spectral component corresponding to the sinusoid frequency is not present in the calculated amplitude spectrum because of the number of periods, which is not complete, and the low resolution. Its energy is distributed around the true frequency value (spectral leakage effect).

EXERCISE 5.6.

Consider the following two digital signals: $x=[0, 0.25, 0.5, 0.75, 1]$ and $h=[0, 0.5, 1, 0.5, 0]$. Calculate the product of their DFTs and then come back to the time domain by inverse DFT. Compare the obtained result to the convolution product of the two signals and conclude.

```
x=[0,0.25,0.5,0.75,1]; h=[0,0.5,1,0.5,0];
X=fft(x,9); H=fft(h,9);Y=X.*H;
y=abs(ifft(Y)); y2=conv(x,h)
figure
subplot(211);
stem(y); xlabel('n'); ylabel('A');
title('Signal obtained with the first method')
subplot(212);
stem(y2); xlabel('n'); ylabel('A');
title('Signal obtained with the second method')
```

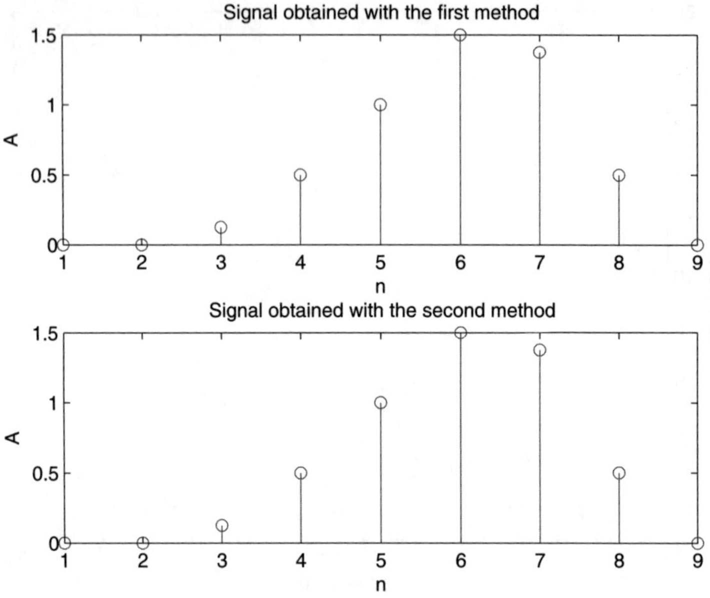

Figure 5.11. *Illustration of the convolution product calculation using two different methods*

EXERCISE 5.7.

Consider the sum of two sinusoids of 50 Hz and 120 Hz, corrupted by an additive, zero-mean, white noise. Calculate and plot its power spectral density.

```
t=0:0.001:0.8; % T=0.001, so Fs=1 kHz
x=sin(2*pi*50*t)+sin(2*pi*120*t)+2*randn(1,length(t));
subplot(211);
plot(x(1:500));
title('Signal')
X=fft(x,512);
Px=X.*conj(X)/512;
f=1000*(0:255)/512; % f=Fe*(k-1)/N
subplot(212);
plot(f,Px(1:256));
title('Power spectral density')
```

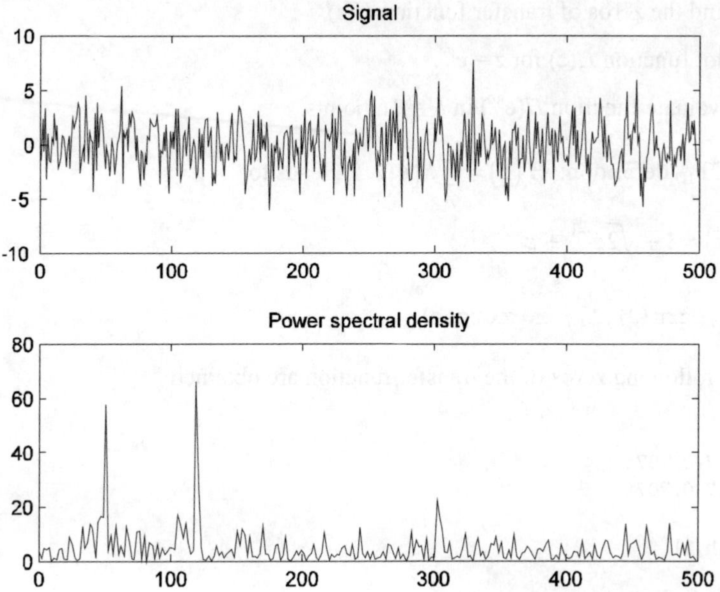

Figure 5.12. *Noisy sinusoid mixture and its power spectral density*

Because of the random nature of the analyzed signal, its power spectral density has to be estimated. The power spectral density estimation methods will be discussed in Chapter 10.

EXERCISE 5.8.

Calculate the DCT_{1D} of the digital signal $x[n] = n + 50\cos(2\pi n/40)$, $n = 1..100$ and determine the percentage of its total energy cumulated by its first three coefficients.

```
x=(1:100)+50*cos(2*pi/40*(1:100)); X=dct(x);
norm([X(1),X(2),X(3)])/norm(X)
```

The percentage of the cumulated energy is thus:

```
ans =
    0.8590
```

EXERCISE 5.9.

Consider the following impulse response of a digital filter: $h = \left[1, \sqrt{2}, 1\right]$.

a. Calculate its z-transform.

b. Find the zeros of transfer function $H(z)$.

c. Plot function $H(z)$ for $z = e^{j\omega}$.

d. Evaluate function $H(e^{j\omega})$ in $N = 16$ points.

a. The ZT_{1D} definition: $H(z) = \sum_n h[n] z^{-n}$, leads to:

$$H(z) = 1 + \sqrt{2} z^{-1} + z^{-2}.$$

b. `h=[1,sqrt(2),1]; r=roots(h)`

The following zeros of the transfer function are obtained:

```
r =
  -0,707+0,707i
  -0,707-0,707i
```

```
zplane(h,1)
```

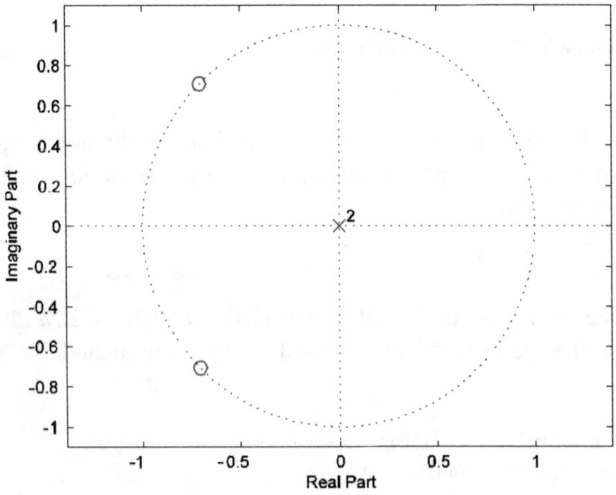

Figure 5.13. *Zeros of the transfer function*

c.

```
[H,w]=freqz(h,1)
figure
plot(w/(2*pi),abs(H))
xlabel('Normalized frequency')
ylabel('Amplitude')
```

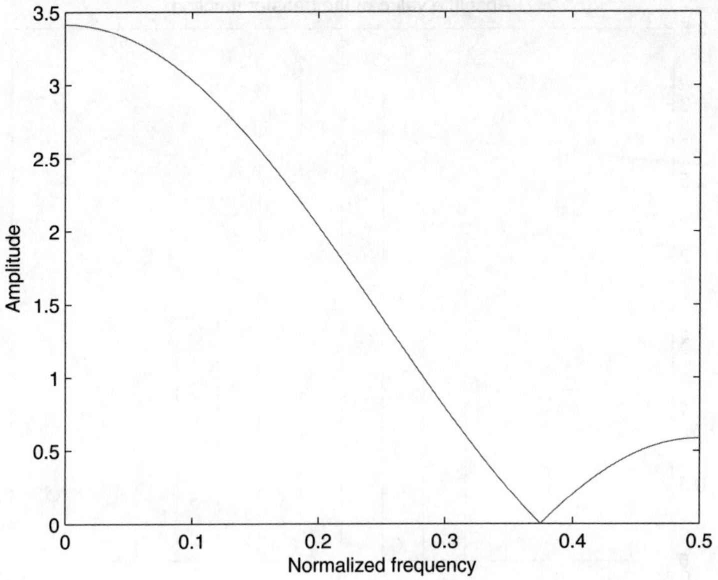

Figure 5.14. *Transfer function magnitude representation*

d.

```
H = fftshift(fft(h,16)); N = length(H);
fqv = [-0.5:1/N:0.5-1/N];
stem(fqv,abs(H))
title('Absolute value of the transfer function');
xlabel('Normalized frequency');ylabel('Amplitude')
```

or:

```
Hz=czt([h,zeros(1,13)]);
fqv = [-0.5:1/N:0.5-1/N];
stem(fqv,abs(H))
title('Absolute value of the transfer function');
xlabel('Normalized frequency');ylabel('Amplitude')
```

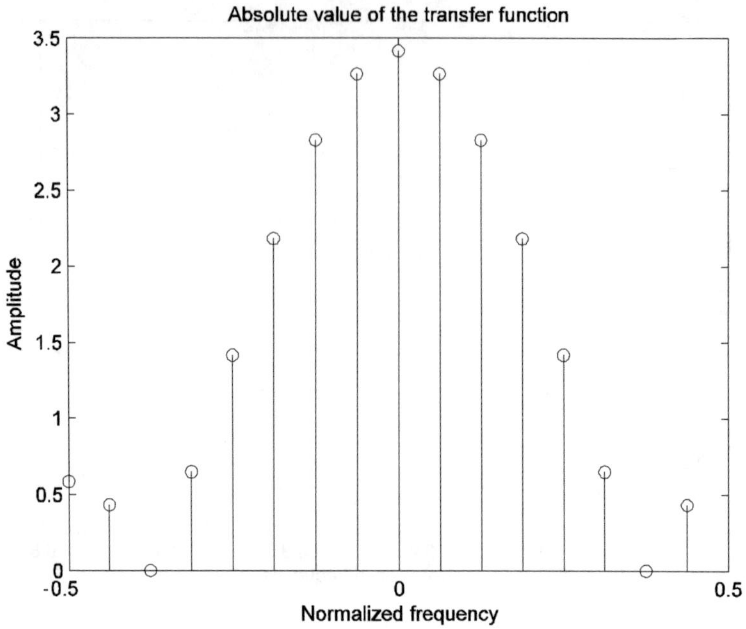

Figure 5.15. *Absolute value of the Fourier transform of the filter pulse response*

EXERCISE 5.10.

Calculate the DFT_{2D} of the following matrix using two DFT_{1D}:

$$x = \begin{bmatrix} x_{00} & x_{01} \\ x_{10} & x_{11} \end{bmatrix}$$

The DFT_{2D} definition: $X_{k_1,k_2} = \sum_{n_1=0}^{1} W_2^{n_1 k_1} \left[\sum_{n_2=0}^{1} W_2^{n_2 k_2} x_{n_1,n_2} \right]$, results in:

$$\begin{bmatrix} x_{00} & x_{01} \\ x_{10} & x_{11} \end{bmatrix} \underset{\text{(linewise)}}{\overset{TFD_{1D}}{\Longrightarrow}} \begin{bmatrix} W^0 x_{00} + W^0 x_{01} & W^0 x_{00} + W^1 x_{01} \\ W^0 x_{10} + W^0 x_{11} & W^0 x_{10} + W^1 x_{11} \end{bmatrix}$$

$$\xrightarrow[\text{(columnwise)}]{\text{TFD}_{1D}} \begin{bmatrix} W^0\left(W^0 x_{00}+W^0 x_{01}\right) & W^0\left(W^0 x_{00}+W^1 x_{01}\right) \\ +W^0\left(W^0 x_{10}+W^0 x_{11}\right) & +W^0\left(W^0 x_{10}+W^1 x_{11}\right) \\ W^0\left(W^0 x_{00}+W^1 x_{01}\right) & W^0\left(W^0 x_{00}+W^1 x_{01}\right) \\ +W^1\left(W^0 x_{10}+W^0 x_{11}\right) & +W^1\left(W^0 x_{10}+W^1 x_{11}\right) \end{bmatrix}$$

For example, in the case of the matrix $x = \begin{bmatrix} 1 & 2 \\ 2 & 3 \end{bmatrix}$, the following MATLAB code:

```
x=[1,2;2,3];
for n1=1:2; X1(n1,1:2)=fft(x(n1,1:2)); end
for n2=1:2; X(1:2,n2)=fft(X1(1:2,n2)); end
```

yields: $X_1 = \begin{bmatrix} 3 & -1 \\ 5 & -1 \end{bmatrix}$ and then $X = \begin{bmatrix} 8 & -2 \\ -2 & 0 \end{bmatrix}$.

The same result can be obtained with:

`X=fft(fft(x).').'` or `X=fft2(x)`

Verify that: `X(1,1)=sum(sum(x));`

EXERCISE 5.11.

The following matrix:

```
x=[1 1 1 1 1
   1 0 0 0 1
   1 0 0 0 1
   1 0 0 0 1
   1 1 1 1 1];
```

represents a binary digital image in the form of a black square on a white background.

Plot this image. Then calculate and plot the DFT_{2D} and the DCT_{2D} of this image.

```
Xf=fft2(x);Xfsh=fftshift(Xf); Xc=dct2(x);
subplot(221);
imagesc(x); title('Original image')
subplot(222);
```

```
imagesc(log(abs(Xf)));
title('Magnitude of the image spectrum')
subplot(223);
imagesc(log(abs(Xfsh)));
title('Magnitude of the image symmetric spectrum')
subplot(224);
imagesc(log(abs(Xc)));
title('Magnitude of the image cosine transform')
colormap(gray)
```

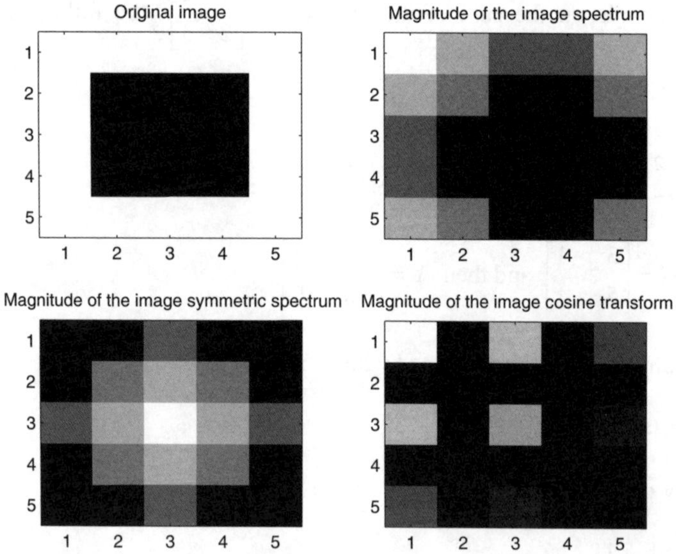

Figure 5.16. *Binary image and its Fourier and cosine transform*

EXERCISE 5.12.

DFT calculation with a digital computer requires truncating the signal because the number of samples has to be finite (generally a power of 2 in the case of the fast Fourier transform). The signal is truncated by default with a rectangular window, but several other weighting windows can also be used (Hamming, Hanning, Blackman, etc.). Its choice depends on the spectral and dynamic resolutions required for a given application.

The objective of this exercise is to compare different weighting windows in two important cases: two closely spaced sinusoids having the same amplitude and two distant sinusoids with very different amplitudes.

1. Generate a sinusoid on $N = 32$ points, with the amplitude 1 V and a frequency of 100 Hz, sampled at 256 Hz. Calculate its DFT on 1,024 points using successively the rectangular, triangular, Hamming, Hanning and Blackman windows to truncate the signal. Conclude about the spectral resolution provided by each weighting window.

```
t=(1:32);f1=50;
Fe=256;Nfft=1024;
y1=sin(2*pi*f1/Fe*t);
sig=y1.*boxcar(32)';
y_rect=abs(fftshift((fft(sig,Nfft))));
sig=y1.*triang(32)';
y_tria=abs(fftshift((fft(sig,Nfft))));
sig=y1.*hamming(32)';
y_hamm=abs(fftshift((fft(sig,Nfft))));
sig=y1.*hanning(32)';
y_hann=abs(fftshift((fft(sig,Nfft))));
sig=y1.*blackman(32)';
y_blac=abs(fftshift((fft(sig,Nfft))));
f=[-Fe/2:Fe/Nfft:(Fe/2-Fe/Nfft)];
subplot(511)
semilogy(f(513:1024),y_rect(513:1024));
axis([0 128 1e-3 100]);grid
legend('rectangular window',-1)
subplot(512);
semilogy(f(513:1024),y_tria(513:1024));
axis([0 128 1e-3 100]);grid
legend('triangular window',-1)
subplot(513);
semilogy(f(513:1024),y_hamm(513:1024));
grid;ylabel('Spectral amplitude')
axis([0 128 1e-4 100]);
legend('Hamming window',-1)
subplot(514);
semilogy(f(513:1024),y_hann(513:1024));
axis([0 128 1e-3 100]);grid ;
legend('Hanning window',-1)
subplot(515);
semilogy(f(513:1024),y_blac(513:1024));
axis([0 128 1e-3 100]);grid;
legend('Blackman window',-1)
xlabel('frequency [Hz]')
```

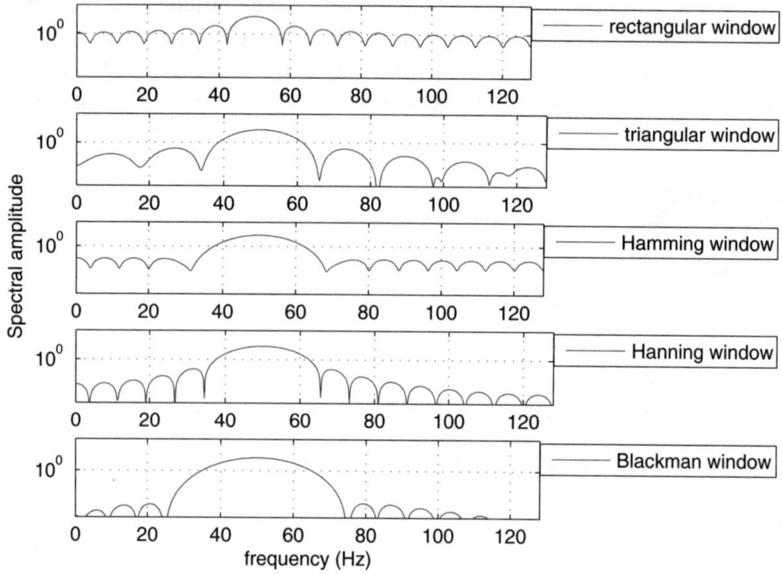

Figure 5.17. *Spectral representation of the weighting windows*

The simplest spectral window is the rectangular one. It provides the best spectral resolution among all the weighting windows because it has the narrowest main lobe. However, the maximum level of its sidelobes is also the highest (about 13 dB), so it provides the worse dynamical resolution.

The other spectral windows improve the dynamical resolution, but degrade the spectral resolution. Thus, it is necessary to achieve a trade-off between the following elements:

– first sidelobe level (or maximum sidelobe level),

– mainlobe width,

– sidelobe attenuation rate.

2. Generate a mixture of two sinusoids with the amplitude 1 and the frequencies 100 Hz and 94 Hz. Consider $N = 32$ samples for the signal and calculate its DFT on 1,024 points.

```
t=(1:32); f1=94; f2=100; Fe=256;Nfft=1024;
y1= sin(2*pi*f1/Fe*t)+ sin(2*pi*f2/Fe*t);
sig=y1.*boxcar(32)';
y_rect=abs(fftshift((fft(sig,Nfft)))); sig=y1.*triang(32)';
```

```
y_tria=abs(fftshift((fft(sig,Nfft)))); sig=y1.*hanning(32)';
y_hann=abs(fftshift((fft(sig,Nfft)))); sig=y1.*hamming(32)';
y_hamm=abs(fftshift((fft(sig,Nfft)))); sig=y1.*blackman(32)';
y_blac=abs(fftshift((fft(sig,Nfft))));
f=[-Fe/2:Fe/Nfft:(Fe/2-Fe/Nfft)];
subplot(511)
plot(f(513:1024),y_rect(513:1024)); grid; axis([0 128 0 20])
legend('rectangular window',-1)
subplot(512);
plot(f(513:1024),y_tria(513:1024));grid; axis([0 128 0 7])
legend('triangular window',-1)
subplot(513);
plot(f(513:1024),y_hann(513:1024)); grid; axis([0 128 0 7])
legend('Hanning window',-1)
ylabel('spectral amplitude')
subplot(514);
plot(f(513:1024),y_hamm(513:1024)); grid; axis([0 128 0 7])
legend('Hamming window',-1)
subplot(515);
plot(f(513:1024),y_blac(513:1024)); grid; axis([0 128 0 6])
legend('Blackman window',-1); xlabel('frequency [Hz]')
```

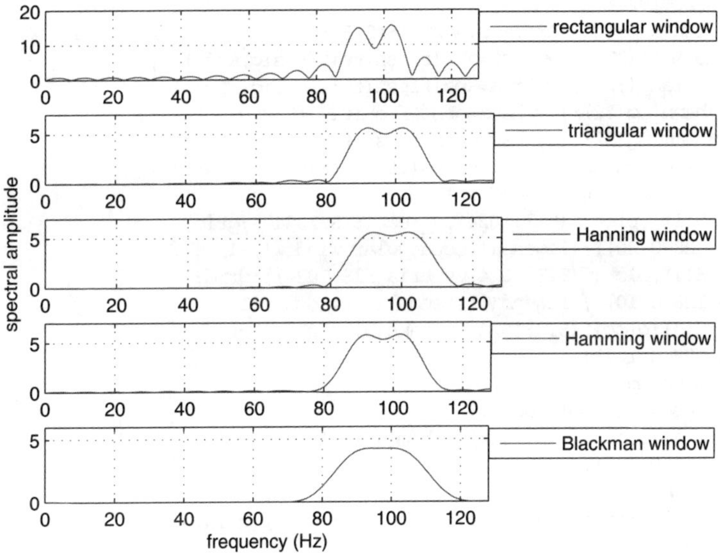

Figure 5.18. *Spectral representation of two closely spaced sinusoids*

The spectral accuracy should be not taken for the spectral resolution, which has the capability of resolving two closely spaced spectral components. It is generally defined as the frequency difference between two sinusoids having the same amplitude which allow us to obtain on their sum spectrum a gap of minimum 3 dB between their maxima.

The time support limited to N points leads to the lobe occurrence in the sinusoid spectrum. For a rectangular weighting window the width of the main lobe is equal to $2/N$. Thus, if $x[n]$ contains two sinusoids whose frequencies are separated by at least $1/N$, the corresponding main lobes are considered separable. If $f_s = 1/T_s$ stands for the sampling frequency, the spectral resolution in Hz is equal to f_s/N, i.e. the inverse of the total analysis time.

3. Finally generate the following signal on 32 points:

$$x[n] = \sin\left(2\pi\frac{100}{256}n\right) + 0.1\sin\left(2\pi\frac{74}{256}n\right)$$

Plot its spectrum obtained for different weighting windows and comment upon this.

```
t=(1:32);f1=74;f2=100;
Fe=256;Nfft=1024;
y1= 0.1*sin(2*pi*f1/Fe*t)+ sin(2*pi*f2/Fe*t);
sig=y1.*boxcar(32)'; y_rect=abs(fftshift(fft(sig,Nfft)));
sig=y1.*triang(32)'; y_tria=abs(fftshift(fft(sig,Nfft)));
sig=y1.*hanning(32)'; y_hann=abs(fftshift(fft(sig,Nfft)));
sig=y1.*hamming(32)'; y_hamm=abs(fftshift(fft(sig,Nfft)));
sig=y1.*blackman(32)'; y_blac=abs(fftshift(fft(sig,Nfft)));
f=[-Fe/2:Fe/Nfft:(Fe/2-Fe/Nfft)];
subplot(511); plot(f(513:1024),y_rect(513:1024));grid;
axis([0 128 0 20]); legend('rectangular window',-1)
subplot(512);plot(f(513:1024),y_tria(513:1024));grid;
axis([0 128 0 10]); legend('triangular window',-1)
subplot(513);plot(f(513:1024),y_hann(513:1024));grid;
axis([0 128 0 10]);
legend('Hanning window',-1)
ylabel('spectral amplitude')
subplot(514);plot(f(513:1024),y_hamm(513:1024));grid;
axis([0 128 0 10]);
legend('Hamming window',-1)
subplot(515);plot(f(513:1024),y_blac(513:1024));grid;
axis([0 128 0 8])
legend('Blackman window',-1);
xlabel('frequency [Hz]')
```

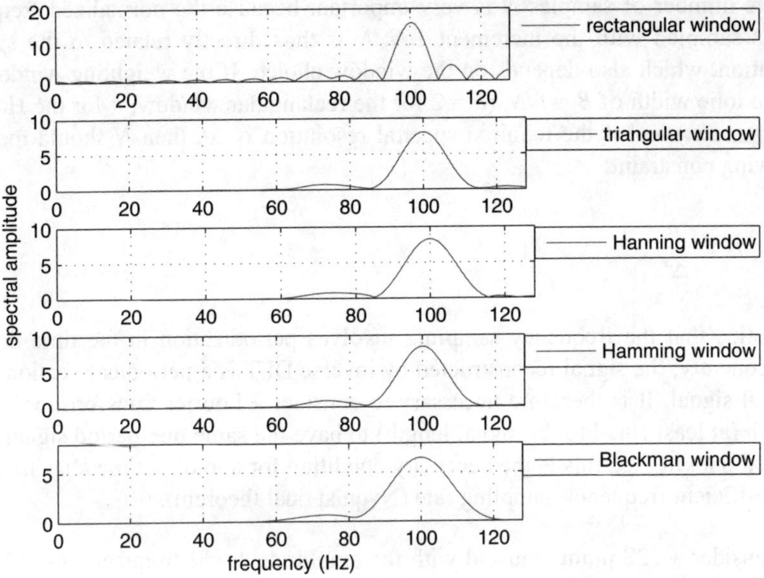

Figure 5.19. *Spectral representation of a mixture of two sinusoids with different amplitudes*

It can be seen that the two components are not resolved when the rectangular window is used. The weak sinusoid is hidden by the sidelobes of the weighting window spectrum, centered on the high amplitude sinusoid frequency.

The Hamming window provides a lower resolution than the rectangular window for an amplitude ratio of 0 dB, but it becomes more interesting for an amplitude ratio of 30 dB.

EXERCISE 5.13.

The last step of the DFT calculation is the frequency sampling. In fact, a continuous frequency cannot be considered when using a digital computer. The frequency axis is therefore sampled and the Fourier transform is calculated for harmonic frequencies:

$$f_n = \frac{nF_s}{N}$$

where $n = -N/2..N/2-1$ and F_s denotes the sampling frequency.

The number of samples N is very important because the normalized frequency axis is sampled with the increment $1/N$. N is thus directly related to the spectral resolution, which also depends on the window choice. If the weighting window has a main lobe width of $B = b/N$, ($b = 2$ for the rectangular window, 4 for the Hanning window, etc.) and if the required spectral resolution is Δf, then N should meet the following constraint:

$$N \geq \frac{b}{\Delta f}$$

Notice that the frequency sampling involves periodization in the time domain. Consequently, the signal reconstructed by inverse DFT is a periodical version of the original signal. It is therefore necessary to consider a Fourier transform order high enough (at least equal to the signal length) to have the same one-period signal as the original one. In fact, this is the necessary condition for avoiding time aliasing due to an insufficient frequency sampling rate (Nyquist dual theorem).

Consider a 128 point sinusoid with the amplitude 1 and frequency $f = 12.5$ Hz. This signal is sampled at $F_s = 64$ Hz or $F_s = 128$ Hz. Its DFT is calculated on 128 or 256 points.

Write a MATLAB code for illustrating the spectral representation of the signals corresponding to the following three cases:

a. $F_s = 64$ Hz, $N_{fft} = 128$

b. $F_s = 128$ Hz, $N_{fft} = 128$

c. $F_s = 128$ Hz, $N_{fft} = 256$

```
t=(1:128); f1=12.5 ;Fs=64;
y1= sin(2*pi*f1/Fs*t); sig=y1; Nfft=128;
y=abs(((fft(sig,Nfft))));
f0=[0:Fs/Nfft:(Fs-Fs/Nfft)];
f1=12.5;Fs=128;
y1= sin(2*pi*f1/Fs*t); sig=y1; Nfft=128 ;
y_rect=abs(fftshift((fft(sig,Nfft))));
Nfft1=256;
y_rect1=abs(fftshift((fft(sig,Nfft1))));
f=[-Fs/2:Fs/Nfft:(Fs/2-Fs/Nfft)];
f1=[-Fs/2:Fs/Nfft1:(Fs/2-Fs/Nfft1)];
subplot(311); stem(f0,y);
grid; axis([0 64 0 80]);
title('Fs=64 kHz, Nfft=128 points');
ylabel('Spectral amplitude')
subplot(312)
stem(f(Nfft/2+1:Nfft),y_rect(Nfft/2+1:Nfft));
```

```
grid; axis([0 64 0 80]);
title('Fs=128 kHz, Nfft=128 points');
ylabel('Spectral amplitude')
subplot(313);
stem(f1(Nfft1/2+1:Nfft1),y_rect1(Nfft1/2+1:Nfft1));
grid; axis([0 64 0 80])
xlabel('frequency [Hz]');
ylabel('Spectral amplitude');
title('Fs=128 kHz, Nfft=256 points');
```

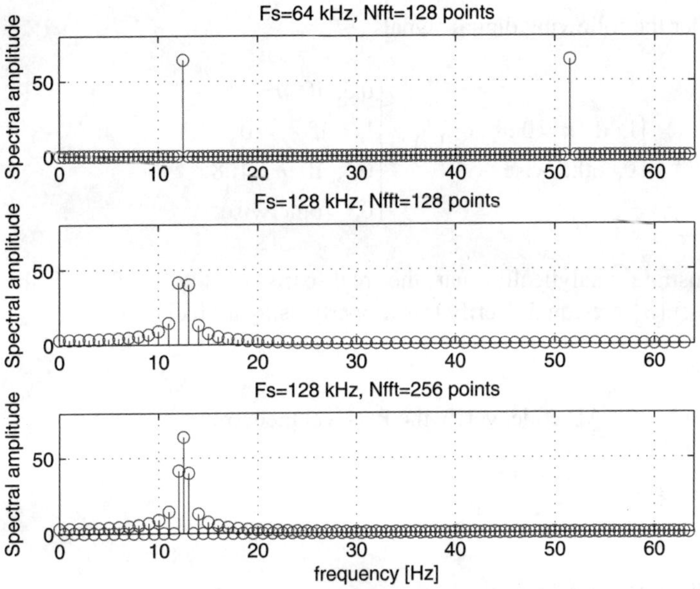

Figure 5.20. *Influence of the FFT parameters*

In the first case, the spectral resolution is equal to 0.5 Hz and the analyzed frequency is a harmonic one. The spectrum thus has the aspect of a Dirac pulse centered on 12.5 Hz. Another spectral component occurs at 64–12.5 Hz because the spectrum is represented on an interval of width F_s.

In the second case, the spectral resolution is equal to 1 Hz and the sinusoid frequency is not analyzed by the DFT (spectral leakage). For the frequencies which are not equal to or a multiple of F_s/N_{fft}, where N_{fft} is the number of points used by the FFT, a pure sinusoid appears in the form of several non-zero values around the true frequency. Without an *a priori* model, it is not possible to detect spectral components closer than $1/N_{fft}$ in normalized frequency.

In the third case, when the zero-padding is used, the spectral accuracy is equal to 0.5 Hz, while the spectral resolution is 1 Hz. The spectral representation is sampled with an increment of 0.5 Hz and contains the main lobe centered on the true frequency and the associated sidelobes.

5.3. Exercises

EXERCISE 5.14.

Consider the following digital signals:

$$x[n]=\begin{cases}1, \text{ if } n=0..8\\0, \text{ otherwise}\end{cases}, \quad y[n]=\begin{cases}0.5, \text{ if } n=-8..-1\\1, \text{ if } n=0\\0.5, \text{ if } n=1..8\\0, \text{ otherwise}\end{cases}$$

Demonstrate analytically that the real parts of the DFT_{1D} of the signals $x[n]$ and $y[n]$ are equal. Verify this property using a MATLAB code.

EXERCISE 5.15.

Using a MATLAB code, verify the Parseval theorem:

$$\sum_{n=0}^{N-1}|x[n]|^2=\frac{1}{N}\sum_{k=0}^{N-1}|X[k]|^2$$

for the following digital signal:

$$x[n]=\begin{cases}n, \text{ if } n=0..7\\0, \text{ otherwise}\end{cases}$$

EXERCISE 5.16.

The function $\rho^i \bmod N,\ \forall i\in[1,N-1]$ performs the permutation of the sequence $\{1, 2, 3, ..., N-1\}$ into the sequence $\{\rho^1,\rho^2,\rho^3,...,\rho^{N-1}\}$ if N is a prime number and ρ is a primitive root of N. For example, the permutation of the sequence {1, 2, 3, 4} into the sequence {2, 4, 3, 1} is obtained for $N = 5$, as it is illustrated in the table below.

i =	1	2	3	4
$\rho^i (\text{mod}\, 5) =$	2	4	3	1

The sequence $\{1, 2, 3, ..., N-1\}$ can be also permutated into the sequence $\{\rho^i \bmod N \text{ with } i = 0..N-2\}$. Thus, in the case $N = 5$ and $\rho = 2$, the following permutation is obtained:

i =	0	1	2	3
$\rho^i (\text{mod}\, 5) =$	1	2	4	3

In the Galois field GF(5), the element $\rho = 2$ is a primitive root, while $\rho^{-1} = 3$ is the inverse element, because $((\rho \cdot \rho^{-1}))_N = ((2 \cdot 3))_5 = 1$. In this case, the sequence $i = \{1, 2, 3, 4\}$ or $i = \{0, 1, 2, 3\}$ is permutated in the following manner:

i =	0	1	2	3
$\rho^i (\text{mod}\, N) = [\rho^{-1}]^i (\text{mod}\, N) = 3^{-i} (\text{mod}\, 5) =$	1	3	4	2

Write a MATLAB code for performing these permutations.

Chapter 6

Linear and Invariant Discrete-Time Systems

6.1. Theoretical background

A discrete-time system transforms an input digital signal into an output one according to its transfer function. A discrete linear time-invariant system (LTI) can be mathematically described by an operator $\aleph\{x\}$, so that $x \to y = \aleph\{x\}$, where x and y are the input and output signals respectively.

Digital filtering is one of the most important signal processing functions and the digital filters are the most commonly used LTI. The filtering algorithm can be implemented in the form of a "software solution", as a routine of a more general program running in the memory of a digital signal processor, or in the form of a "hardware solution", as a dedicated electronic circuit.

6.1.1. *LTI response calculation*

The output signal y of a LTI is calculated as the convolution product of its impulse response h with the input signal x, which is considered here as a finite energy signal:

$$\begin{aligned} x &= \{x[n] \ , \ n = 0..N-1\} = \{x_n, \ n = 0..N-1\} \\ h &= \{h[n] \ , \ n = 0..L-1\} = \{h_n, \ n = 0..L-1\} \end{aligned} \qquad [6.1]$$

$$y[n]=(x*h)[n]=(h*x)[n]=\sum_{m=0}^{N-1}x[m]h[n-m]=\sum_{m=0}^{L-1}h[m]x[n-m]=$$
$$=\sum_{m=n-N+1}^{n}h[m]x[n-m]=\sum_{m=\max(n+1-L,0)}^{\min(n,N-1)}x[m]h[n-m],\ n=0..N+L-2 \qquad [6.2]$$

Notice that the output signal y is the sum of partial products obtained through three main operations:

– time-reversing of the input signal $x[-m]$,

– time-shifting of the reversed input signal $x[n-m]$,

– multiplication of $h[m]$ by $x[n-m]$.

If $x[n]$ and $h[n]$ are represented by the polynomials:

$$X(z)=\sum_{i=0}^{N-1}x_i z^i \quad \text{with} \quad \text{order}\{X(z)\}=N-1$$
$$H(z)=\sum_{i=0}^{L-1}h_i z^i \quad \text{with} \quad \text{order}\{H(z)\}=L-1 \qquad [6.3]$$

then the following polynomial is associated with the output signal $y = x * h$:

$$Y(z)=\sum_{i=0}^{N+L-2}y_i z^i = X(z)H(z) \text{ with order}\{Y(z)\}=N+L-2 \qquad [6.4]$$

In the case of two 2D signals, x and h:

$$x=\{x_{n_1,n_2} \quad \text{with} \quad n_1=0..N'-1 \quad \text{and} \quad n_2=0..N''-1\}$$
$$h=\{h_{n_1,n_2} \quad \text{with} \quad n_1=0..L'-1 \quad \text{and} \quad n_2=0..L''-1\} \qquad [6.5]$$

the linear convolution can be defined:

– as a matrix $y = x**h = h**x$ having the following elements:

$$\begin{cases} y_{n_1,n_2}=\sum_{r_1=0}^{N'-1}\sum_{r_2=0}^{N''-1}h_{n_1-r_1,n_2-r_2}x_{r_1,r_2} \\ n_1=0..N'+L'-2 \quad \text{with} \quad n_2=0..N''+L''-2 \end{cases} \qquad [6.6]$$

– as a polynomial convolution:

$$Y_{n_1}(z) = \sum_{r_1=0}^{N'-1} H_{n_1-r_1}(z) X_{r_1}(z) \text{ with } n_1 = 0..N'+L'-2 \quad [6.7]$$

– as a polynomial product:

$$Y(z_1,z_2) = H(z_1,z_2)X(z_1,z_2) \quad [6.8]$$

6.1.2. *LTI response to basic signals*

The impulse response of a LTI is defined by:

1D: $h[n] = \aleph_{1D}\{\delta[n]\}$

2D: $h[n_1,n_2] = \aleph_{2D}\{\delta[n_1,n_2]\}$

The indicial response of a LTI is defined by:

1D: $g[n] = \aleph_{1D}\{u[n]\} = \sum_{m=0}^{\infty} h[n-m]$

2D: $g[n_1,n_2] = \aleph_{2D}\{u[n_1,n_2]\} = \sum_{m_1=0}^{\infty} \sum_{m_2=0}^{\infty} h[n_1-m_1,n_2-m_2]$

LTI response to complex exponential signals

1D: if $x[n] = z^n$ with $z \in \mathbf{C}$, then the $\mathrm{LTI_{1D}}$ response to this type of signal is:

$$y[n] = z^n H(z)\big|_{z=\exp(j\omega)} \rightarrow \mathrm{e}^{j\omega n} H\left(\mathrm{e}^{j\omega}\right) \quad [6.9]$$

On the unit circle, for $z = \exp(jk2\pi/N)$, $k = 1..N$, this results in:

$$y[n] = \exp\left(jk\frac{2\pi}{N}n\right) H\left(\exp\left(jk\frac{2\pi}{N}\right)\right) \quad [6.10]$$

2D: if $x[n_1,n_2] = z_1^{n_1} z_2^{n_2}$, then the $\mathrm{LTI_{2D}}$ response is given by:

$$y[n_1,n_2] = z_1^{n_1} z_2^{n_2} H(z_1,z_2)\Big|_{z_{1,2}=\exp(j\omega_{1,2})} \rightarrow \mathrm{e}^{j(\omega_1 n_1+\omega_2 n_2)} H\left(\mathrm{e}^{j\omega_1},\mathrm{e}^{j\omega_2}\right) \quad [6.11]$$

On the unit sphere, for $z_{1,2} = \exp(jk_{1,2}2\pi / N_{1,2})$, $k_{1,2} = 1..N_{1,2}$ this results in:

$$y[n_1,n_2] = \mathrm{e}^{j\left(k_1\frac{2\pi}{N_1}n_1 + k_2\frac{2\pi}{N_2}n_2\right)} H\left(\mathrm{e}^{jk_1\frac{2\pi}{N_1}}, \mathrm{e}^{jk_2\frac{2\pi}{N_2}}\right) \qquad [6.12]$$

LTI response to periodical signals

1D: if $x[n] = x[n+N] = \frac{1}{N}\sum_{k=0}^{N-1} X[k]\exp(jkn2\pi / N)$, then the $\mathrm{LTI}_{1\mathrm{D}}$ response is the periodical signal:

$$y[n] = y[n+N] = \frac{1}{N}\sum_{k=0}^{N-1} X[k]\mathrm{e}^{jk\frac{2\pi}{N}n} H\left(\mathrm{e}^{jk\frac{2\pi}{N}}\right) \qquad [6.13]$$

2D: if:

$$x[n_1,n_2] = x[n_1+N_1, n_2+N_2] =$$

$$= \frac{1}{N_1N_2}\sum_{k_1=0}^{N_1-1}\sum_{k_2=0}^{N_2-1} X[k_1,k_2]\mathrm{e}^{jk_1\frac{2\pi}{N_1}n_1}\mathrm{e}^{jk_2\frac{2\pi}{N_2}n_2}$$

then the $\mathrm{LTI}_{2\mathrm{D}}$ response is the periodical signal:

$$y[n_1,n_2] = y[n_1+N_1, n_2+N_2] =$$

$$= \frac{1}{N_1N_2}\sum_{k_1=0}^{N_1-1}\sum_{k_2=0}^{N_2-1} X[k_1,k_2]e^{jk_1\frac{2\pi}{N_1}n_1}e^{jk_2\frac{2\pi}{N_2}n_2} H\left(e^{jk_12\pi/N_1}, e^{jk_22\pi/N_2}\right) \qquad [6.14]$$

LTI response to a general signal

1D: let us consider the signal $y[n] = \aleph_{1\mathrm{D}}\{x[n]\}$, which meet the following condition:

$$y[n] = \frac{1}{a_0}\left\{-\sum_{m=1}^{N_a} a_m y[n-m] + \sum_{m=0}^{N_b} b_m x[n-m]\right\} \qquad [6.15]$$

or the 1D convolution relationship:

$$y[n] = x[n] * h[n] \qquad [6.16]$$

The LTI_{1D} can thus be calculated using one of the following equations:

$$\begin{aligned} y[n] &= \mathrm{ZT}_{1D}^{-1}\{H(z)X(z)\} \\ y[n] &= \mathrm{DTFT}_{1D}^{-1}\left\{H\left(e^{j\omega}\right)X\left(e^{j\omega}\right)\right\} \\ y[n] &= \mathrm{DFT}_{1D}^{-1}\{H[k]X[k]\} \end{aligned} \qquad [6.17]$$

2D: let us consider the signal $y[n_1,n_2] = \aleph_{2D}\{x[n_1,n_2]\}$, which meets the following condition:

$$y[n_1,n_2] = \frac{1}{a_{0,0}}\left\{-\sum_{m_1=1}^{N_{a1}}\sum_{m_2=1}^{N_{a2}} a_{m_1,m_2}\, y[n_1-m_1,n_2-m_2] + \right. \\ \left. + \sum_{m_1=0}^{N_{b1}}\sum_{m_2=0}^{N_{b2}} b_{m_1,m_2}\, x[n_1-m_1,n_2-m_2]\right\} \qquad [6.18]$$

or the 2D convolution relationship:

$$y[n_1,n_2] = x[n_1,n_2] * *h[n_1,n_2] \qquad [6.19]$$

The LTI_{2D} can thus be calculated using one of the following equations:

$$\begin{aligned} y[n_1,n_2] &= \mathrm{ZT}_{2D}^{-1}\{H(z_1,z_2)X(z_1,z_2)\} \\ y[n_1,n_2] &= \mathrm{DTFT}_{2D}^{-1}\{H(\omega_1,\omega_2)X(\omega_1,\omega_2)\} \\ y[n_1,n_2] &= \mathrm{DFT}_{2D}^{-1}\{H[k_1,k_2]X[k_1,k_2]\} \end{aligned} \qquad [6.20]$$

6.2. Solved exercises

EXERCISE 6.1.

1. Write a MATLAB code to verify the linearity of system $\aleph$, represented by the transfer function:

$$H(z) = \frac{2.2403 + 2.4908z^{-1} + 2.2403z^{-2}}{1 - 0.4z^{-1} + 0.75z^{-2}}$$

Consider for this two different signals x and y and demonstrate the following relationship: $\aleph\{x[n]+y[n]\}=\aleph\{x[n]\}+\aleph\{y[n]\}$.

```
clf; n = 0:40; a = 2;b = -3;
x1 = cos(2*pi*0.1*n); x2 = cos(2*pi*0.4*n); x = a*x1 + b*x2;
num = [2.2403 2.4908 2.2403]; den = [1 -0.4 0.75];
ic = [0 0]; % set the initial conditions to zero
y1 = filter(num,den,x1,ic); %Calculation of the output signal y1[n]
y2 = filter(num,den,x2,ic); %Calculation of the output signal y2[n]
y = filter(num,den,x,ic); %Calculation of the output signal y[n]
yt = a*y1 + b*y2; d = y - yt; %Calculation of the error signal d[n]
% Plotting the error and the output signals
subplot(3,1,1); stem(n,y); ylabel('Amplitude');
title('Output corresponding to: a\cdot x_{1}[n] + b\cdot x_{2}[n]');
subplot(3,1,2); stem(n,yt); ylabel('Amplitude');
title('Output calculated as: a\cdot y_{1}[n] + b\cdot y_{2}[n]');
subplot(3,1,3); stem(n,d); xlabel('Time index n');
ylabel('Amplitude'); title('Error signal');
```

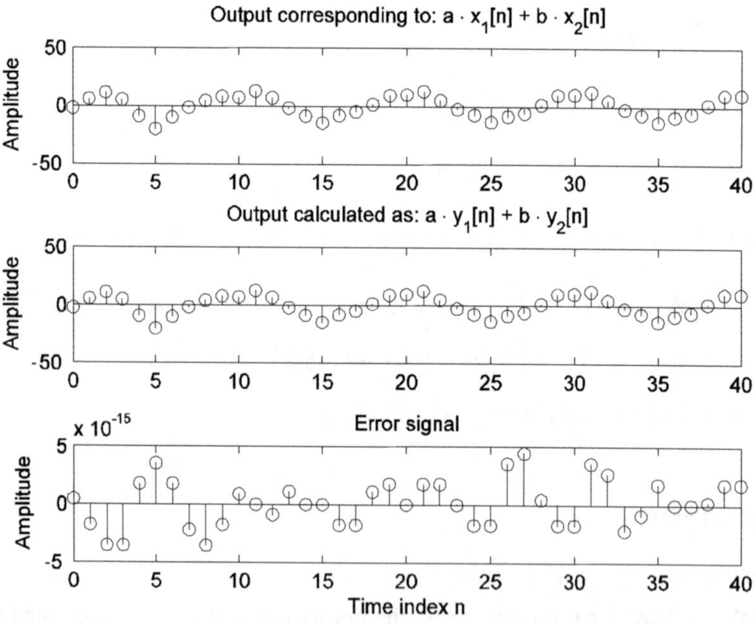

Figure 6.1. *Linearity of the convolution operator*

Note that the error signal amplitude is very close to zero, so the two output signals are equivalent.

2. Verify that this system is also time invariant.

```
clf; n = 0:40; D = 10; a = 3.0;b = -2;
x = a*cos(2*pi*0.1*n) + b*cos(2*pi*0.4*n); xd = [zeros(1,D) x];
num = [2.2403 2.4908 2.2403]; den = [1 -0.4 0.75];
ic = [0 0]; % set the initial conditions to zero
% Calculation of the output y[n]
y = filter(num,den,x,ic);
% Calculation of the output yd[n]
yd = filter(num,den,xd,ic);
% Calculation of the error signal d[n]
d = y - yd(1+D:41+D);
% Plotting the outputs
subplot(3,1,1); stem(n,y); ylabel('Amplitude');
title('Output y[n]'); grid;
subplot(3,1,2); stem(n,yd(1:41)); ylabel('Amplitude');
title(['Output corresponding to the shifted input x[n -', num2str(D),']']);
grid;
subplot(3,1,3); stem(n,d); xlabel('time index n');
ylabel('Amplitude');title('Error signal'); grid;
```

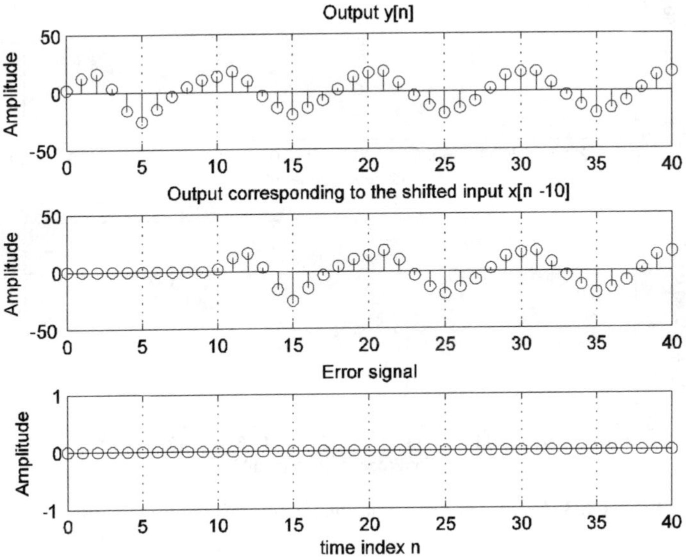

Figure 6.2. *Time invariance of the convolution operator*

It can be easily seen from the example above that when the input signal of a time invariant linear system is shifted, its output is also time delayed by the same amount.

EXERCISE 6.2.

Using a MATLAB code, demonstrate that the execution time associated with the function `conv.m` is proportional to N^2 where N is the number of points.

```
maxlengthlog2=5; m=9;
for k=1:maxlengthlog2
     i=k+m;
     disp(['The convolution length is: ' num2str(2^i) ]);
     h=[1:2^i]; dep=cputime; y=conv(h,h);
     timeexe(k)=cputime-dep;
end
longvec=[2.^[m+1:maxlengthlog2+m]];
subplot(211); plot(longvec,timeexe,'o'); grid
xlabel('convolution length') ;
ylabel('CPU time [s]');
subplot(212); plot(log2(longvec),log2(timeexe),'o') ;grid
xlabel('log2(convolution length)'); ylabel('log2(CPU time [s])');
```

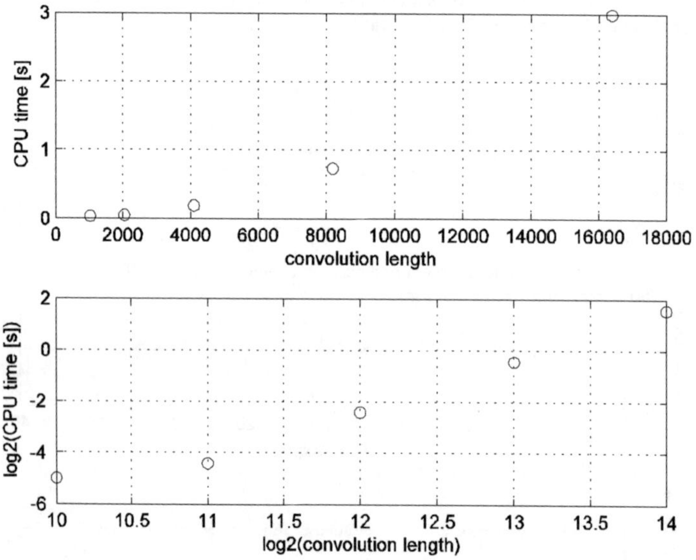

Figure 6.3. *Execution time associated with the convolution operator*

EXERCISE 6.3.

Calculate the 1D linear convolution for the following signals:

$$x = \{\ x_0, x_1, x_2\} \Rightarrow X(z) = x_0 + x_1 z + x_2 z^2, \text{ with } \quad N = 3$$

$$h = \{\ h_0, h_1\} \Rightarrow H(z) = h_0 + h_1 z, \text{ with } \quad L = 2$$

According to the definition: $y_n = \sum_{m=0}^{N-1} x_m h_{n-m}$. This results in:

$$y = \{y_0, y_1, y_2, y_3\} = \{x_0h_0, x_0h_1 + x_1h_0,\ x_1h_1 + x_2h_0,\ x_2h_1\}$$

The same result is obtained using the polynomial representations of the two signals, $X(z)$ and $H(z)$:

$$Y(z) = X(z)H(z) = \left(x_0 + x_1z + x_2z^2\right)\left(h_0 + h_1z\right) =$$
$$= x_0h_0 + (x_0h_1 + x_1h_0)z + (x_1h_1 + x_2h_0)z^2 + x_2h_1z^3$$

The MATLAB function, conv.m, for the linear convolution calculation, in fact performs the product of the polynomials $X(z)$ and $Y(z)$. For example, for $x = \{1, 2, 3\}$ and $h = \{1, 1\}$ the following result is obtained:

```
x = [1,2,3] ; h = [1,1] ; [y] = conv(x,h)

y = 1 3 5 3
```

The inverse operator is the deconvolution and performs the polynomial division. Thus, the MATLAB function deconv.m leads to:

```
[q,r]=deconv(y,x)

q = 1 1      r = 0 0 0 0
```

EXERCISE 6.4.

Write a MATLAB code to illustrate Plancherel's theorem using the following signals:

$$x_1(n) = 2n - 1,\ n = 1..9$$
$$x_2(n) = 1,\ -2,\ 3,\ -2,\ 1,\ 0,\ 0,\ ...$$

```
w = -pi:2*pi/255:pi;
x1 = [1 3 5 7 9 11 13 15 17];
x2 = [1 -2 3 -2 1];
y = conv(x1,x2);
h1 = freqz(x1, 1, w);
h2 = freqz(x2, 1, w);
hp = h1.*h2;
h3 = freqz(y,1,w);
subplot(2,2,1);
plot(w/(2*pi),abs(hp));grid
title('Product of the amplitude spectra')
```

```
subplot(2,2,2);
plot(w/(2*pi),abs(h3));grid
title('Amplitude spectrum of the convolution result')
subplot(2,2,3);
plot(w/(2*pi),angle(hp));grid
title('Sum of the phase spectra')
subplot(2,2,4);
plot(w/(2*pi),angle(h3));grid
title('Phase spectrum of the convolution result')
```

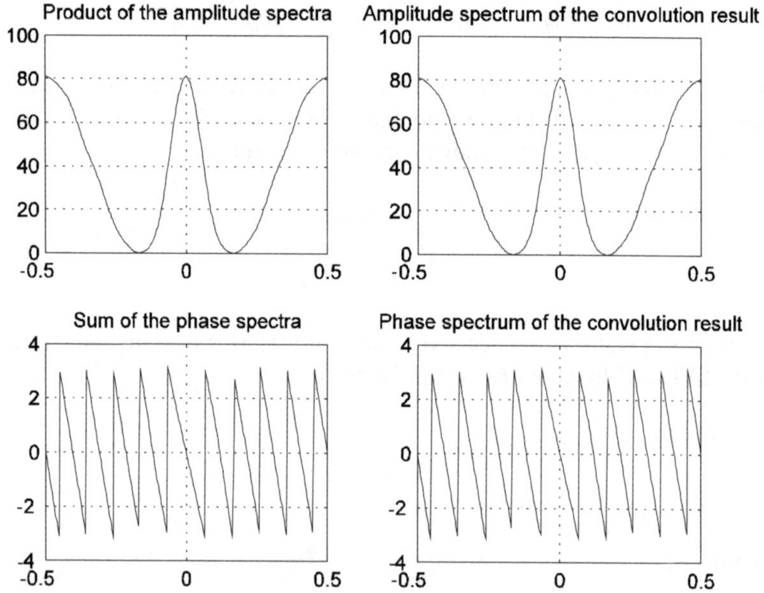

Figure 6.4. *Plancherel's theorem illustration*

EXERCISE 6.5.

Calculate sequentially the convolution of two discrete-time signals. The origin of the two functions is assumed to be $n = 0$.

```
x=ones(3,1); h=exp(-[1:10]);
% Convolution length calculation
Ly=length(x)+length(h)-1;
lh=length(h);
lx=length(x);
y=zeros([Ly,1]);
% x inversion
xf=fliplr(x);
disp(['Type any key to begin']);
```

```
for i=1:1:Ly
     indlo=max(0,i-lx);
     indhi=min(i-1,lh-1);
     for j=indlo:indhi
          y(i)=h(j+1)*x(i-j)+y(i);
     end;
     subplot(311);
     stem(-lx+i:1:i-1,xf);
     ylabel('Reversed x');
     lim1= max(lh,lx)  ;
     lim2= min(min(min(x),min(h)),0)  ;
     lim3= max(max(x),max(h))  ;
     axis([-lx,lim1,lim2,lim3]);
     subplot(312);
     stem(0:1:lh-1,h);
     ylabel('h');
     axis([-lx,lim1,lim2,lim3]);
     subplot(313);
     stem(0:1:Ly-1,y);
     ylabel('y');
     pause;
end
```

The first and last step of the convolution is illustrated in Figures 6.5 and 6.6. The rectangular signal is reversed and is then shifted along the exponential signal.

The convolution value corresponding to each point is finally obtained as the sum of the partial point to point products of the two signals.

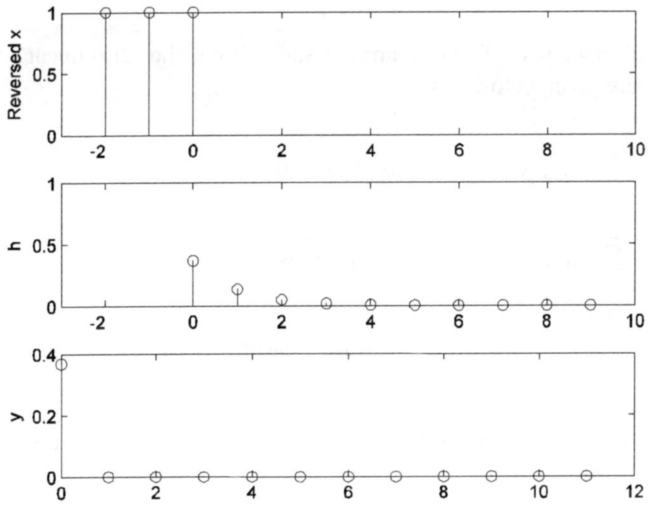

Figure 6.5. *First step of the convolution of two discrete-time signals*

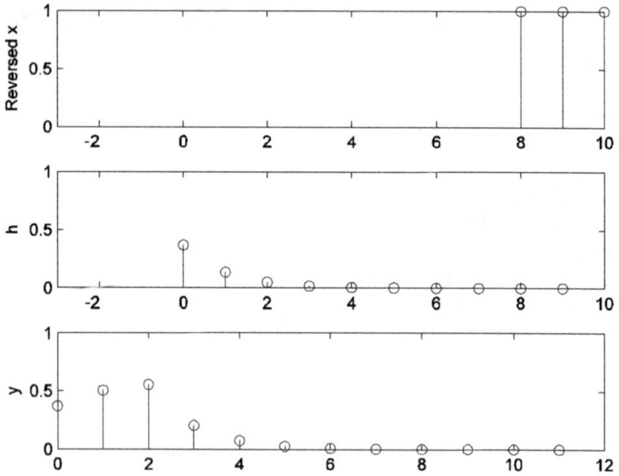

Figure 6.6. *Last step of the convolution of two discrete-time signals*

EXERCISE 6.6.

Calculate the 2D linear convolution for the following discrete-time signals:

$$x = \left\{ x_{n_1,n_2}, n_1 = 0, 1;\ n_2 = 0, 1, 2 \right\} = \left[0, 1, 2; 3, 4, 5 \right]$$

$$x = \left\{ h_{n_1,n_2}, n_1 = 0, 1;\ n_2 = 0, 1, 2 \right\} = \left[1, 0, 1; 1, 0, 1 \right]$$

The first elements of the matrix issued from the 2D linear convolution $y = h**x$, are given below:

$$y_{0,0} = \sum_{r_1=0}^{1} \sum_{r_2=0}^{2} h_{0-r_1,0-r_2} x_{r_1,r_2} = h_{0,0}x_{0,0} = 0$$

$$y_{0,1} = \sum_{r_1=0}^{1} \sum_{r_2=0}^{2} h_{0-r_1,1-r_2} x_{r_1,r_2} = h_{0,1}x_{0,0} + h_{0,0}x_{0,1} = 0 + 1 = 1$$

$$y_{0,2} = \sum_{r_1=0}^{1} \sum_{r_2=0}^{2} h_{0-r_1,2-r_2} x_{r_1,r_2} = h_{0,2}x_{0,0} + h_{0,1}x_{0,1} + h_{0,0}x_{0,2} = 0 + 0 + 1 \cdot 2 = 2$$

The final result can be obtained using the following MATLAB code:

```
x=[0,1,2;3,4,5];
h=[1,0,1;1,0,1];
```

```
y=conv2(x,h)

y=0 1  2 1 2
  3 5 10 5 7
  3 4  8 4 5
```

The same result can be obtained if the convolution is calculated as a product of polynomials.

$$Y_0(z) = \sum_{r_1=0}^{1} H_{0-r_1}(z)X_{r_1}(z) = H_0(z)X_0(z) = z + 2z^2 + z^3 + 2z^4$$

$$Y_1(z) = \sum_{r_1=0}^{1} H_{1-r_1}(z)X_{r_1}(z) = H_1X_0 + H_0X_1 = 3 + 5z + 10z^2 + 5z^3 + 7z^4$$

$$Y_2(z) = \sum_{r_1=0}^{1} H_{2-r_1}(z)X_{r_1}(z) = H_1(z)X_1(z) = 3 + 4z + 8z^2 + 4z^3 + 5z^4$$

Each 2D signal can also be represented as a polynomial depending on two variables:

$$\begin{aligned} X(z_1, z_2) &= 0z_1^0z_2^0 + 1z_1^0z_2^1 + 2z_1^0z_2^2 + 3z_1^1z_2^0 + 4z_1^1z_2^1 + 5z_1^1z_2^2 \\ &= z_2 + 2z_2^2 + 3z_1 + 4z_1z_2 + 5z_1z_2^2 \\ H(z_1, z_2) &= 1z_1^0z_2^0 + 0z_1^0z_2^1 + 1z_1^0z_2^2 + 1z_1^1z_2^0 + 0z_1^1z_2^1 + 1z_1^1z_2^2 \\ &= 1 + z_2^2 + z_1 + z_1z_2^2 \end{aligned}$$

The 2D convolution thus takes the form of such a polynomial:

$$\begin{aligned} Y(z_1, z_2) &= H(z_1, z_2)X(z_1, z_2) \\ &= \left(z_2 + 2z_2^2 + 3z_1 + 4z_1z_2 + 5z_1z_2^2\right)\left(1 + z_2^2 + z_1 + z_1z_2^2\right) \end{aligned}$$

EXERCISE 6.7.

Write a MATLAB code to demonstrate that the filtering operator is in fact a convolution operator. Consider for this the following signal:

$x(n)$ = 1, -2, 3, -4, 3, 2, 1, 0,...

and the system with the finite impulse response given below:

$h(n)$ = 3, 2, 1, -2, 1, 0, -4, 0, 3, 0, ...

```
h = [3 2 1 -2 1 0 -4 0 3];
x = [1 -2 3 -4 3 2 1];
y = conv(h,x) ; n = 0:14 ;
subplot(2,1,1);stem(n,y);
xlabel('Time index n');
ylabel('Amplitude');
title('Output obtained by convolution');
grid;
x1 = [x zeros(1,8)];
y1 = filter(h,1,x1);
subplot(2,1,2);stem(n,y1);
xlabel('Time index n');
ylabel('Amplitude');
title('Output obtained by filtering');
grid;
```

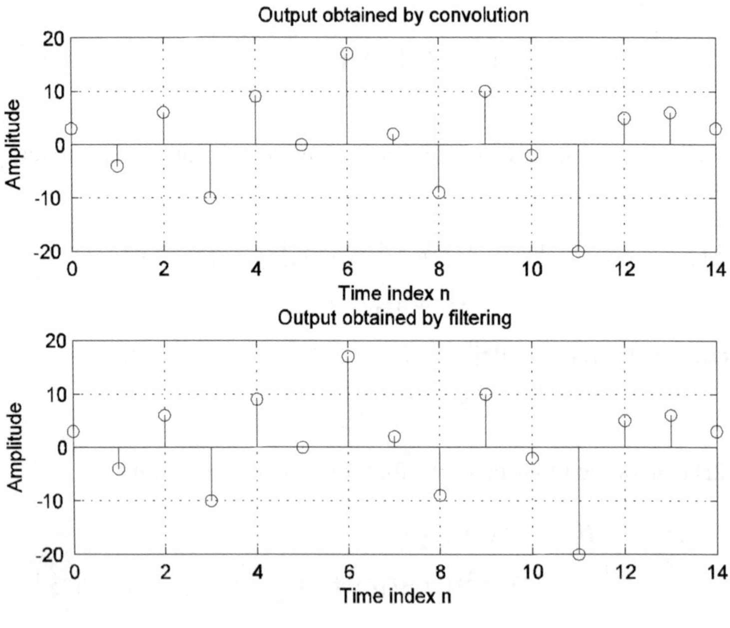

Figure 6.7. *Equivalence of the convolution and the filtering operators*

EXERCISE 6.8.

Consider a LTI_{1D}, characterized in the discrete-time domain by the following equation:

$$y[n] - 0.334y[n-1] = 0.334x[n] + 0,334x[n-1]$$

or equivalently by the transfer function:

$$H(z) = \frac{Y(z)}{X(z)} = \frac{0.334 + 0.334z^{-1}}{1 - 0.334z^{-1}}$$

Because of the MATLAB rule about the vector and matrix indices, the transfer function coefficients are denoted as indicated below:

$$H(z) = \frac{b(1) + b(2)z^{-1} + ... + b(nb+1)z^{-nb}}{a(1) + a(2)z^{-1} + ... + a(na+1)z^{-na}}$$

If $a(1) \neq 1$, the coefficients are normalized so that $a(1) = 1$.

The following MATLAB code calculates and plots the complex gain, the poles and the zeros of a LTI_{1D}.

```
b=[0.334,0.334]; a=[1.0,-0.334];
figure(1); zplane(b,a); figure(2); freqz(b,a)
```

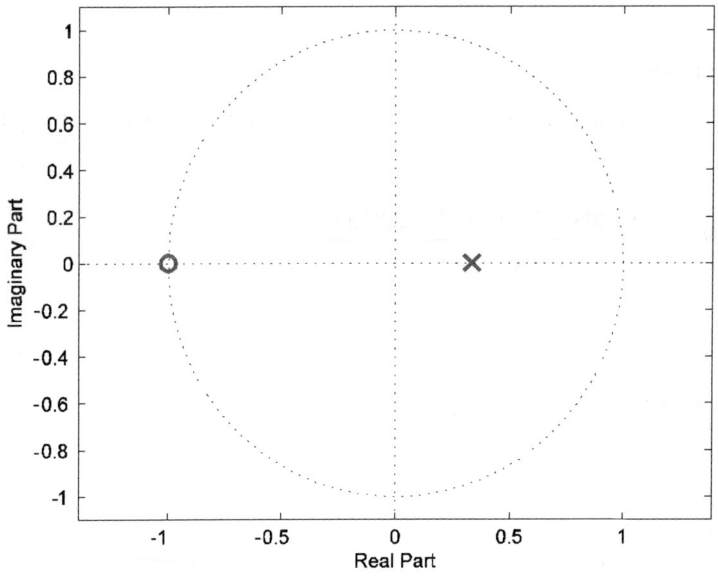

Figure 6.8. *Localization of the transfer function poles and zeros*

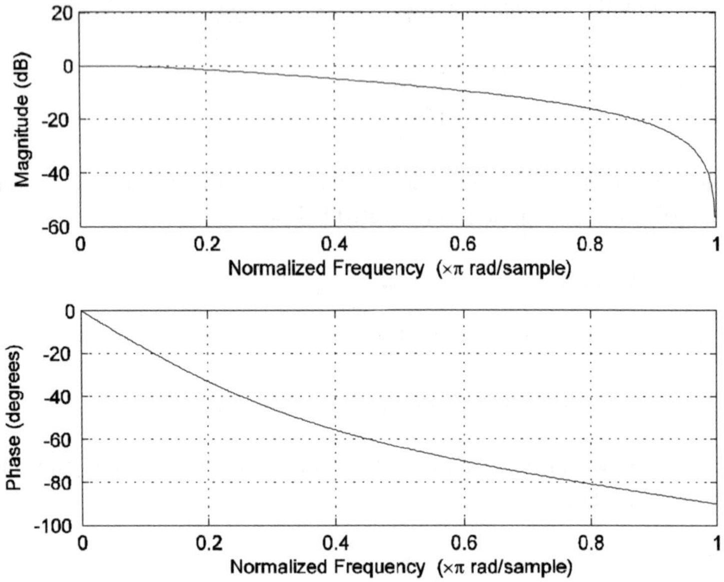

Figure 6.9. *Amplitude and phase responses of the system*

EXERCISE 6.9.

Calculate the impulse response of the system defined by the following transfer function:

$$H(z) = \frac{2.2403 + 2.4908z^{-1} + 2.2403z^{-2}}{1 - 0.4z^{-1} + 0.75z^{-2}}$$

```
clf;
N = 40;
num = [2.2403 2.4908 2.2403];
den = [1 -0.4 0.75];
y = impz(num,den,N);
% Impulse response representation
stem(y);
xlabel('Time index n');
ylabel('Amplitude');
title('Impulse response');
grid;
```

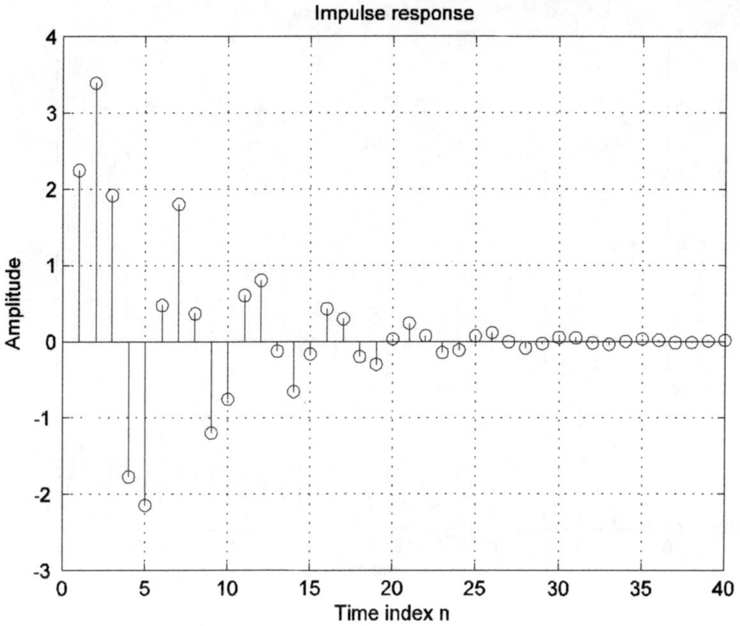

Figure 6.10. *System impulse response*

If initial conditions are not zero, the previously defined functions are not able to take them into account. In this case the function `filter.m` should be used.

A simple calculation indicates that for a first order system, the initial conditions can be written in the form:

$$z0 = [b1x(-1) - ay(-1)]$$

For a 2nd order system, they become:

$$z0 = [b1x(-1) + b2x(-2) - a1y(-1) - a2y(-2),\;\; b1x(-1) - a2y(-1)]$$

Consider for example the previous system with $y(-1) = 2$, $y(-2) = 1$ and $x(n) = u(n)$.

```
n=0:34 ; x=ones(1,35);
num = [2.2403 2.4908 2.2403]; den = [1 -0.4 0.75];
z0=[0.4*2-0.75*1,-0.75*2] ;
y=filter(num,den,x,z0);
stem(y); xlabel('Time index n'); ylabel('Amplitude');
title('Indicial response'); grid;
```

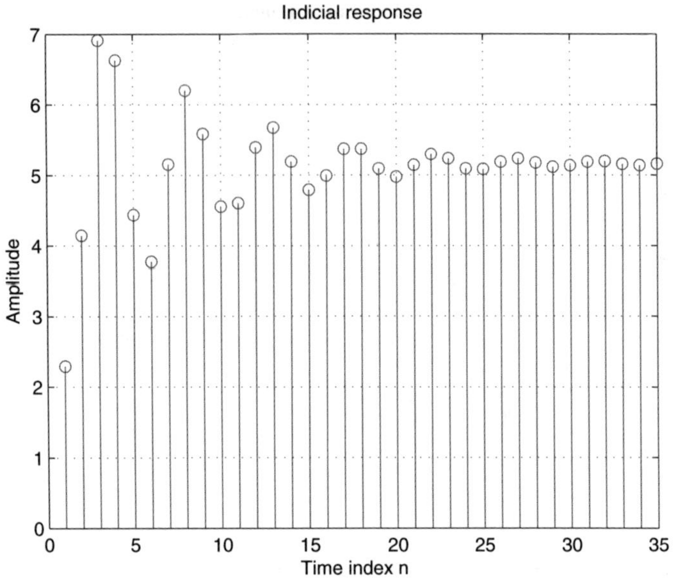

Figure 6.11. *System indicial response*

EXERCISE 6.10.

Compare the impulse responses of two 5th order lowpass Butterworth and Cauer filters, whose normalized cutoff frequency is equal to $W_n = 0.5$.

```
[b,a]=butter(5,0.5);
imp=[1;zeros(49,1)];  % Dirac pulse
h=filter(b,a,imp);
subplot(211)
stem(h);
title('butterworth')
[b,a]=ellip(5,1,20,0.5);
h=filter(b,a,imp);
subplot(212)
stem(h);
title('cauer')
```

The same result can be obtained using the MATLAB command:

```
impz(b,a)
```

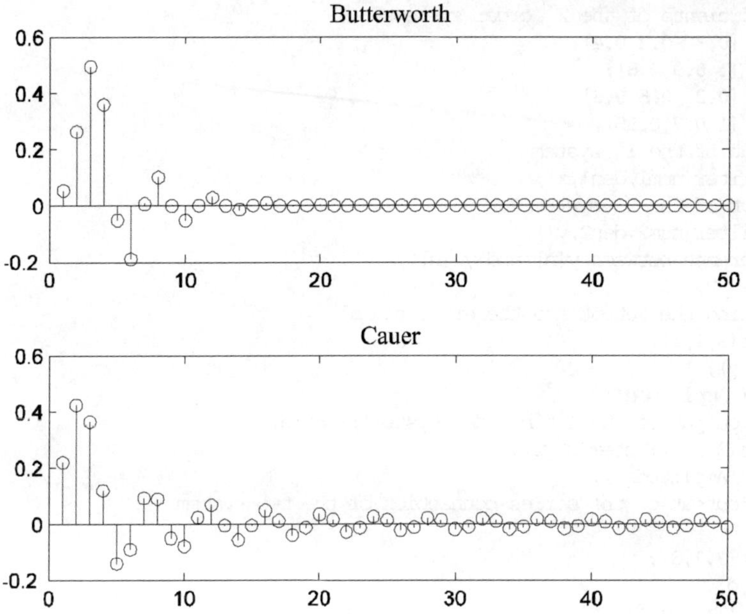

Figure 6.12. *Impulse response of two systems corresponding to Butterworth (top) and Cauer (bottom) models*

EXERCISE 6.11.

Consider two 2nd order systems defined by the following transfer functions:

$$H_1(z) = \frac{0.3 - 0.2z^{-1} + 0.4z^{-2}}{1 + 0.9z^{-1} + 0.8z^{-2}}$$

$$H_2(z) = \frac{0.2 - 0.5z^{-1} + 0.3z^{-2}}{1 + 0.7z^{-1} + 0.85z^{-2}}$$

Find the transfer function equivalent to that of the series connection of the two systems.

```
clf;
x = [1 zeros(1,40)]; n = 0:40;
% 4th order system coefficients
den = [1 1.6 2.28 1.325 0.68];
num = [0.06 -0.19 0.27 -0.26 0.12];
% 4th order system output
y = filter(num,den,x);
```

```
% Coefficients of the 2nd order systems
num1 = [0.3 -0.2 0.4];
den1 = [1 0.9 0.8];
num2 = [0.2 -0.5 0.3];
den2 = [1 0.7 0.85];
% Output of the 1st system
y1 = filter(num1,den1,x);
% Output of the 2nd system
y2 = filter(num2,den2,y1);
% Difference between y[n] and y2[n]
d = y - y2;
% Plotting the output and the error signal
subplot(3,1,1);
stem(n,y);
ylabel('Amplitude');
title('output of the 4^t^h order system'); grid;
subplot(3,1,2); stem(n,y2);
ylabel('Amplitude');
title('output of the series connection of the two systems');
grid;
subplot(3,1,3);
stem(n,d)
xlabel('Time index n');
ylabel('Amplitude');
title('Error signal'); grid;
```

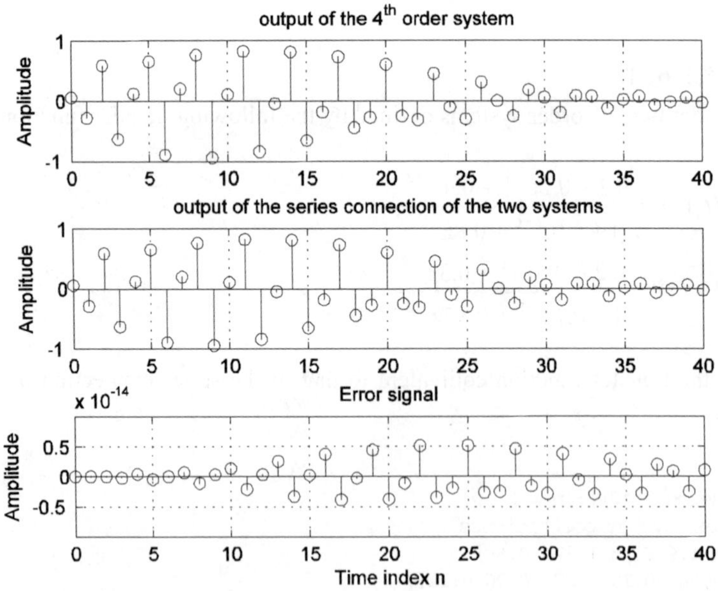

Figure 6.13. *Influence of the series connection of two digital systems*

Notice that this result is also valid for analog filters only if the series connection does not change their features, i.e. if the input impedance of the second filter is much larger than the output impedance of the first. This condition is always met in the case of digital filters.

EXERCISE 6.12.

Demonstrate that the system with the transfer function below is stable using time domain and z domain tests:

$$H(z) = \frac{1 - 0.8z^{-1}}{1 + 1.5z^{-1} + 0.9z^{-2}}$$

```
clf;
num = [1 -0.8];
den = [1 1.5  0.9];
N = 200;
% Impulse response calculation
h = impz(num,den,N+1);
parsum = 0;
for k = 1:N+1;
     parsum = parsum + abs(h(k));
     if abs(h(k)) < 10^(-6),
          break,
     end
end
% Plotting the impulse response
n = 0:N;
stem(n,h);
xlabel('Time index n');
ylabel('Amplitude');
% Plotting abs(h(k))
figure;
zplane(num,den);
disp('Value =');
disp(abs(h(k)));
```

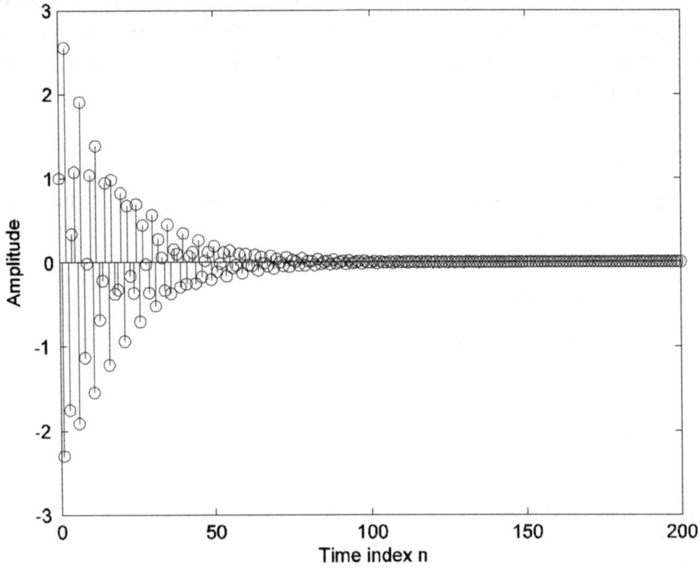

Figure 6.14. *Impulse response of a stable system*

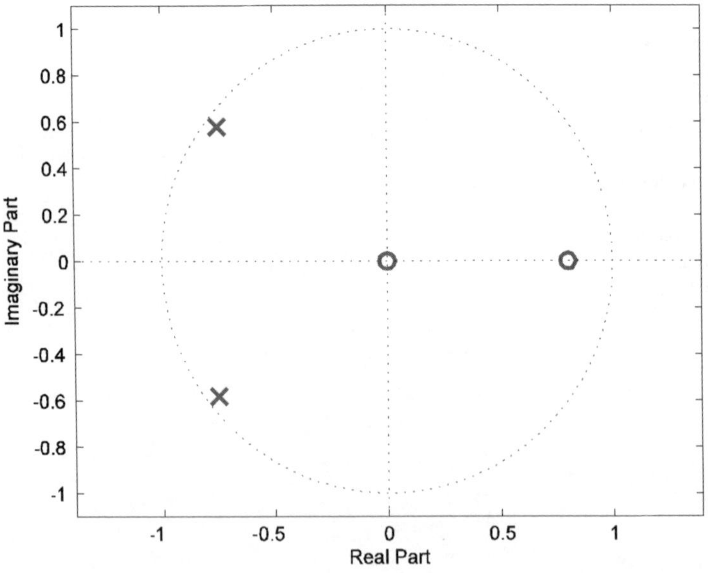

Figure 6.15. *Poles and zeros localization on the unit circle*

```
Value =
  1.6761e-005
```

In the z-domain the stability condition requires that all the poles of the transfer function, which are the roots of its denominator, have a magnitude less than 1.

```
% Stability test
ki = poly2rc(den);
disp('The denominator roots are the poles of the transfer function and are
equal to: '); disp(ki);
```

```
The denominator roots are the poles of the transfer function and are
equal to:
    0.7895
    0.9000
```

EXERCISE 6.13.

The poles and the zeros of a LTI_{1D}, in the Z plane are:

```
z=[exp(j*pi/5);exp(-j*pi/5)]; p=[0.9*z];
```

while the coefficients of its transfer function can be obtained by:

```
b=poly(z);
a=poly(p);
```

Calculate the transfer function of this LTI_{1D} using the function `freqz.m` and conclude about its type.

The residues, poles and direct terms of the transfer function are related by the following relationship:

$$H(z)=\frac{B(z)}{A(z)}=\frac{r(1)}{1-p(1)z^{-1}}+\ldots+\frac{r(n)}{1-p(n)z^{-1}}+k(1)+k(2)z^{-1}+\ldots$$

With MATLAB they are obtained using the function `residue.m`:

```
[r,p,k]=residue(b,a);
```

The system impulse response can thus be calculated in the form:

$$h[n]=r_1 p_1^n + r_2 p_2^n + k\delta[n]$$

Compare the two impulse responses coded below:

```
n=(0:100)';
h1=r(1)*p(1).^n+r(2)*p(2).^n;
h1(1)=h1(1)+k;
h2=impz(b,a,n);
subplot(211);
stem(h1);
xlabel('Time index n');
ylabel('Impulse response');
title('First method')
subplot(212);
stem(h2)
xlabel('Time index n');
ylabel('Impulse response');
title('Second method')
```

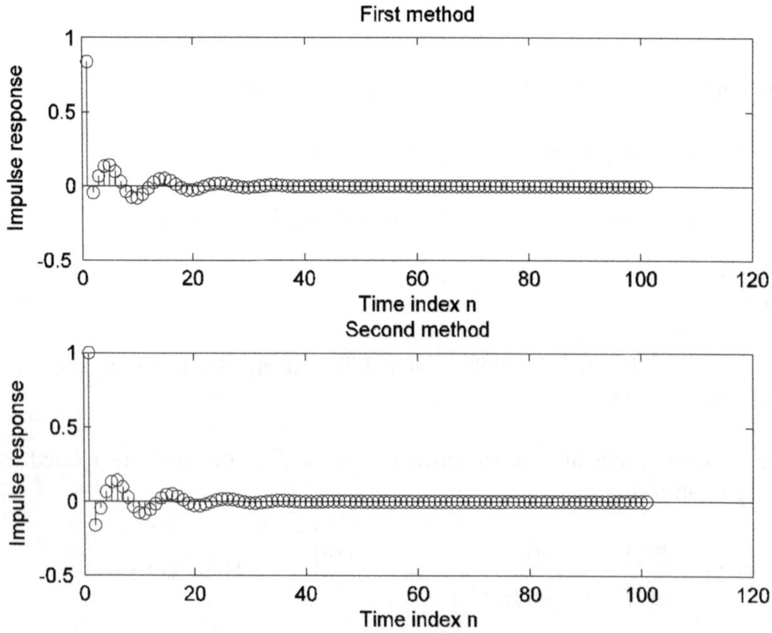

Figure 6.16. *Impulse response obtained with the residue method (top) and by direct calculation (bottom)*

Find out the system response for the following two sinusoids and their sum:

```
x1=sin(pi/5*n);
y1=filter(b,a,x1);
```

```
x2=sin(pi/6*n);
y2=filter(b,a,x2);
subplot(221)
plot(n,y1);
xlabel('Time index n');
ylabel('Amplitude');
title('System response for the 1^s^t sinusoid');
subplot(222)
plot(n,y2)
xlabel('Time index n');
ylabel('Amplitude');
title('System response for the 2^n^d sinusoid');
x=x1+x2;y=y1+y2;
subplot(223)
plot(n,x)
xlabel('Time index n');
ylabel('Amplitude');
title('Sum of the two sinusoids');
subplot(224)
plot(n,y)
xlabel('Time index n');
ylabel('Amplitude');
title('System response for the sum of the 2 sinusoids');
```

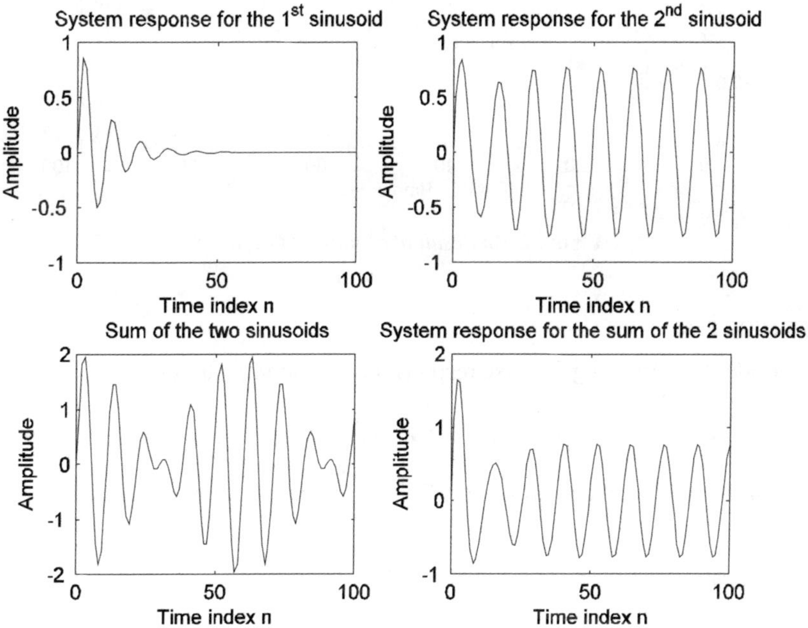

Figure 6.17. *Filtering of the sum of two sinusoids*

Finally, evaluate the transfer function for $z = \exp(j\pi/6)$ and then calculate the stationary response of the system to the signal $\sin(\pi n/6)$.

```
H=polyval(b,exp(j*pi/6))/polyval(a,exp(j*pi/6));
ys=abs(H)*sin(pi/6*n+angle(H));
plot(n,ys)
xlabel('Time index n');
ylabel('Amplitude');
```

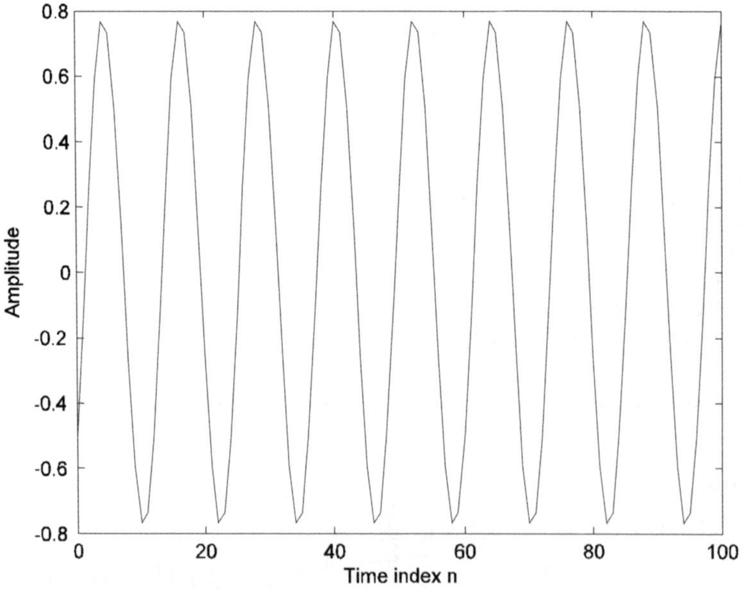

Figure 6.18. *Result of a sinusoid filtering*

EXERCISE 6.14.

Consider the following impulse responses corresponding to four LTI_{1D}:

```
h1=[1,2,1]; h2=[1,2,2,1]; h3=[1,0,-1]; h4=[1,2,0,-2,-1];
```

Calculate and plot their zeros in the Z plane, using the following code lines:

```
figure;
subplot(221);
zplane(roots(h1));
title('1^s^t system')
subplot(222);
zplane(roots(h2));
```

```
title('2^n^d system')
subplot(223);
zplane(roots(h3));
title('3^r^d system')
subplot(224);
zplane(roots(h4));
title('4^t^h system')
```

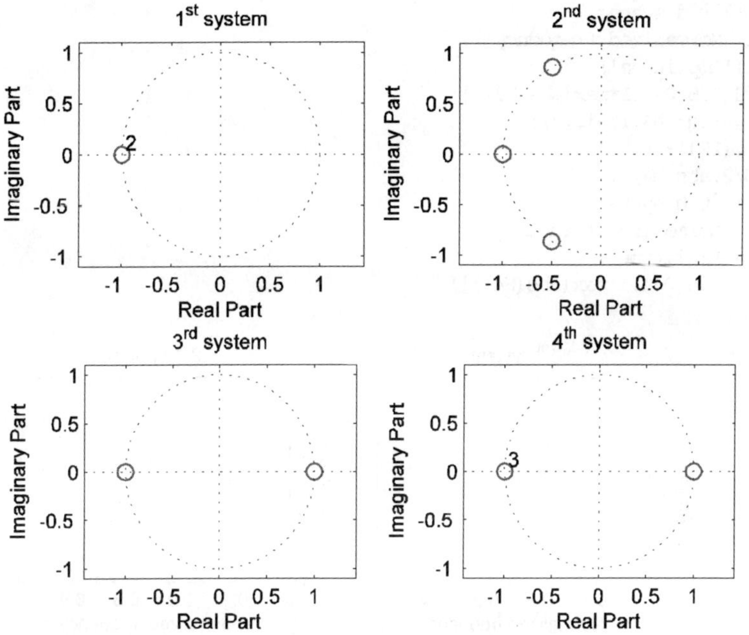

Figure 6.19. *Zeros of the four transfer functions provided in exercise 6.15*

The MATLAB code below can be used for plotting the magnitude of the transfer functions $\left|H_i\left(e^{j\omega}\right)\right|$:

```
figure;
[H1,W]=freqz(h1,1,512,2);
subplot(221);
plot(W/2,abs(H1));
title('1^s^t system')
xlabel('Normalized frequency');
ylabel('Amplitude')
axis([0 0.5 0 1.1*max(abs(H1))])
[H2,W]=freqz(h2,1,512,2);
subplot(222);
```

```
plot(W/2,abs(H2));
title('2^n^d system')
xlabel('Normalized frequency');
ylabel('Amplitude')
axis([0 0.5 0 1.1*max(abs(H2))])
[H3,W]=freqz(h3,1,512,2);
subplot(223);
plot(W/2,abs(H3));
title('3^r^d system')
xlabel('Normalized frequency');
ylabel('Amplitude')
axis([0 0.5 0 1.1*max(abs(H3))])
[H4,W]=freqz(h4,1,512,2);
subplot(224);
plot(W/2,abs(H4));
title('4^t^h system')
xlabel('Normalized frequency');
ylabel('Amplitude')
axis([0 0.5 0 1.1*max(abs(H4))])
```

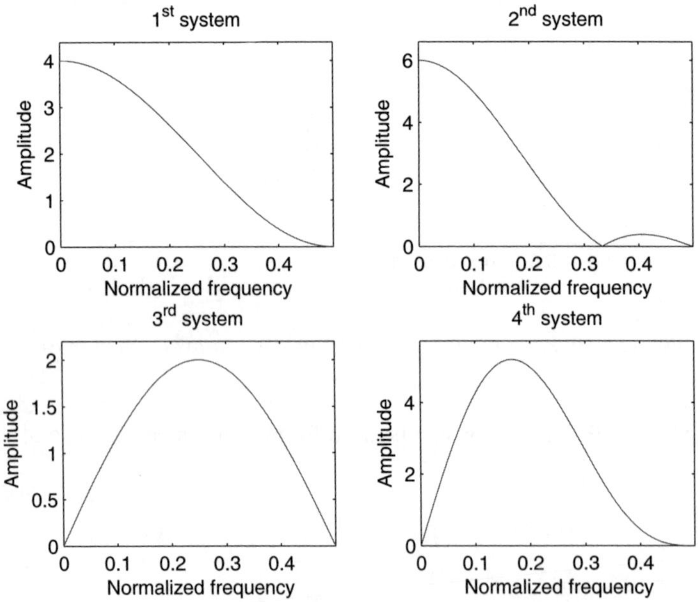

Figure 6.20. *Spectral characterization of the four systems described in exercise 6.15*

The first system has a double zero in $z = -1$, so it has a lowpass filter behavior. This is the same for the second system, whose zeros are: –1, –0.5 + 0.866i and –0.5 – 0.866i.

The two last systems have the same zeros, 1 and −1, but the second zero is triple in the case of the 4th system. They have therefore a bandpass filter behavior.

EXERCISE 6.15.

Consider a LTI_{2D} having the transfer function:

$$H(z_1, z_2) = B(z_1, z_2) = \sum_{m_1=0}^{1} \sum_{m_2=0}^{1} b_{m_1,m_2} z_1^{-m_1} z_2^{-m_2}$$

The coefficient matrix b_{m_1,m_2} has the form:

$$b = \begin{bmatrix} 1 & 1 \\ 1 & 0 \end{bmatrix} = \begin{bmatrix} b_{1,1} & b_{1,2} \\ b_{2,1} & b_{2,2} \end{bmatrix}$$

The MATLAB code below calculates the impulse response, the indicial response and the response to a complex exponential of this LTI_{2D}, on 50×50 points.

```
b=[1,1;1,0];
dirac2D=zeros(50);
dirac2D(1,1)=1;
h=conv2(b,dirac2D);
subplot(221)
mesh(dirac2D);
title('2D Dirac pulse');
subplot(223)
mesh(h);
title('impulse response')
subplot(222)
unit2D=ones(50);
mesh(unit2D);
title('2D Heaviside function');
subplot(224)
g=conv2(b,unit2D);
mesh(g);
title('indicial response')
for n1=1:64
    for n2=1:64
        e(n1,n2)=exp(j*n1*2*pi/64)*exp(j*n2*2*pi/64);
    end
end
y=conv2(b,e);
figure;
subplot(211)
mesh(real(e));
title('real part of the 2D exponential sequence')
```

```
subplot(212)
mesh(real(y));
title('real part of the system response')
```

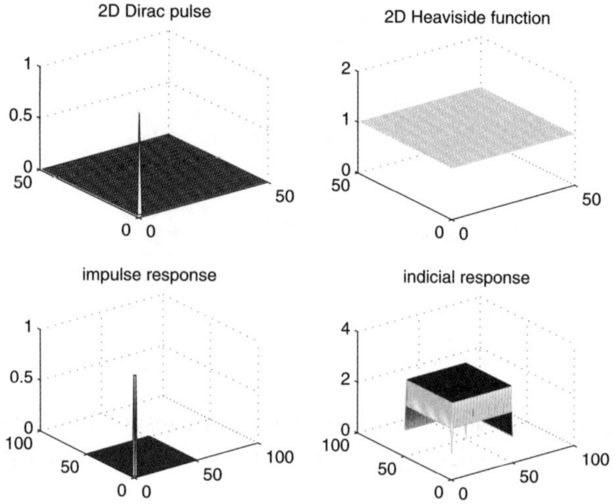

Figure 6.21. *Impulse and indicial responses of a LTI_{1D}*

EXERCISE 6.16.

The impulse response of a LTI_{2D} is given by:

$$h[n_1, n_2] = \delta[n_1 + 1, n_2] + \delta[n_1 - 1, n_2] + \delta[n_1, n_2 + 1] + \delta[n_1, n_2 - 1]$$

The transfer function of this system can thus be calculated as follows:

$$H(\omega_1, \omega_2) = \sum_{n_1=-\infty}^{+\infty} \sum_{n_2=-\infty}^{+\infty} h[n_1, n_2] \exp(-j\omega_1 n_1 - j\omega_2 n_2) = 2(\cos\omega_1 + \cos\omega_2)$$

The following MATLAB code calculates and plots this transfer function:

```
h=[0,1,0;1,0,1;0,1,0];
freqz2(h,[64 64]);
title('Transfer function');
```

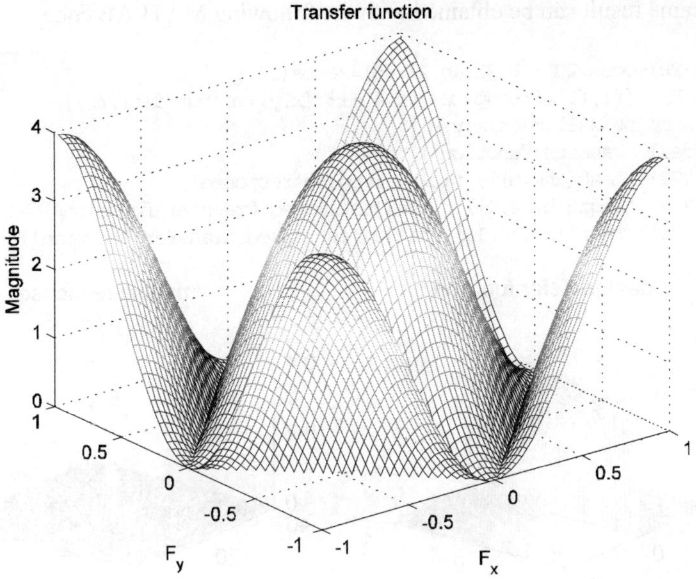

Figure 6.22. *2D transfer function*

The code lines below lead to the same result:

```
for w1=1:64
      for w2=1:64
            H(w1,w2)=2*cos(w1*2*pi/64)+2*cos(w2*2*pi/64);
      end
end
mesh(abs(H))
```

EXERCISE 6.17.

Calculate the impulse response $h[n_1, n_2]$ of a LTI_{2D}, whose transfer function is given by:

$$H(\omega_1, \omega_2) = \begin{cases} 1, & |\omega_1| \le a < \pi, \ |\omega_2| \le b < \pi \\ 0, & \text{otherwise} \end{cases}$$

According to the definition:

$$h[n_1, n_2] = \frac{1}{4\pi^2} \int_{-a}^{+a} \int_{-b}^{+b} \exp(j\omega_1 n_1 + j\omega_2 n_2)\, d\omega_1 d\omega_2 = \frac{\sin a n_1}{\pi n_1} \frac{\sin b n_2}{\pi n_2}$$

The same result can be obtained with the following MATLAB code:

```
H(1:20,1:20)=ones(20); H(21:40,21:40)=zeros(20);
h=fsamp2(H); [f1,f2]=freqspace([40,40]); [x,y]=meshgrid(f1,f2);
figure; subplot(221),mesh(x,y,H);
title('ideal transfer function')
subplot(222); mesh(abs(h)); title('impulse response')
subplot(223);freqz2(h,[4,4]);title('estimated transfer function')
subplot(224);freqz2(h,[16,16]);title('estimated transfer function')
```

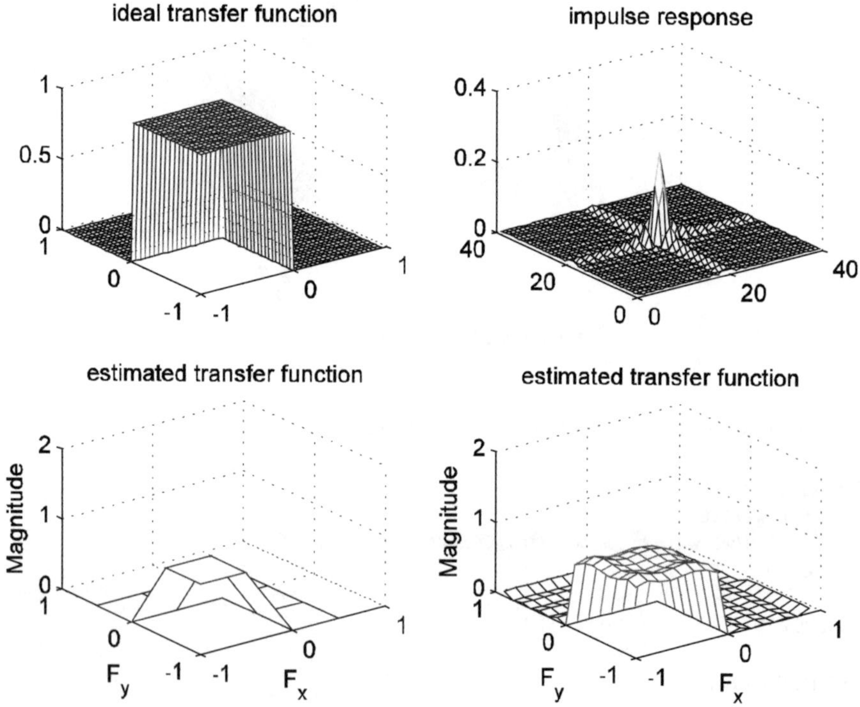

Figure 6.23. *Approximation of a 2D transfer function*

Notice that the impulse response is similar to the "2D sinc" function. The transfer function estimates make use of 4×4 and 16×16 impulse response coefficients. The approximation is obviously better when the number of coefficients increases.

6.3. Exercises

EXERCISE 6.18.

Consider the following transfer functions:

$$H_1(z)=1-2z^{-1}+z^{-2}$$

$$H_2(z)=1+2z^{-1}+z^{-2}$$

$$H_3(z)=1-z^{-1}+0.5z^{-2}$$

$$H_4(z)=\frac{\left(1-z^{-1}\right)^2}{1-z^{-1}+0.5z^{-2}}$$

$$H_5(z)=\frac{\left(1+z^{-1}\right)^2}{1-z^{-1}+0.5z^{-2}}$$

$$H_6(z)=\frac{1}{1-z^{-1}+0.5z^{-2}}$$

a. Plot the poles and zeros of these functions using the command `zplane.m`.

b. Represent the above transfer functions, using the command `freqz.m`, and indicate the type of corresponding systems.

c. Determine the indicial response of the system represented by $H_5(z)$.

EXERCISE 6.19.

A LTI_{1D} is characterized by the following discrete-time equation:

$$y[n]-1.8\cos\left(\frac{\pi}{16}\right)y[n-1]+0.81y[n-2]=x[n]+\frac{1}{2}x[n-1]$$

a. Determine the poles of the system transfer function which are the roots p_k of the polynomial: $A(z)=1+\sum_{m=1}^{Na} a_m z^m$ (use the MATLAB command `roots.m`). If these roots are complex conjugated, the system response will have the form of a complex exponential. Represent the real and imaginary parts of signals $p_k^n u[n]$.

b. As the transfer function of the system is of 2nd order, its impulse response has the form $h[n]=\left(\alpha p_1^n+\beta p_2^n\right)u[n]$. Verify this statement using the system impulse response calculated using the MATLAB function `impz.m`

c. Find constants α and β from the expression of $h[n]$.

d. Calculate the system stationary response to the complex exponential signal $x[n]=e^{j\pi/4n}$, $n=0..30$, using the MATLAB function `filter.m`.

EXERCISE 6.20.

Consider the following transfer function:

$$H(z) = \frac{(z+0.5)\,(z^2+4)}{(z-0.8)\,(z^2-z+0.64)}$$

a. Calculate and plot its poles and zeros.

b. Plot the system phase function and conclude about its linearity.

c. Write the transfer function *H*(*z*) as the product between a minimum phase function $H_{MP}(z)$ and an all-pass transfer function $H_A(z)$. Plot the phase functions of the two systems characterized by $H_{MP}(z)$ and $H_A(z)$.

d. Calculate and plot the phase shifting introduced by the systems corresponding to the following transfer functions: $H(z)$, $H_{MP}(z)$ and $H_A(z)$, if the input signal is the sinusoid $x[n] = \sin(n\pi/2)$.

EXERCISE 6.21.

Analyze the effect of the poles and the zeros of a transfer function $H_{1D}(z)$ on its magnitude $\left|H\left(e^{j\omega}\right)\right|$.

a. Consider the transfer function $H(z) = 1-(j/2)z^{-1}$, having only one zero: $z = j/2$. The polynomial coefficients are introduced as follows:

```
b1=[1,-0.5*j];
```

In order to analyze the variation of $\left|H\left(e^{j\omega}\right)\right|$ when the zero of the transfer function approaches the unit circle, the following additional polynomials are also defined:

```
b2=[1,-0.7*j]; b3=[1,-0.9*j]; b4=[1,-j];
```

Then use the following code line, with *k* taking values from 1 to 4:

```
figure(k), zplane (bk),
```

b. Consider the transfer function $H(z)=\dfrac{1}{1-(j/2)z^{-1}}$, having only one pole $z = j/2$. The polynomial coefficients are therefore introduced using:

```
a1=[1,-0.5*j];
```

In order to analyze the variation of $\left|H\left(e^{j\omega}\right)\right|$ when the pole of the transfer function approaches the unit circle, repeat the previous procedure, but using the command:

```
zplane(1,a1)
```

EXERCISE 6.22.

Verify if the systems characterized by the following impulse responses are stable or not:

$$h_1[n]=\cos(2\pi n/6)$$

$$h_2[n]=0.9^n\cos(2\pi n/6)$$

$$h_3[n]=1, 2, 3, 4, 5, 6, 7, 8, 9, 0, 0, 0, \ldots$$

Chapter 7

Infinite Impulse Response Filters

7.1. Theoretical background

7.1.1. *Transfer function and filter specifications for infinite impulse response (IIR) filters*

The infinite impulse response (IIR) filters, also called recursive filters due to the implementation method, are defined by the finite difference equation below:

$$y[n]+\sum_{i=1}^{N} b_i \cdot y[n-i]=\sum_{i=0}^{M} a_i \cdot x[n-i] \tag{7.1}$$

This equation results in the following rational transfer function in the z-domain:

$$H(z)=\frac{\sum_{i=0}^{M} a_i \cdot z^{-i}}{1+\sum_{i=1}^{N} b_i \cdot z^{-i}}=\frac{P(z)}{Q(z)} \tag{7.2}$$

where at least one of the b_i coefficients is non-zero and $M < N$.

The IIR filters are quite similar to the analog filters. Indeed, in both cases the filter specifications are defined in the form of margin constraints on its frequency response (magnitude, phase and group propagation time). An example is provided in the figure below for the transfer function magnitude of a lowpass analog (a) and IIR digital (b) filter.

$\delta_0(\Delta_0)$ stands for the maximum accepted passband ripple, $\delta_b(\Delta_b)$ is the minimum required stopband rejection (or attenuation), $\left[\omega_p, \omega_b\right]\left(\left[\Omega_p, \Omega_b\right]\right)$ defines the transition band and $\omega_c(\Omega_c)$ represents the cutoff frequency.

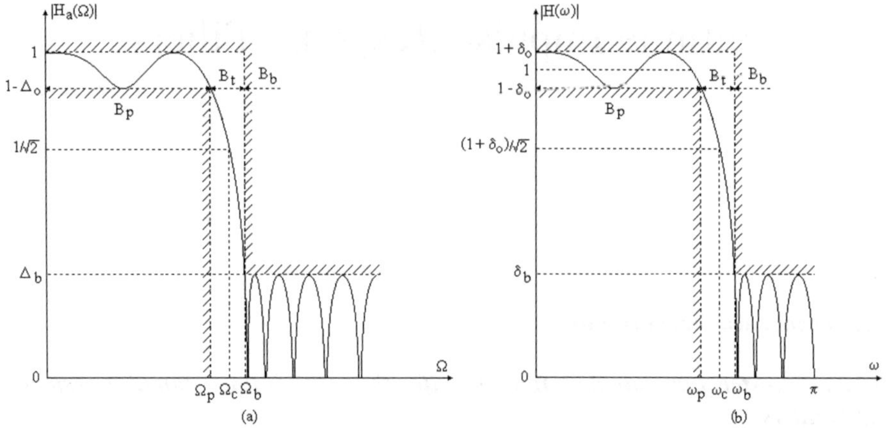

Figure 7.1. *Filter specifications for the transfer function magnitude of an analog (a) and an IIR digital lowpass filter (b)*

The specifications of the two types of filters are linked by the following relationships:

$$\begin{cases} \Delta_0 = \dfrac{2 \cdot \delta_0}{1+\delta_0} \\ \Delta_b = \dfrac{\delta_b}{1+\delta_b} \end{cases} \quad \text{or} \quad \begin{cases} \Delta a_M = -20 \cdot \log\left(1-\Delta_0\right) = 20 \cdot \log \dfrac{1+\delta_0}{1-\delta_0} \\ \Delta a_m = -20 \cdot \log \Delta_b = -20 \cdot \log \delta_b \end{cases} \qquad [7.3]$$

7.1.2. *Design methods for IIR filters*

Different digital filter design methods have been conceived. Some of them are specific to the digital filters, others are very similar to those used for analog filter design.

An IIR filter design involves three main steps:

a. define the filter specifications;

b. calculate the filter parameters to meet these specifications,

b1. direct design by CAD (computer-aided-design) procedures,

b2. analog prototype based design:

– filter specification transformation (according to section 7.1.3),

– analog prototype design,

– transfer function sampling,

– frequency transformation (possibly);

c. set up the IIR filter using a specific structure.

Direct filter design

The direct filter design aims to adjust its coefficients in order to optimize a given criterion. The filter structure (the numerator and denominator orders, M and N) is supposed fixed prior to the optimization procedure.

The following criterion is usually considered in the case when the filter specifications are relative to the transfer function magnitude:

$$Q\left[(a_k)_{k=1..M},(b_k)_{k=1..N}\right]=\sum_{i=1}^{L} W(\nu_i)\cdot\left\| H_0(\nu_i)\right|-\left|H(\nu_i)\right\|^{2q} \qquad [7.4]$$

where: $H_0(\nu_i)$ is the ideal frequency response for $\nu=\nu_i$,

$H(\nu_i)$ is the actual filter response for $\nu=\nu_i$ and a given set of coefficients $\{a_k\}_{k=1..M}$ and $\{b_k\}_{k=1..N}$,

$W(\nu_i)\geq 0$ is a weighting function,

$2q$ is the criterion order, with q a positive integer ($2q = 2$: quadratic criterion).

The optimization algorithm looks for the best set of the transfer function coefficients $\{a_k\}_{k=1..M}$ and $\{b_k\}_{k=1..N}$, so that Q is minimized.

Iterative methods ("Fletcher and Powell" and others) are used to solve this non-linear problem. It can be also useful to combine several criteria: magnitude response, phase response, group propagation time, etc.

Analog prototype based design

In this case, the coefficients of the corresponding analog prototype transfer function:

$$H_a(s) = \frac{C(s)}{D(s)} = \frac{\sum_{i=0}^{M} c_i \cdot s^i}{\sum_{i=0}^{N} d_i \cdot s^i}, \quad N \geq M \qquad [7.5]$$

are considered already calculated by one of the methods specific to the analog filter design (Butterworth, Chebychev, Cauer, etc.).

In all the cases, the following equation is used for separating the analog prototype transfer function $H_a(s)$:

$$H_a(s) \cdot H_a(-s) = \frac{C(s) \cdot C(-s)}{D(s) \cdot D(-s)} = \left|H_a(j\Omega)\right|^2\Big|_{\Omega^2=-s^2} = \frac{E(-s^2)}{G(-s^2)} \qquad [7.6]$$

This separation is performed through the two steps described below:

– the poles of $H_a(s)$ are represented by the zeros of $G\left(-s^2\right)$, localized in the left side of the s-plane (for insuring the stability),

– the zeros of $H_a(s)$ are obtained from the zeros of $E\left(-s^2\right)$ after distributing them between $H_a(s)$ and $H_a(-s)$, so that the complex conjugated zero couples are not separated (the solution is thus not unique).

The analog prototype transfer function sampling should meet two basic requirements:

1. stability conservation;
2. transfer function amplitude and phase conservation.

These conditions are equivalent to the following two requirements:

1. transformation of the s-plane left side into the z-plane unit circle inside;
2. linear conversion of the s-plane axis $j\Omega$ into the z-plane unit circle contour ($z = e^{j\omega}$).

Two methods are currently used to perform the transformation $H_a(s) \to H(z)$:

a. the impulse invariance method;

b. the bilinear transformation.

The first method creates a digital filter whose impulse response is equal to the sampled impulse response of the analog filter (T_e is the sampling period):

$$h_{ae}(t) = h_a(t) \cdot \delta_{T_e}(t) \Rightarrow H_{ae}(\Omega) = \frac{1}{T_e} \sum_{k=-\infty}^{\infty} H_a(\Omega + k \cdot \Omega_e) \qquad [7.7]$$

According to the Nyquist theorem:

$$H_a(\Omega) = 0 \text{ for } |\Omega| \geq \Omega_M \text{ and } \Omega_e \geq 2\Omega_M \qquad [7.8]$$

This results in:

$$T_e \cdot H_{ae}(\Omega) = H_a(\Omega) \text{ for } \Omega \in \left[-\frac{\Omega_e}{2}, \frac{\Omega_e}{2}\right] \qquad [7.9]$$

Consequently:

$$H_a(\Omega) = H(\omega) \Leftrightarrow H(\omega) = T_e \cdot H_{ae}(\Omega)$$

$$\Leftrightarrow \sum_{n=-\infty}^{\infty} h[n] \cdot e^{-j\omega n} = \sum_{n=-\infty}^{\infty} T_e \cdot h_a(nT_e) \cdot e^{-jn\Omega T_e}$$

$$\Rightarrow \begin{cases} h[n] = T_e \cdot h_a(nT_e) \\ \omega = \Omega T_e \end{cases} \qquad [7.10]$$

The following relationship is therefore obtained:

$$H(\omega) = H_a\left(\frac{\omega}{T_e}\right) \text{ for } \omega \in [-\pi, \pi] \qquad [7.11]$$

This means that the digital filter obtained with this method has exactly the same frequency response as the corresponding analog prototype.

The main steps of the algorithm used for implementing the impulse invariance method are given below:

1. Decompose $H_a(s)$ in the form:

$$H_a(s) = \sum_{k=1}^{N} \frac{A_k}{s + c_k} \qquad [7.12]$$

2. Calculate:

$$h_a(t) = \sum_{k=1}^{N} A_k \cdot e^{-c_k \cdot t} \cdot u(t) \qquad [7.13]$$

3. Derive:

$$h[n] = T_e \cdot h_a(nT_e) = T_e \cdot \sum_{k=1}^{N} A_k \cdot e^{-c_k \cdot n \cdot T_e} \cdot u[n] \qquad [7.14]$$

4. Determine:

$$H(z) = \sum_{n=0}^{\infty} h[n] \cdot z^{-n} = \sum_{k=1}^{N} \frac{T_e \cdot A_k}{1 - e^{-c_k \cdot T_e} \cdot z^{-1}} \qquad [7.15]$$

The link between the s-plane and the z-plane is represented by the transformation below and is illustrated in Figure 7.2.

$$z = e^{s \cdot T_e} = e^{\sigma \cdot T_e} \cdot e^{j\Omega \cdot T_e} \qquad [7.16]$$

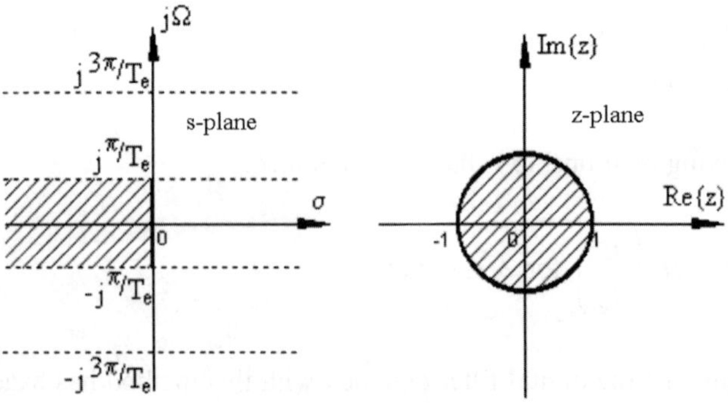

Figure 7.2. *Connection between the s-plane and the z-plane from the impulse invariance method*

Although the two previously mentioned constraints are fulfilled by the impulse invariance method, its application is not very convenient. In fact, when the frequency $\Omega_e = 2\pi / T_e$ is increased to avoid aliasing, the poles $z_i = e^{s_i \cdot T_e}$ approach the unit circle and may lead to unstable system behavior after the quantification of its transfer function denominator coefficients $\{b_k\}_{k=1..N}$.

The bilinear transformation (Figure 7.3) is performed using the following relationship:

$$s = \frac{2}{T_e} \cdot \frac{1 - z^{-1}}{1 + z^{-1}} \Leftrightarrow z = \frac{\frac{2}{T_e} + s}{\frac{2}{T_e} - s} \qquad [7.17]$$

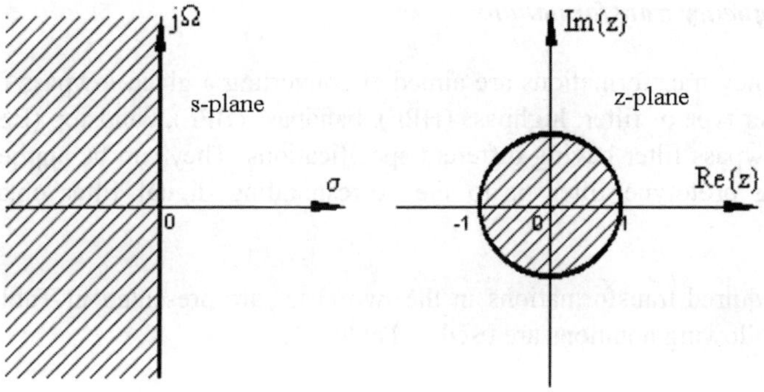

Figure 7.3. *Connection between the s-plane and the z-plane from the bilinear method*

Figure 7.3 shows that the bilinear transformation meets the stability constraint, but axis $j\Omega$ of the s-plane is not linearly converted to the contour of the unit circle in the z-plane:

$$j\frac{\Omega T_e}{2} = \frac{1 - e^{-j\omega}}{1 + e^{-j\omega}} \Leftrightarrow \omega = 2 \cdot \tan^{-1}\left(\frac{\Omega T_e}{2}\right) \qquad [7.18]$$

However, in the low frequency domain $\omega/2 \ll \pi/2 \Leftrightarrow \tan(\omega/2) \cong \omega/2$ this condition is satisfied:

$$\Omega T_e / 2 = \omega / 2 \Leftrightarrow \omega = \Omega T_e \qquad [7.19]$$

Furthermore, even in the high frequency domain, the non-linearity of the transformation can be corrected using a predistortion technique.

In the last case, the application of the bilinear transformation method involves the following steps:

1. define the digital filter specifications;

2. perform the frequency distortion using the equation:

$$\Omega = (2/T_e) \cdot \tan\left(\omega/2\right) \qquad [7.20]$$

3. calculate $H(s)$ using standard methods;

4. sample the obtained transfer function using equation [7.17].

7.1.3. *Frequency transformations*

Frequency transformations are aimed at converting a given lowpass filter (LPF) into another type of filter: highpass (HPF), bandpass (BPF), bandstop (BSF), or into another lowpass filter having different specifications. They can be applied either to the analog prototype filter or to the corresponding digital filter obtained after conversion.

The required transformations in the two cases are presented in Tables 7.1 and 7.2. The following notations are used in Table 7.1:

$$s_n = \frac{s}{\Omega_0} \quad , \quad s_n^{'} = \frac{s^{'}}{\Omega_0^{'}} \quad , \quad \delta = \frac{B}{\Omega_0^{'}} \qquad [7.21]$$

where Ω_0 is the cutoff frequency of the initial lowpass filter, while Ω'_0 stands for the cutoff frequency of the transformed filter if this is a lowpass or highpass filter and is defined by the following relationship in the case of a bandpass or a bandstop filter:

$$\Omega_0^{'2} = \Omega_H \cdot \Omega_B \qquad [7.22]$$

Ω_B and Ω_H denote the lower and upper edges of the passband or stopband filter and:

$$B = \Omega_H - \Omega_B \qquad [7.23]$$

In Table 7.2, ω_c is the cutoff frequency of the initial lowpass filter, while ω_B and ω_H stand for the lower and upper edges of the passband or stopband of the transformed filters.

Conversion	Normalized frequency transformation	Frequency transformation
LPF ⇨ LPF	$s_n = s_n'$	$s_n = \dfrac{\Omega_0}{\Omega_0'} s_n'$
LPF ⇨ HPF	$s_n = \dfrac{1}{s_n'}$	$s_n = \dfrac{\Omega_0 \cdot \Omega_0'}{s_n'}$
LPF ⇨ BPF	$s_n = \dfrac{s_n'^2 + 1}{s_n' \delta}$	$s_n = \dfrac{\Omega_0 \cdot s_n'^2 + \Omega_0 \cdot \Omega_0'^2}{B \cdot s_n'}$
LPF ⇨ BSF	$s_n = \dfrac{s_n' \delta}{s_n'^2 + 1}$	$s_n = \dfrac{\Omega_0 \cdot B \cdot s_n'}{s_n'^2 + \Omega_0'^2}$

Table 7.1. *Frequency transformations in the analog domain*

Conversion	Transformation	Parameter expression
LPF ⇨ LPF	$z^{-1} = \dfrac{z^{-1} - \alpha}{1 - \alpha z^{-1}}$	$\alpha = \dfrac{\sin\left(\frac{\omega_c - \omega_B}{2}\right)}{\sin\left(\frac{\omega_c + \omega_B}{2}\right)}$
LPF ⇨ HPF	$z^{-1} = -\dfrac{z^{-1} + \alpha}{1 + \alpha z^{-1}}$	$\alpha = -\dfrac{\cos\left(\frac{\omega_c + \omega_H}{2}\right)}{\cos\left(\frac{\omega_c - \omega_H}{2}\right)}$
LPF ⇨ BPF	$z^{-1} = -\dfrac{z^{-2} - \left(\frac{2\alpha k}{k+1}\right) z^{-1} + \frac{k-1}{k+1}}{\frac{k-1}{k+1} z^{-2} - \frac{2\alpha k}{k+1} z^{-1} + 1}$	$\alpha = \dfrac{\cos\left(\frac{\omega_H + \omega_B}{2}\right)}{\cos\left(\frac{\omega_H - \omega_B}{2}\right)}$ $k = \cot\left(\dfrac{\omega_H - \omega_B}{2}\right) \tan\dfrac{\omega_c}{2}$
LPF ⇨ BSF	$z^{-1} = \dfrac{z^{-2} - \left(\frac{2\alpha}{1+k}\right) z^{-1} + \frac{1-k}{1+k}}{\frac{1-k}{1+k} z^{-2} - \frac{2\alpha}{1+k} z^{-1} + 1}$	$\alpha = \dfrac{\cos\left(\frac{\omega_H + \omega_B}{2}\right)}{\cos\left(\frac{\omega_H - \omega_B}{2}\right)}$ $k = \tan\left(\dfrac{\omega_H - \omega_B}{2}\right) \tan\dfrac{\omega_c}{2}$

Table 7.2. *Frequency transformations in the discrete-time domain*

7.2. Solved exercises

EXERCISE 7.1.

Define in the Laplace domain the transfer function of an anti-aliasing filter, which attenuates with 0.5 dB at the frequency $\nu_p = 3{,}500$ Hz and with 30 dB at the frequency $\nu_a = 4{,}500$ Hz.

```
Fp = 3500;Fs = 4500;
Wp = 2*pi*Fp; Ws = 2*pi*Fs;
[N, Wn] = buttord(Wp, Ws, 0.5, 30,'s');
[b,a] = butter(N, Wn, 's');
wa = 0:(3*Ws)/511:3*Ws;
h = freqs(b,a,wa);
plot(wa/(2*pi), 20*log10(abs(h)));grid
xlabel('Frequency [Hz]');
ylabel('Gain [dB]');
title('Frequency response');
axis([0 3*Fs -60 5]);
```

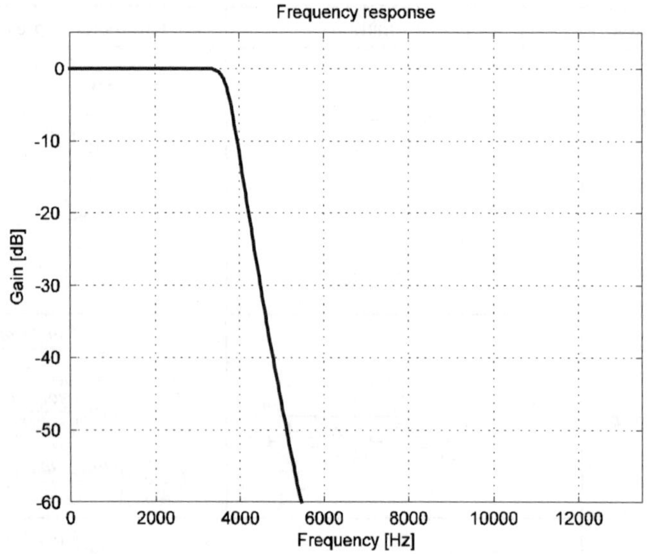

Figure 7.4. *Frequency response of an analog Butterworth lowpass filter*

EXERCISE 7.2.

Consider a system with the impulse response $h[n] = 0.9^n$. Plot the impulse response of this system sampled at 1 Hz for values of n between 0 and 50. Demonstrate that its time constant is equal to 10.

```
n=[0:50]; h=(0.9).^n;
subplot(211); stem(n,h)
grid; xlabel('n'); ylabel('h(n)')
title('Impulse response')
h1=exp(-n/10);
subplot(212)
plot(n,h1,'+',n,h,'-r')
grid; xlabel('n');
legend('exp(-n/10)','h(n)')
```

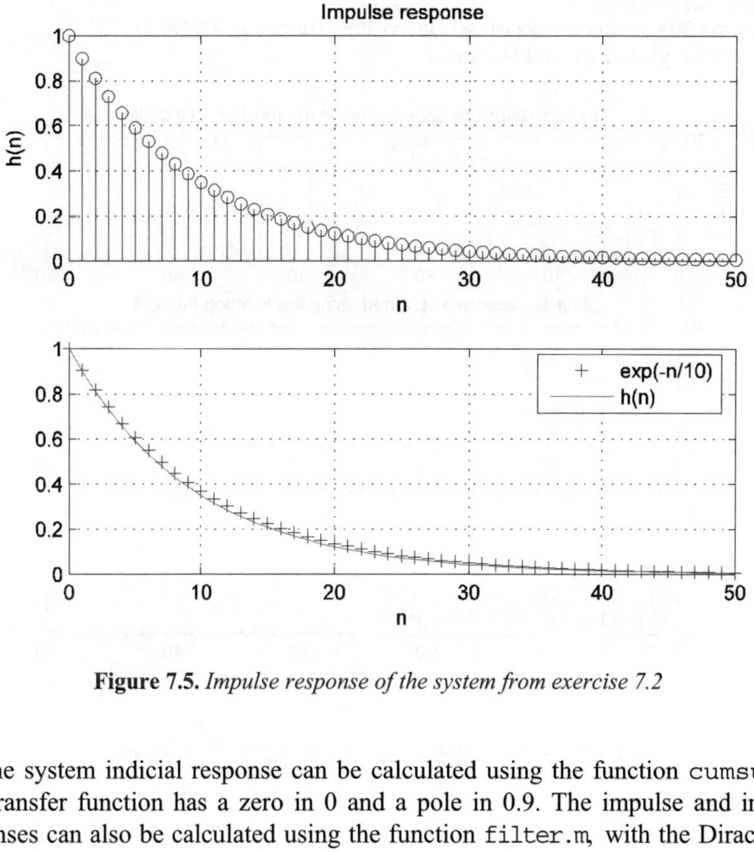

Figure 7.5. *Impulse response of the system from exercise 7.2*

The system indicial response can be calculated using the function cumsum.m. The transfer function has a zero in 0 and a pole in 0.9. The impulse and indicial responses can also be calculated using the function filter.m, with the Dirac pulse and the step function as input signals.

The following MATLAB code evaluates the indicial response $s(n)$ of the system having the impulse response $h(n)$.

```
s=cumsum(h); n1=[0:length(s)-1];
subplot(311); stem(n,s)
title('Indicial response obtained using the function CUMSUM');
```

```
ylabel('s(n)'); grid
u=ones(1,100); b=[1]; a=[1 -0.9];
s1=filter(b,a,u);
subplot(312) ;
stem(n,s1(1:length(n)))
title('Indicial response obtained using the function FILTER');
ylabel('s1(n)'); grid
delta=zeros(1,100); delta(1)=1;
h1=filter(b,a,delta);
subplot(313);
stem(n,h1(1:length(n)))
title('Impulse response obtained using the function FILTER');
xlabel('n'); ylabel('h1(n)'); grid
```

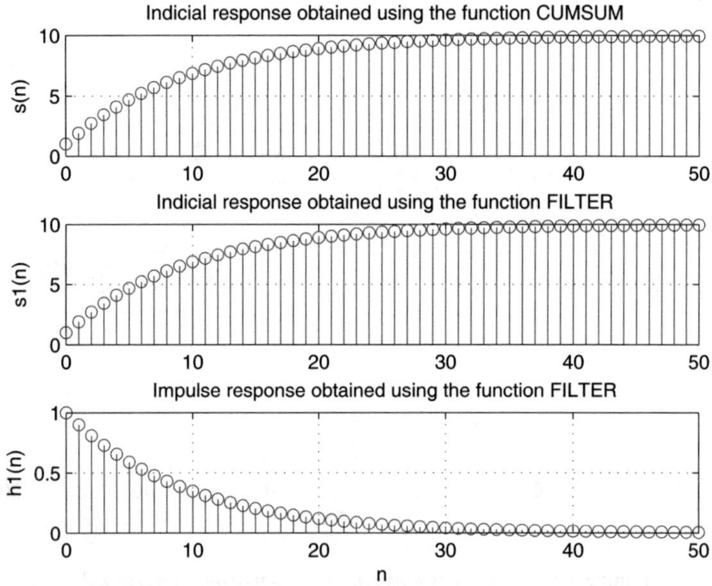

Figure 7.6. *Two methods for calculating the indicial and impulse response of the system from exercise 7.2*

EXERCISE 7.3.

Use the function `filter.m` to calculate the impulse response $h(n)$ of the system defined by the following equation:

$$y(n) = 1.7654\, y(n-1) - 0.81\, y(n-2) + x(n) + 0.5\, x(n-1)$$

Plot $h(n)$ between -10 and 60. Compare the obtained result with the theoretical result.

```
a=[1 -1.7654 0.81]
b=[1 0.5]
delta=[1 zeros(1,99)];
h=filter(b,a,delta);
n=[-10:59];
h1=[zeros(1,10) h];
stem(n,h1(1:70));
xlabel('n');
ylabel('h(n)');
grid
```

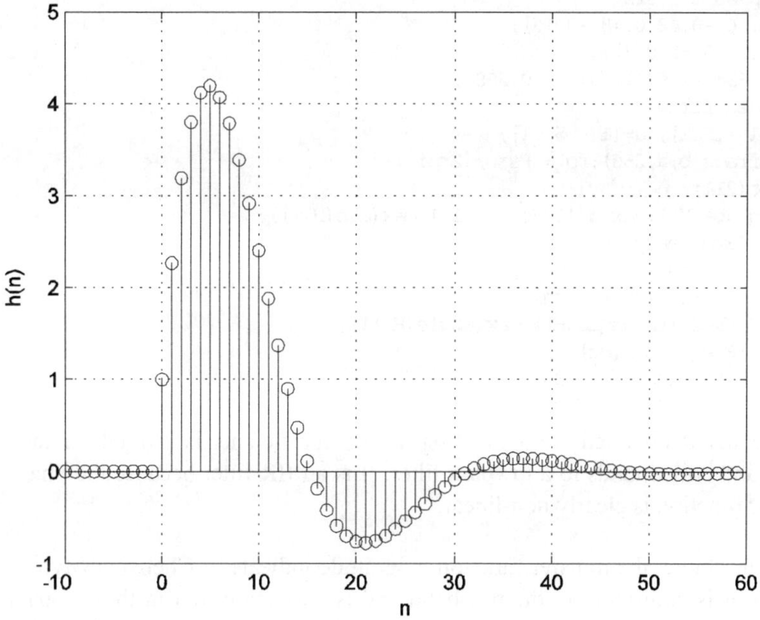

Figure 7.7. *Impulse response of the system from exercise 7.3*

EXERCISE 7.4.

Identify the type of filter which corresponds to the following difference equations:

1. $y[n] = x[n] + 0.5x[n-1] + 1.8\cos[\pi/16]y[n-1] - 0.81y[n-2]$

2. $y[n] = 0.634x[n] + 0.634x[n-1] - 0.268y[n-1]$

3. $y[n]=0.16x[n]-0.48x[n-1]+0.48x[n-2]-0.16x[n-3]$
$-0.13y[n-1]-0.52y[n-2]-0.3y[n-3]$

4. $y[n]=0.634x[n]-0.634x[n-1]+0.268y[n-1]$

5. $y[n]=0.1x[n]-0.5x[n-1]+x[n-2]+0.5y[n-1]-0.1y[n-2]$

The sampling frequency is 20 kHz in all cases.

```
Fs = 20e3;
% Lowpass filter
b=[1 0.5]; a=[1 -1.8*cos(pi/16) 0.81];
% b=[0.634 0.634]; a=[1 0.268];
% Highpass filters
% b=[0.16 -0.48 0.48 -0.16];
a=[1 0.13 0.52 0.3];
% b=[0.634 -0.634]; a=[1 -0.268];
% Allpass filter
% b=[.1 -.5 1]; a=[1 -.5 .1];
[H,w]=freqz(b,a,256);fq = Fs*w/(2*pi);
subplot(211);
plot(fq,abs(H)); axis([0 Fs/2 0 1.1*max(abs(H))]);
xlabel('Frequency [Hz]');
ylabel('|H(\nu)|'); grid
subplot(212); plot(fq,angle(H))
axis([0 Fs/2 min(angle(H)) max(angle(H))]);
xlabel('Frequency [Hz]');
ylabel('\Phi(\nu)'); grid
```

The result obtained from the first set of coefficients is plotted on the figure below and corresponds to a lowpass filter. It is an IIR filter because the phase of its transfer function is clearly non-linear.

Furthermore, the transfer function magnitude indicates a Chebychev type I filter, because it is equiripple in the passband and is maximally flat in the stopband. The filter order (2 in this case) can also be derived from the corresponding difference equation.

Perform the same type of analysis for the other four cases.

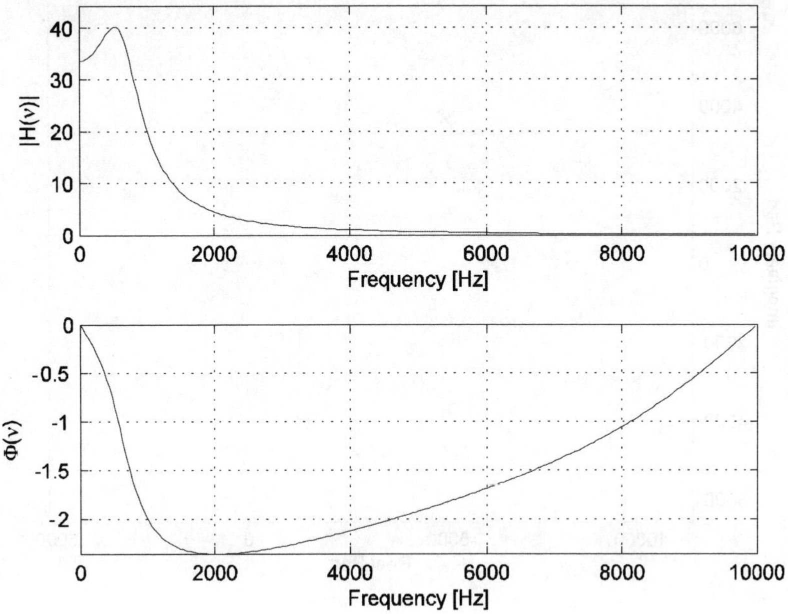

Figure 7.8. *Amplitude and phase response of the system described by the first difference equation*

EXERCISE 7.5.

Calculate the poles of a 5th order analog Butterworth filter with the cutoff frequency 1 kHz.

```
% Prototype filter design
N=5;
[zc,pc,kc]=buttap(N);
% Frequency transformation
Omega=2*pi*10^3;
pc=pc*Omega;
% Plotting the poles
zplane(zc,pc)
```

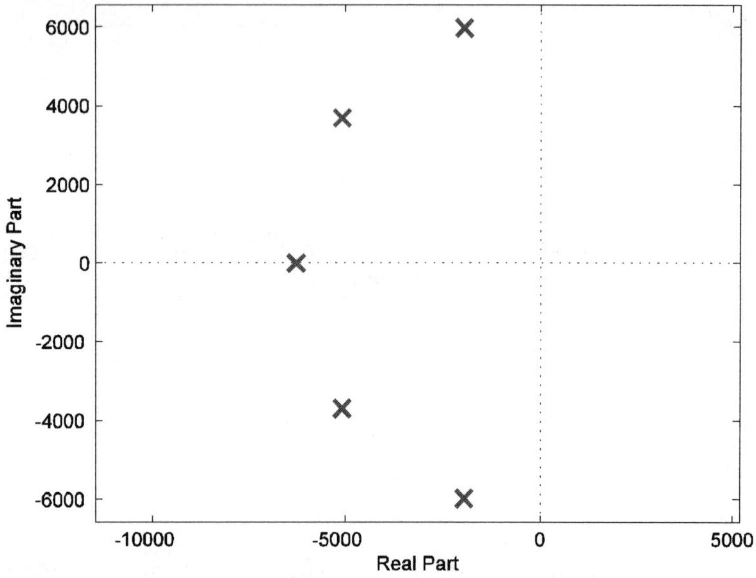

Figure 7.9. *Poles of a 5th order analog Butterworth filter*

EXERCISE 7.6.

Design an IIR lowpass filter for a sampling frequency of 1 Hz, with an attenuation of 0.75 dB in the frequency band [0 0.1306] Hz, an attenuation in the stopband of at least 20 dB and a lower stopband edge frequency of 0.201 Hz. Calculate the poles of the associated analog filter using a Butterworth model.

```
% wp: upper passband edge frequency [rad]
% ws: lower stopband edge frequency [rad]
% atmax: passband ripple or maximum permissible passband loss [dB]
% atmin: minimum stopband attenuation [dB]
atmax=0.75
atmin=20
Wp=0.1306*2*pi
Ws=0.201*2*pi
% Butterworth analog filter design
N=ceil(log10((10^(0.1*atmin)-1)/(10^(0.1*atmax)-1))/Wnorm);
Wnorm=2*log10(Ws/Wp);Whp=Wp/((10^(.1*atmax)-1)^(1/(2*N)));
Omega=Whp;[zc,pc,kc]=buttap(N);
% Frequency transformation
pc=pc*Omega;
% Plotting the poles
zplane(zc,pc)
```

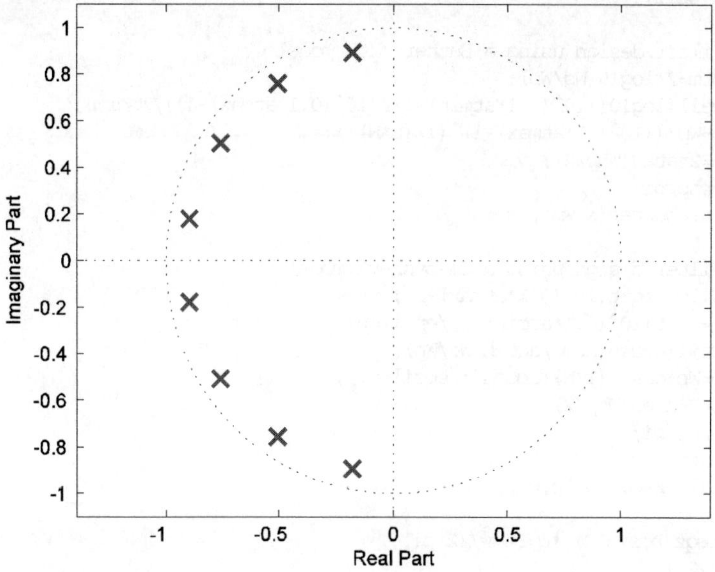

Figure 7.10. *Poles of the filter designed in exercise 7.6*

EXERCISE 7.7.

Design a digital filter which meets the following specifications using the bilinear method:

– sampling frequency: 1 Hz,

– 0 dB attenuation of the DC component,

– maximum attenuation of 1 dB at 0.1 Hz,

– minimum attenuation of 15 dB at 0.15 Hz.

Consider a Butterworth and then a Chebychev model for the filter transfer function.

```
% FS: sampling frequency [Hz]
% wp: upper passband edge frequency [rad]
% ws: lower stopband edge frequency [rad]
% atmax: maximum attenuation in the passband[dB]
% atmin: minimum  attenuation in the stopband[dB]
% fil: filter type (1 = Butterworth, 0 = Chebyshev)
%[b,a]: numerator/denominator coefficients
Fs = 1; atmax = 1; atmin = 15; wp = 0.1*2*pi;
ws=0.15*2*pi; fil=0; %(or fil=1 according to the filter type)
Ws=2*tan(ws/2);
```

```
Wp=2*tan(wp/2);
if fil==1
    % Filter design using a Butterworth model
    Wnorm=2*log10(Ws/Wp);
    N=ceil(log10((10^(0.1*atmin)-1)/(10^(0.1*atmax)-1))/Wnorm);
    Whp=Wp/((10^(.1*atmax)-1)^(1/(2*N)));
    whp=2*atan(Whp/2);
    wn=whp/pi;
     [b,a]=butter(N,wn);
else
    % Filter design using a Chebyshev1 model
    epsilon=sqrt(10^(0.1*atmax)-1);
    num=sqrt(10^(0.1*atmin)-1)/epsilon;
    N=ceil(acosh(num)/acosh(Ws/Wp))
    Whp=Wp*cosh((1/N)*acosh(1/epsilon));
    whp=2*atan(Whp/2);
    wn=whp/pi;
    Rp=atmax;
     [b,a]=cheby1(N,Rp,wn);
end
[H,w]=freqz(b,a,512);fq=Fs*w/(2*pi);
subplot(221)
zplane(b,a)
subplot(222)
mag=abs(H);
plot(fq,mag)
ylabel('Amplitude');
xlabel('Frequency [Hz]');
axis([0 0.5 0 1.1*max(mag)]); grid
thresh1 = [-atmax*ones(1,length(mag))];
thresh2 = [-atmin*ones(1,length(mag))];
subplot(223)
plot(fq,20*log10(mag),w/pi,thresh1,w/pi,thresh2)
ylabel('Amplitude [dB]');
xlabel('Frequency [Hz]'); grid
axis([0 .5 -1.1*atmin 1]);
xlabel('Frequency [Hz]'); grid
phase_function = unwrap(angle(H));
subplot(224)
plot(fq,phase)
axis([0 .5 1.1*min(phase_function) 1]);
ylabel('Phase [rad]');
xlabel('Frequency [Hz]'); grid
```

The example of the Chebychev filter is shown in the figure below.

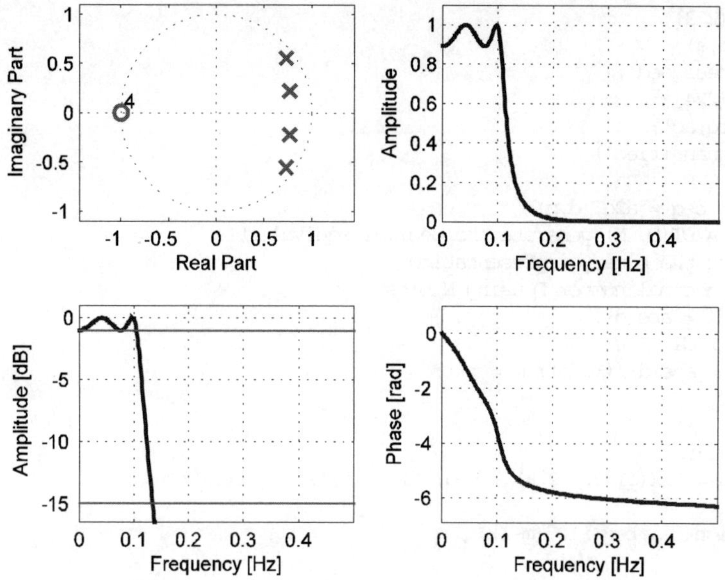

Figure 7.11. *Poles, zeros and transfer function for a digital lowpass Chebychev type I filter*

EXERCISE 7.8.

Implementing a digital filter on a digital signal processor requires the quantification of its coefficients. Thus, the actual filter characteristics are slightly modified with respect to the designed filter.

Consider a 6th order elliptic filter having a cutoff normalized frequency of 0.2, a ripple of 0.05 dB in the passband and a minimum attenuation of 60 dB in the stopband.

Write a MATLAB code to illustrate the effect of the quantification on 5 bits on the filter characteristics (transfer function, poles and zeros).

```
[b,a] = ellip(6,0.05,60,2*0.2);
w = 0:pi/255:pi;
h = freqz(b,a,w); g = 20*log10(abs(h));
bq = a2dT(b,5); aq = a2dT(a,5);
h = freqz(bq,aq,w); gq = 20*log10(abs(h));
subplot(211);
plot(w/(2*pi),g,'b',w/(2*pi),gq,'r--');
axis([0 0.5 -80 10]); grid;
xlabel('Normalized frequency');
ylabel('Gain [dB]');
```

```
legend('designed', 'quantified');
subplot(223);
zplane(b,a);
title('designed')
subplot(224);
zplane(bq,aq);
title('quantified')

function beq = a2dT(d,n)
% BEQ = A2DT(D, N) generate the decimal equivalent
%  beq of the binary representation
%  of a decimal number D using N bits
m = 1; d1 = abs(d);
while fix(d1) > 0
     d1 = abs(d)/(10^m); m = m+1;
end
beq = 0;
for k = 1:n
     beq = fix(d1*2)/(2^k) + beq; d1 = (d1*2) - fix(d1*2);
end
beq = sign(d).*beq.*10^(m-1);
```

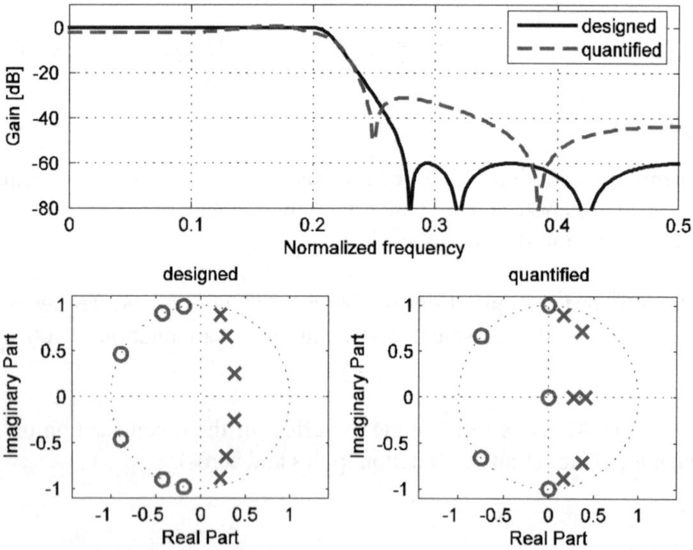

Figure 7.12. *Effect of the quantification of a filter's coefficients on its characteristics*

EXERCISE 7.9.

Write a MATLAB code to illustrate the effect of the quantification of a filter's coefficients when it is formed by the series connection of 2nd order cells.

```
[z,p,k] = ellip(6,0.05,60,2*0.2);
[b,a] = zp2tf(z,p,k);
w = 0:pi/255:pi;
h = freqz(b,a,w);
g = 20*log10(abs(h));
sos = zp2sos(z,p,k);
sosq = a2dR(sos,6);
R1 = sosq(1,:); R2 = sosq(2,:); R3 = sosq(3,:);
b1 = conv(R1(1:3),R2(1:3));
bq = conv(R3(1:3),b1);
a1 = conv(R1(4:6),R2(4:6));
aq = conv(R3(4:6),a1);
h = freqz(bq,aq,w);
gg = 20*log10(abs(h));
subplot(211);
plot(w/(2*pi),g,'b-', w/(2*pi),gq,'r--');
axis([0 0.5 -80 10]);grid;
xlabel('Normalized frequency');
ylabel('Gain [dB]');
legend('designed','quantified');
subplot(223);
zplane(b,a);
title('designed');
subplot(224) ;
zplane(bq,aq);
title('quantified');

function beq = a2dR(d,n)
% BEQ = A2DR(D, N) generate the decimal equivalent
% beq of the binary representation
% of a decimal quantity D using N bits (rounded)
m = 1; d1 = abs(d);
while fix(d1) > 0
     d1 = abs(d)/(10^m); m = m+1;
end
beq = 0;
d1 = d1 + 2^(-n-1);
for k = 1:n
     beq = fix(d1*2)/(2^k) + beq;
    d1 = (d1*2) - fix(d1*2);
end
beq = sign(d).*beq.*10^(m-1);
```

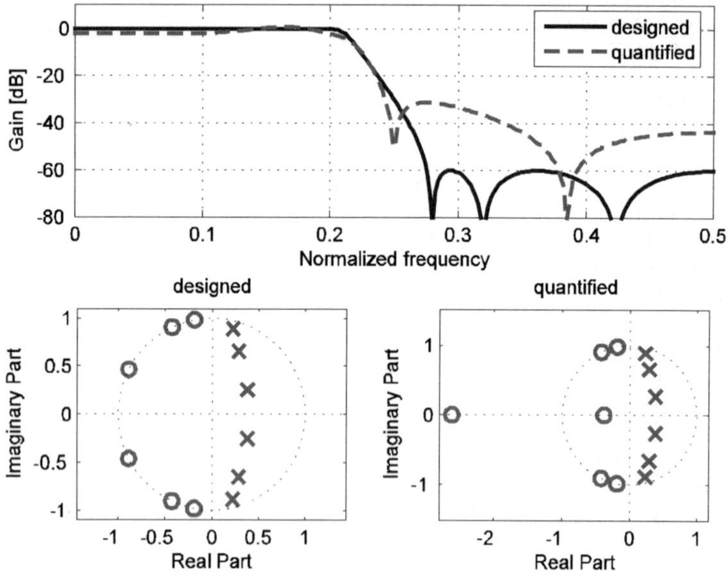

Figure 7.13. *Effect of the coefficient quantification for a filter formed by the series connection of 2nd order cells*

7.3. Exercises

EXERCISE 7.10.

a. Calculate the coefficients of a digital IIR 4th order lowpass filter for a sampling frequency of 10 kHz and a cutoff frequency of 3 kHz.

b. Plot the amplitude and phase of its frequency response.

c. Plot the poles and zeros of the filter transfer function.

d. Plot the filter impulse response.

e. Repeat the exercise for a Chebyshev type I filter, having the same specifications and a maximum passband ripple of 0.5 dB. Then compare the two filters.

EXERCISE 7.11.

Design a lowpass filter with the following specifications: sampling frequency 2 Hz, passband upper edge 0.28 Hz, lower stopband edge 0.32 Hz, bandpass maximum ripple 0.1 dB and stopband minimum attenuation 30 dB.

a. Find out the order of the Butterworth, Chebychev type I and II, and elliptic filters, which meet these specifications.

b. Why is the elliptic filter order always the lowest for the same required specifications?

EXERCISE 7.12.

A linear sampled filter, with initial conditions equal to zero, is defined by the following recurrent relationship:

$$y[nT] = x[nT] + 2x[(n-1)T] + 3x[(n-2)T] + 4x[(n-3)T] \\ + 3x[(n-4)T] + 2x[(n-5)T] + x[(n-6)T]$$

a. Calculate the filter transfer function and impulse response.

b. Plot the magnitude of the transfer function. Indicate the type of corresponding filter.

c. Find the zeros of the transfer function. What is their effect on the filter frequency response?

Chapter 8

Finite Impulse Response Filters

8.1. Theoretical background

Unlike IIR filters, finite impulse response (FIR) filters are specific to the discrete-time domain and cannot be obtained by direct transposition of analog filters. The linear phase of the transfer function makes them useful for many applications.

8.1.1. *Transfer function and properties of FIR filters*

For this type of filter, also called non-recursive due to the implementation method, the transfer function is defined by:

$$H(z) = \sum_{n=0}^{N-1} h[n] \cdot z^{-n} \Rightarrow H(\omega) = \sum_{n=0}^{N-1} h[n] \cdot e^{-j \cdot \omega \cdot n} \qquad [8.1]$$

where $h[n]$ stands for the (finite) impulse response of the filter.

In order to avoid the phase discontinuities of the transfer function, $H(\omega)$ is expressed in the following form:

$$H(\omega) = H_0(\omega) \cdot e^{j\theta(\omega)} \qquad [8.2]$$

where the zero-phase transfer function $H_0(\omega)$ and the phase function $\theta(\omega)$ are real and continuous.

In practice, the impulse response $h[n]$ is required to be real, which means that the real part of $H(\omega)$ is even, while its imaginary part is odd. The linearity constraint for the phase function $\theta(\omega)$ leads to the four FIR filter classes, whose properties are summarized in Table 8.1.

FIR	N	$h[n]$	$H_0(\omega)$	$\theta(\omega)$	$H(\omega)$ or $H_0(\omega)$ $\omega = 0$	$H(\omega)$ or $H_0(\omega)$ $\omega = \pi$
1	odd	Symmetric $h[n] = h[N-1-n]$	$\sum_{n=0}^{\frac{N-1}{2}} a_n \cdot \cos[n\omega]$	$-\frac{N-1}{2}\omega$	Without constraints	Without constraints
2	even	Symmetric $h[n] = h[N-1-n]$	$\sum_{n=1}^{\frac{N}{2}} b_n \cdot \cos\left[\left(n - \frac{1}{2}\right)\omega\right]$	$-\frac{N-1}{2}\omega$	Without constraints	0
3	odd	anti-symmetric $h[n] = -h[N-1-n]$ $h\left(\frac{N-1}{2}\right) = 0$	$\sum_{n=0}^{\frac{N-1}{2}} a_n \cdot \sin[n\omega]$	$\frac{\pi}{2} - \frac{N-1}{2}\omega$	0	0
4	even	anti-symmetric $h[n] = -h[N-1-n]$	$\sum_{n=1}^{\frac{N}{2}} b_n \cdot \sin\left[\left(n - \frac{1}{2}\right)\omega\right]$	$\frac{\pi}{2} - \frac{N-1}{2}\omega$	0	Without constraints

Table 8.1. *Properties of FIR filters*

Coefficients a_i and b_i in Table 8.1 can be obtained using the following equations:

$$\begin{cases} a_0 = h\left(\dfrac{N-1}{2}\right) \\ a_n = 2h\left(\dfrac{N-1}{2} - n\right), \quad n = 1..\dfrac{N-1}{2} \\ b_n = 2h\left(\dfrac{N}{2} - n\right), \quad n = 1..\dfrac{N}{2} \end{cases} \qquad [8.3]$$

The first two classes' filters are usually involved in pure filtering applications, while the others are useful for phase shifting filtering applications (integrator operator, derivative operator, Hilbert transformer). The following relationship can easily be obtained using equation [8.1] and the symmetry/anti-symmetry property of a FIR impulse response (see Table 8.1):

$$H\left(z^{-1}\right) = \pm z^{N-1} \cdot H(z) \qquad [8.4]$$

According to equation [8.4], if z_i is a zero of $H(z)$, then $1/z_i$ also becomes a zero. On the other hand, as $h[n]$ is real, z_i^* and $1/z_i^*$ are also zeros of $H(z)$. Thus, the typical distribution of FIR filter zeros is shown in Figure 8.1.

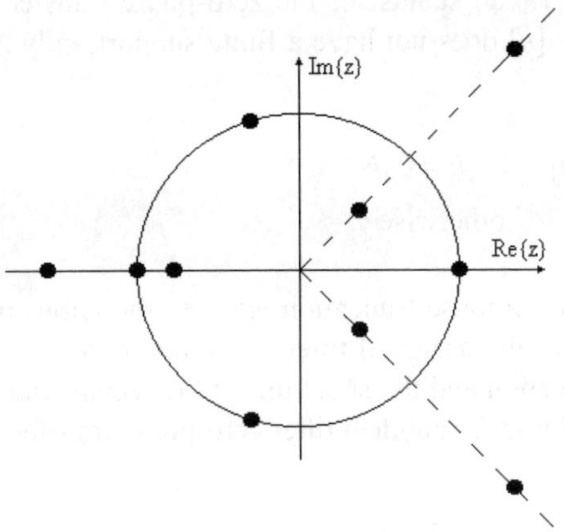

Figure 8.1. *Typical distribution of FIR filter zeros*

$z = -1$ and $z = 1$ are always zero for a 2nd class and 4th class FIR filter respectively, while they are both always zero for a 3rd class filter (see Table 8.1).

8.1.2. *Design methods*

The main methods which can be used for designing digital FIR filters are given below:

1. window method,
2. frequency sampling method,
3. optimization methods.

Window method

The basic idea of this method is to fix the ideal frequency response:

$$H_{\infty}(\omega) = H_d(\omega) \cdot e^{-j\frac{N-1}{2}\omega} \qquad [8.5]$$

and then to invert it to obtain the filter impulse response:

$$h_{\infty}[n] = \frac{1}{2\pi} \int_{2\pi} H_{\infty}(\omega) e^{j\omega n} d\omega \qquad [8.6]$$

In equation [8.5] $H_d(\omega)$ stands for the zero-phase transfer function of the ideal filter. As sequence $h_{\infty}[n]$ does not have a finite support, only N samples are kept to obtain a FIR filter:

$$h[n] = \begin{cases} h_{\infty}[n] & \text{if } n = 0..N-1 \\ 0 & \text{otherwise} \end{cases} \qquad [8.7]$$

The ideal impulse response truncation leads to the Gibbs phenomenon, because the transfer function of the designed filter is actually obtained by the convolution of the target transfer function and the spectrum of the rectangular window of length N. This statement is still true for an ideal filter zero-phase transfer function:

$$H_0(\omega) = \frac{1}{2\pi} H_d(\omega) * W_0(\omega) \qquad [8.8]$$

$W_0(\omega)$ can be seen as the analog domain equivalent of the function sinc(ω):

$$W_0(\omega) = \frac{\sin\left(\frac{N}{2}\omega\right)}{\sin\left(\frac{\omega}{2}\right)} = S_a(\omega) \qquad [8.9]$$

The effects of the infinite impulse response truncation (passband and stopband ripple non-zero transition band), are illustrated in Figure 8.2.

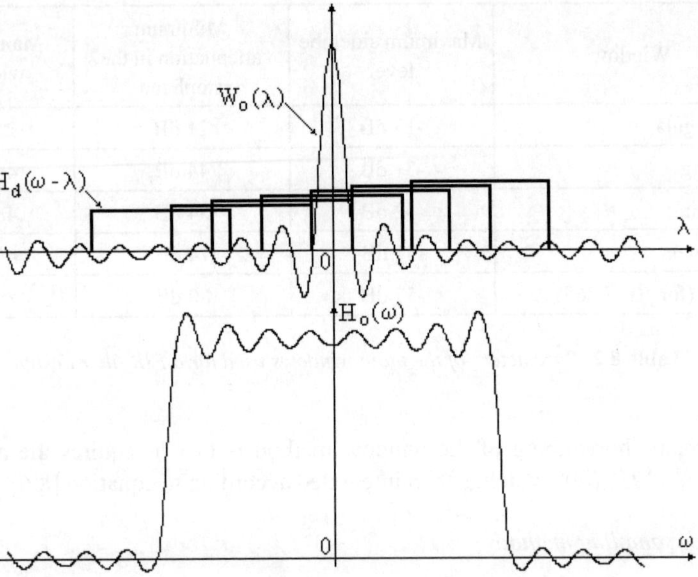

Figure 8.2. *Effect of the ideal impulse response truncation on the transfer function of a FIR filter*

The rectangular window, which is used by default, has to be replaced by another type of window to damp out these effects. The spectrum of this new window should meet the following conditions:

a. narrow mainlobe to obtain a narrow transition band;

b. the mainlobe should concentrate the main part of the window energy to obtain low sidelobes level;

c. the energy should be uniformly distributed over the window sidelobes to obtain a uniform ripple in the filter passband and stopband.

The first two constraints are contradictory because the first requires a narrow mainlobe, while the second a large one. Several windows (Hamming, Hanning, Blackman, Kaiser, etc.) have been proposed to perform a trade-off between these constraints.

Table 8.2 is helpful to determine the number of points N for the impulse response of a FIR filter and the appropriate window.

Window	Maximum sidelobe level	Minimum attenuation in the stopband	Mainlobe width
Rectangular	-13 dB	21 dB	2/N
Hanning	-31 dB	44 dB	4/N
Hamming	-42 dB	54 dB	4/N
Blackman	-57 dB	74 dB	6/N
Kaiser (for $\beta = 7.865$)	-57 dB	80 dB	8/N

Table 8.2. *Parameters of the main windows used for a FIR filter design*

The main shortcoming of the window method is that it requires the analytical expression of $H_\infty(\omega)$, which is then integrated according to equation [8.6].

Frequency sampling method

Only N samples of the transfer function $H_\infty(\omega)$ are considered in the case of the frequency sampling method:

$$H(k) = H_\infty(\omega)\Big|_{\omega = k\frac{2\pi}{N}}, \quad k = 0..N-1 \qquad [8.10]$$

The impulse response of the corresponding FIR filter is then calculated as the inverse DFT of this sequence.

The transfer function of the designed filter will thus be expressed as follows:

$$H(\omega) = e^{-j\omega\frac{N-1}{2}} \sum_{k=0}^{N-1} A(\omega,k) \cdot H_d(k) \qquad [8.11]$$

where $H_d(k)$ stand for the zero-phase transfer function samples and:

$$A(\omega,k) = \frac{1}{N} \frac{\sin\left(\frac{N}{2}\left(\omega - k\frac{2\pi}{N}\right)\right)}{\sin\left[\frac{1}{2}\left(\omega - k\frac{2\pi}{N}\right)\right]} = \frac{1}{N} S_a\left(\omega - k\frac{2\pi}{N}\right) \qquad [8.12]$$

The zero-phase transfer function of the FIR filter is then obtained in the form:

$$H_0(\omega) = \frac{1}{N}\sum_{k=0}^{N-1} H_d(k) \cdot S_a\left(\omega - k\frac{2\pi}{N}\right) \qquad [8.13]$$

Note that $H_0(\omega)$ is issued from the interpolation of the ideal sampled transfer function, i.e. each sample $H_d(k)$ is weighted by a shifted version of function $S_a(\omega)$. Consequently, the approximation error is zero for $\omega_k = 2\pi k/N$ and is finite elsewhere.

Optimization methods

In order to increase the approximation quality in the case of the frequency sampling method, several transfer function samples from the passband can be used as additional variables. They are then optimized using specific CAD (computer-aided design) tools to minimize the approximation error.

The LS (least squares) method for the digital FIR filter design is widely used to perform the optimization. This method is also very useful when the transfer function samples are not equally spaced.

Another solution is provided by Remetz's method, which is based on the alternance theorem. This method leads to the best approximation of the transfer function $H_d(\omega)$, in the Chebychev sense.

8.1.3. *General conclusion about digital filter design*

The choice of filter type (IIR or FIR) and its design method depends on the considered application, on the associated constraints and on the available design tools. Thus, if the filter specifications are defined with respect to the magnitude of its frequency response, the IIR filters are much more cost effective than the FIR filters, which are much larger. The IIR filters are also the best choice whenever a very narrow passband, stopband or transition band is required or whenever the stopband attenuation has to be large.

On the other hand, when the filter specifications are defined with respect to the phase of its frequency response, the FIR filters justify the implementation cost overrun. Indeed, unlike the IIR filter, they have the capability of insuring a linear phase due to the symmetry of the impulse response coefficients. They remain stable for any set of coefficients, which make them suitable for implementing adaptive systems. Finally, we should mention that it is easy to obtain high order FIR filters using the series connection of several basic cells.

8.2. Solved exercises

EXERCISE 8.1.

Design a 25th order FIR lowpass filter with a minimum transition band, using the window method. Consider a sampling frequency f_s = 20 kHz and the following lower and upper passband frequency edges: f_1 = 3 kHz and f_2 = 7 kHz. Plot the transfer function, the impulse response and the zeros of the designed filter.

```
N=25; f1=3e3; f2=7e3; fs=2e4;
wn1=2*f1/fs; wn2=2*f2/fs;
b=fir1(N,[wn1 wn2],boxcar(N+1));
[h,w] = freqz(b,1); fq = fs*w/(2*pi);
amp=20*log10(abs(h)); phase=unwrap(angle(h))*180/pi;
figure; subplot(221); plot(fq,amp);
axis([0 max(fq) min(amp) 10])
xlabel('Frequency [Hz]'); ylabel('Amplitude [dB]') ; grid
subplot(223); plot(fq,phase);
axis([0 max(fq) 1.2*min(phase) 1.2*max(phase)])
xlabel('Frequency [Hz]'); ylabel('Phase [°]'); grid
subplot(2,2,2); impz(b,1); xlabel('n')
subplot(2,2,4); zplane(b,1);
```

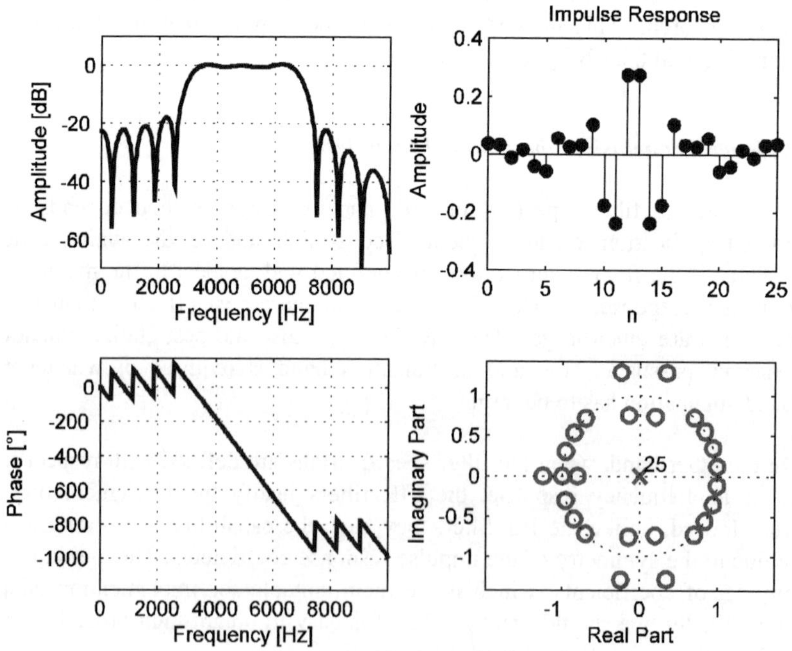

Figure 8.3. *Characteristics of the filter from exercise 8.1*

The rectangular window is the most suitable in this case because it provides the narrowest transition band. Since the impulse response has an even dimension (26) and is symmetric, the designed filter belongs to the 2nd class of FIR filters class.

A common error which should be avoided is to take the filter order (N) for its length or number of coefficients ($N+1$). Also note the presence of the zero $z=-1$, and the fact that the zeros occur as couples of inverse and conjugated values.

EXERCISE 8.2.

Calculate the coefficients of a FIR filter having the same parameters as in the previous exercise, but using a Blackman window instead of the rectangular window. Compare the obtained results in terms of the transition band and the stopband attenuation.

The same MATLAB code as for the previous exercise is used with the exception of the option `boxcar`, which is replaced by `blackman`. The obtained result is provided on the figure below.

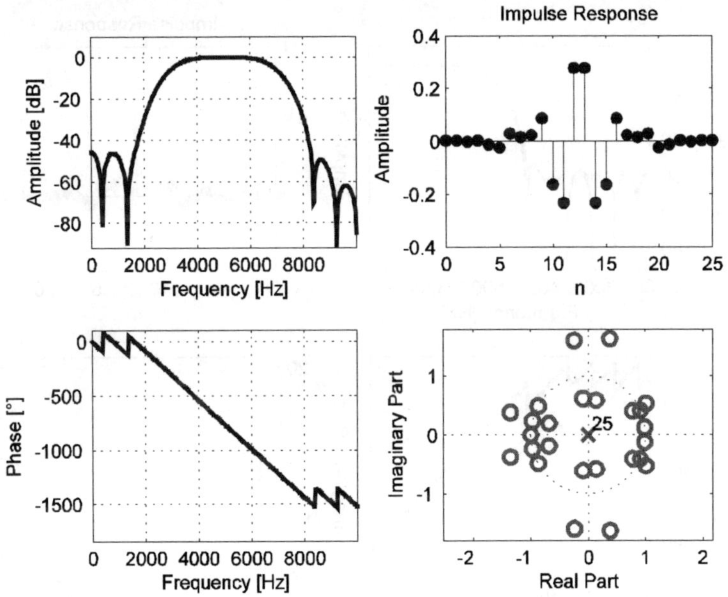

Figure 8.4. *Characteristics of the filter from exercise 8.2*

Note that the transition band is about two times wider than in the previous case. In addition, the stopband attenuation has been increased by about 30 dB. Consequently, the rectangular window should be used whenever a high selectivity is required, while the other windows are more appropriate in the case of a large stopband rejection.

EXERCISE 8.3.

Design a 24th order FIR highpass filter using the frequency sampling method. Consider a cutoff frequency f_c = 6 kHz and a sampling frequency f_s = 20 kHz. Plot the transfer function, the impulse response and the zeros of the designed filter.

```
N=24; fc=6e3; fs=20e3; inc=0.01; wcn=2*fc/fs; ff=[0:inc:1];
hh=[zeros(1,wcn/inc-1) ones(1,(1-wcn)/inc+2)];
b=fir2(N,ff,hh); [h,w] = freqz(b,1); fq = fs*w/(2*pi);
amp=20*log10(abs(h)); phase=unwrap(angle(h))*180/pi;
subplot(221); plot(fq,amp); axis([0 max(fq) min(amp) 10])
xlabel('Frequency [Hz]'); ylabel('Amplitude [dB]'); grid
subplot(223); plot(fq,phase);
axis([0 max(fq) 1.2*min(phase) 1.2*max(phase)])
xlabel('Frequency [Hz]');ylabel('Phase [°]');grid
subplot(2,2,2);impz(b,1);xlabel('n');subplot(2,2,4); zplane(b,1);
```

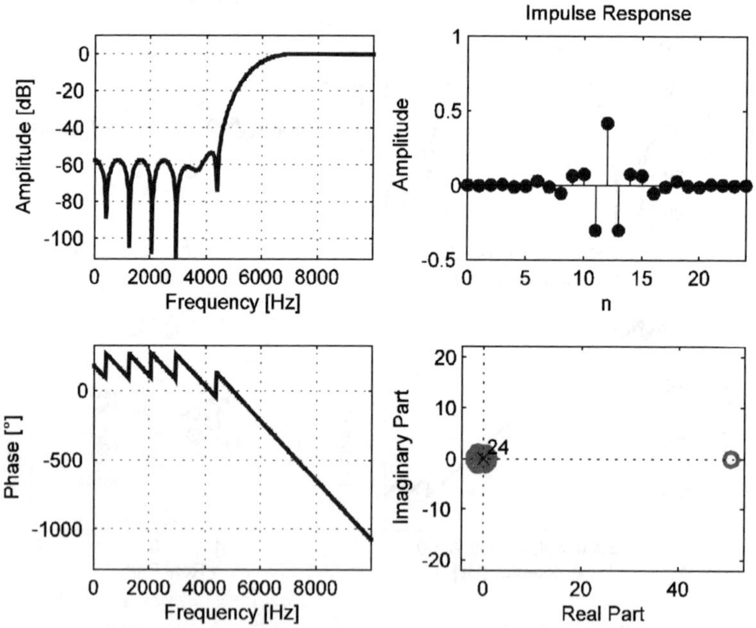

Figure 8.5. *Characteristics of the filter from exercise 8.3*

It can easily be seen from the impulse response shape that the designed filter belongs to the 2nd class FIR filters. In fact, its impulse response is symmetric and its order is even. One of the filter zeros is far from the unit circle because it is the inverse of another one, which is close to the origin. In this case there is no constraint on the zero localization with the exception of the conjugation and the inverse symmetry.

In this case the quality of transfer function approximation is equivalent to that obtained by a window method. Actually, the function fir2 is a combination between the frequency sampling method and the window method (the Hamming window is used by default).

EXERCISE 8.4.

Calculate the coefficients of a 25th order FIR bandstop filter using a LS method. Consider a sampling frequency f_s = 20 kHz and the following lower and upper stopband frequency edges: f_1 = 3 kHz and f_2 = 7 kHz. Plot the transfer function, the impulse response and the zeros of the designed filter. Imagine a possible application for this type of filter.

```
N=25; f1=3e3; f2=7e3; fs=2e4;
wn1=2*f1/fs; wn2=2*f2/fs;
inc=0.01; ff=[0:inc:1-inc];
hh=[ones(1,round((wn1/inc)-1)) zeros(1,round(((wn2-wn1)/inc)+1))];
hh=[hh ones(1,round((1-wn2)/inc))];
b=firls(N,ff,hh);
[h,w] = freqz(b,1); fq = fs*w/(2*pi);
amp=20*log10(abs(h));
phase=unwrap(angle(h))*180/pi;
figure; subplot(221); plot(fq,amp);
axis([0 max(fq) min(amp) 10])
xlabel('Frequency [Hz]');
ylabel('Amplitude [dB]') ; grid
subplot(223); plot(fq,phase);
axis([0 max(fq) 1.2*min(phase) 1.2*max(phase)])
xlabel('Frequency [Hz]');
ylabel('Phase [°]'); grid
subplot(2,2,2); impz(b,1); xlabel('n')
subplot(2,2,4); zplane(b,1);
```

Although the filter is not equiripple in the stopband, it could be helpful to reject some interferences or perturbations present in this band and which corrupt the useful signal component. A better rejection in the stopband can be obtained either with a higher order FIR filter or with an IIR filter.

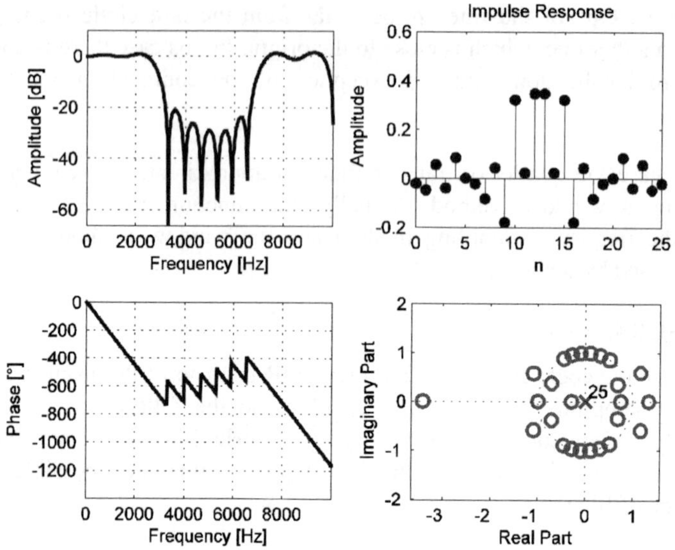

Figure 8.6. *Characteristics of the filter from exercise 8.4*

EXERCISE 8.5.

Design a bandpass FIR filter having the same parameters as in exercise 8.1 using the Remez method. Plot the transfer function, the impulse response and the zeros of the obtained filter and compare with the filter designed in exercise 8.1.

```
f11=2.5e3; f12=3.5e3;
f21=6.5e3; f22=7.5e3;
fs=2e4;
wvect=[f11 f12 f21 f22];
[n,fo,mo,w] = remezord(wvect,[0 1 0],[0.04 0.03 0.05],fs)
b = remez(n,fo,mo,w);
[h,w] = freqz(b,1);
fq = fs*w/(2*pi);
amp=20*log10(abs(h));
phase=unwrap(angle(h))*180/pi;
figure;
subplot(221); plot(fq,amp);
axis([0 max(fq) min(amp) 10])
xlabel('Frequency [Hz]');
ylabel('Amplitude [dB]'); grid
subplot(223); plot(fq,phase);
axis([0 max(fq) 1.2*min(phase) 1.2*max(phase)])
xlabel('Frequency [Hz]');
ylabel('Phase [°]'); grid
subplot(2,2,2);
```

```
impz(b,1); xlabel('n')
subplot(2,2,4);
zplane(b,1);
```

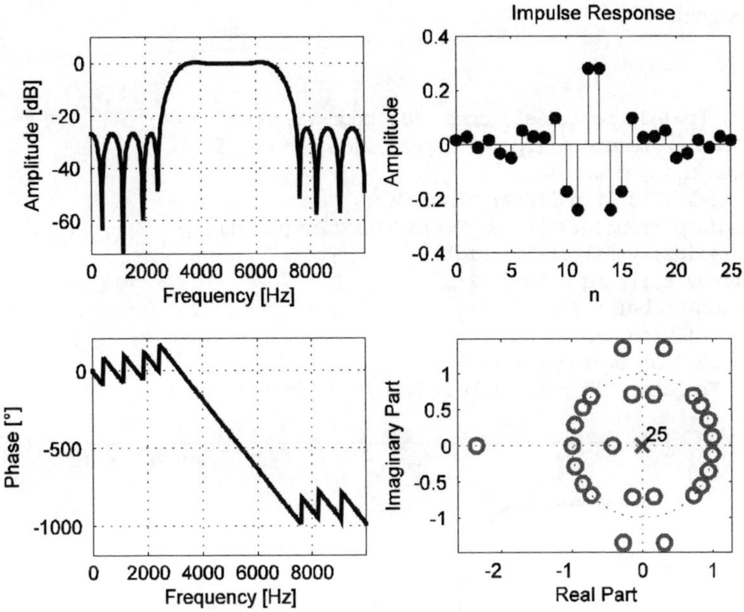

Figure 8.7. *Characteristics of the filter from exercise 8.5*

Recent MATLAB versions have replaced the functions `remezord.m` and `remez.m` with `firpmord.m` and `firpm.m`

Just as with the filter designed in exercise 8.1 the order of the obtained filter is 25. It can also be retrieved from the number of samples of its impulse response or from its number of zeros.

Compared to the filter designed in exercise 8.1 this filter provides a better stopband rejection (the attenuation gain is about 10 dB) and a reduced passband ripple. Furthermore, the filter designed by the Remez method is equiripple in the stopband.

EXERCISE 8.6.

Consider a sum of two sinusoids, whose frequencies are 1 kHz and 1.56 kHz, sampled at 10 kHz. Extract the first sinusoid of this mixture using a lowpass FIR filter, designed using the Remez method:

a. Define the filter specifications, calculate its coefficients and plot its transfer function.

b. Filter the mixture of the two sinusoids using the designed filter and plot the output signal.

a.

```
fs=10e3; fp=1e3; fc=1.56e3; ondp=.01; ondc=.1; ap=1; ac=0;
[n,fo,mo,w] = remezord([fp fc],[ap ac],[ondp ondc],fs);
b = remez(n,fo,mo,w);
f1=1e3;f2=1.56e3; f1n=f1/fs; f2n=f2/fs;
sig1=sin(2*pi*f1n*[0:99]); sig2=sin(2*pi*f2n*[0:99]);
sig=sig1+sig2; y=filter(b,1,sig);
[h,w]=freqz(b,1); fq = fs*w/(2*pi);
amp=20*log10(abs(h));
figure ; plot(fq,amp);
axis([0 max(fq) min(amp) 10])
xlabel('Frequency [Hz]'); ylabel('Amplitude [dB]'); grid
```

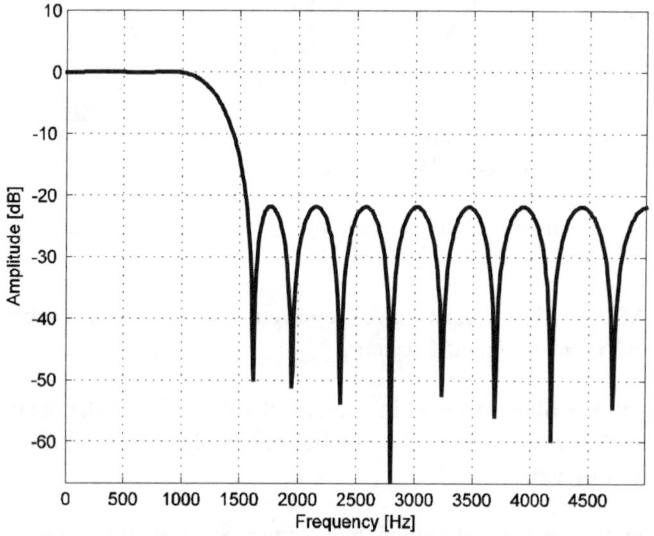

Figure 8.8. *Magnitude of the transfer function of the lowpass filter designed in exercise 8.6*

b.

```
figure ; subplot(2,2,1);
plot(sig1); title('First signal')
subplot(2,2,3); plot(sig2); title('Second signal')
subplot(2,2,2); plot(sig); title('Mixture of the two signals')
subplot(2,2,4); plot(y); title('Filtered signal')
```

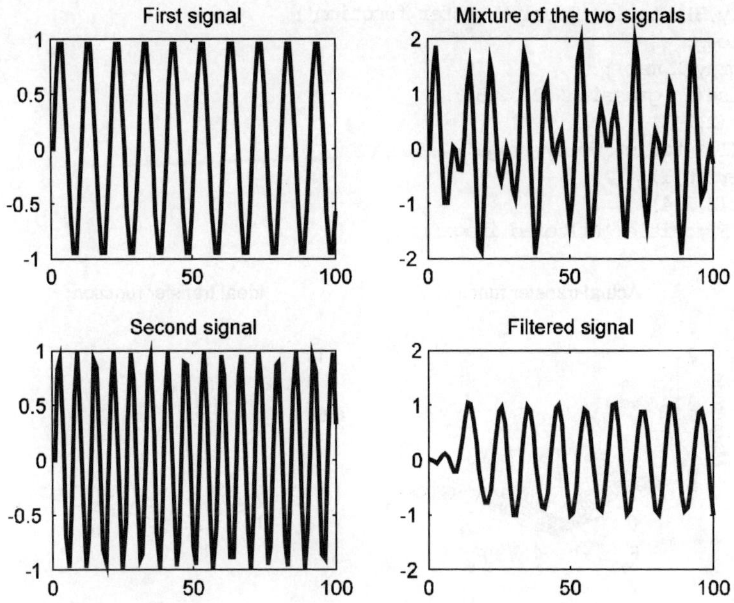

Figure 8.9. *Mixture of two sinusoids and filtered signal*

EXERCISE 8.7.

The methods for designing 2D FIR filters are similar to those used for the 1D filter design. The goal of this exercise is to illustrate the lowpass filtering of an image.

Consider to this end, a mask of 7×7 pixels, the 9 pixels in the middle having the value 1 and the other the value 0. Calculate the coefficients of this filter and plot the corresponding ideal and actual transfer functions.

Using the designed filter then perform the lowpass filtering of the image "clown" from the MATLAB image library, corrupted by a zero-mean white Gaussian noise with the variance 0.05.

```
H = zeros(7,7); H(3:5,3:5) = ones(3,3);
h = fsamp2(H);
figure
subplot(2,2,1);
freqz2(h,32,32);
title('Actual transfer function')
[f1,f2]=freqspace([7 7]);
[x,y]=meshgrid(f1,f2);
subplot(2,2,2)
```

```
mesh(x,y,H); title('Ideal transfer function')
load clown;
I=ind2gray(X,map);
I=imnoise(I,'gaussian',0,0.05);
subplot(2,2,3)
imshow(I); title('Noisy image')
F=filter2(h,I);
subplot(2,2,4)
imshow(F); title('Filtered image')
```

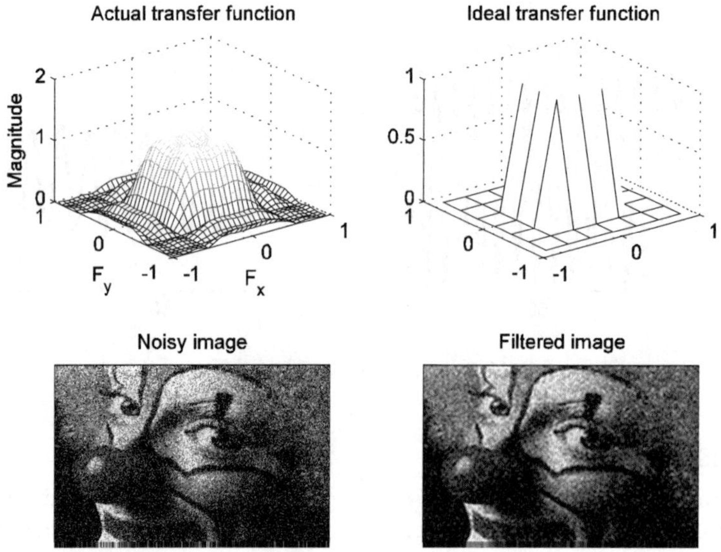

Figure 8.10. *Lowpass filtering image*

Point out the ripple of the transfer function due to the same Gibbs phenomenon as for the 1D filters.

EXERCISE 8.8.

This exercise is aimed at illustrating the effect of the quantification of lowpass FIR filter coefficients. Consider a cutoff normalized frequency of 0.2 and a lower stopband normalized frequency edge of 0.225. Use 19 coefficients coded on 5 bits to design the filter.

```
f = [0 2*0.2  2*0.225 1]; m = [1 1 0 0];
b = remez(19, f, m);
[h,w] = freqz(b,1); g = 20*log10(abs(h));
fq = w/(2*pi); bq = a2dT(b,4);
hq = freqz(bq,1); gq = 20*log10(abs(hq));
```

```
figure
subplot(211); plot(fq,g,'b-',fq,gq,'r-');
axis([0 0.5 -60 10]);grid
xlabel('Normalized frequency');
ylabel('Gain [dB]');
legend('original','quantified');
subplot(223);
zplane(b); title('original');
subplot(224);
zplane(bq); title('quantified');
```

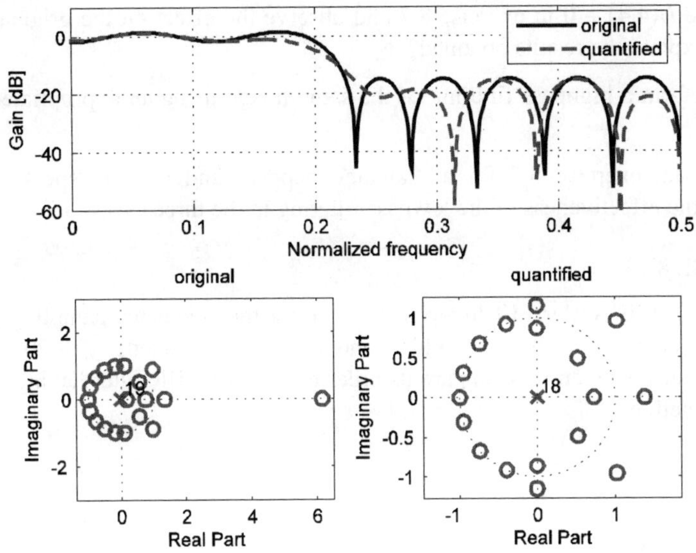

Figure 8.11. *Influence of the coefficient quantification on a FIR filter's characteristics*

8.3. Exercises

EXERCISE 8.9.

Plot the rectangular, triangular, Hanning, Hamming, Blackman and Kaiser windows using MATLAB functions `boxcar`, `bartlett`, `hanning`, `hamming`, `blackman` and `kaiser`. Calculate and plot the magnitude of their spectra. Emphasize the sidelobe level and the bandwidth of these spectra. Conclude about the effect of these windows on a FIR filter design.

EXERCISE 8.10.

Using the properties of the four classes of FIR filters summarized in Table 8.1, find the appropriate applications for each of them from the following list: lowpass filter, bandpass filter, highpass filter, stopband filter, Hilbert transformer and derivative filter.

EXERCISE 8.11.

Calculate the coefficients of a 25th order derivative filter using the LS method, for a sampling frequency f_s = 20 kHz. Plot its transfer function, impulse response and zeros. Imagine a possible application for this filter.

EXERCISE 8.12.

a. Use $H(4,4) = 0$ in exercise 8.7 and observe the effect on the original image. Can you explain the results obtained?

b. Perform a highpass filtering of the same image using an appropriate transfer function.

c. Repeat exercise 8.7 for a "salt and pepper" and then a "speckle" noise. Compare the effectiveness of the lowpass filtering in the three cases.

EXERCISE 8.13.

Design a 120th order FIR highpass filter using the frequency sampling method. Consider a cutoff frequency f_c = 6 kHz, and a sampling frequency f_s = 20 kHz. Plot the filter transfer function. Compare its order to that of an IIR filter having a similar transfer function.

Chapter 9

Detection and Estimation

9.1. Theoretical background

This chapter presents the most important decision models considered in detection or estimation problems.

9.1.1. *Matched filtering: optimal detection of a known noisy signal*

The matched filter is an optimal linear system for the detection of a known signal $s(t)$ corrupted by an independent additive noise. It maximizes the signal-to-noise ratio at the moment when the decision is made. If the noise is white, the impulse response of the matched filter to the signal $s(t)$ has the following expression:

$$g_a(t) = k \cdot s^*(t_0 - t) \qquad [9.1]$$

where k is a constant depending on the filter gain, and t_0 is a time-delay parameter which usually corresponds to the signal duration T.

The output signal of the matched filter driven by the signal $x(t) = s(t) + n(t)$ becomes thus:

$$y(t) = x(t) * g_a(t) = k\Gamma_{ss}(t - t_0) + k\Gamma_{sn}(t - t_0) \qquad [9.2]$$

Consequently, the matched filter has the behavior of a correlator with respect to the signal to be detected.

The signal-to-noise ratio is maximum in $t = t_0$ and depends on the signal energy W_s and the noise power spectral density $N_0/2$. For a given noise and different signals having the same energy W_s, the corresponding matched filters give the same signal-to-noise ratio. However, they may be very different in terms of the time resolution, which means the capability of resolving two closely spaced signals.

In fact, without matched filtering the time resolution is equal to the signal duration T. After matched filtering it is related to the signal correlation length D_τ, which is inversely proportional to the signal bandwidth B_τ. Consequently, the matched filter may greatly improve the time resolution provided that the signal has a large bandwidth (i.e. a narrow autocorrelation function).

In real world applications the signal amplitude is limited by physical systems. Thus, a long signal is often necessary in order to insure the required energy level. Thus, a large BT product is finally the global condition to have both a good signal-to-noise ratio and a high time resolution. The most often used signals with large BT products are the frequency modulated pulse (chirp) or the pseudorandom binary phase modulated signal.

In the case of a colored noise, the matched filter is obtained by the series connection of a whitening filter with the matched filter corresponding to the whitened signal.

9.1.2. *Linear optimal estimates*

The approaches presented in this section look for the optimal estimate over the class of linear filters. The solution is thus the best linear estimate, although an even better non-linear solution could exist. The main advantage of any linear approach is its simplicity of implementation.

The most used linear optimal filters were introduced by Wiener and Kalman. Although the signal presentation mode is different in the two cases, the two filters make use of the same optimization criterion and lead essentially to the same results.

Wiener filter

Norbert Wiener's approach is based on an external signal representation. According to this representation, signals are described by their statistical properties: probability density function, moments, correlation function, etc. Since the considered filters are linear, only the 1st and 2nd order statistics are involved. Actually, Wiener's approach requires the knowledge of the autocorrelation and cross-correlation functions in the time domain or of the spectra and interspectra in the frequency domain.

Let us consider a discrete-time, real, zero-mean and stationary random process $\{X_k\}_{k\in \mathbf{Z}}$ and an infinite series of observations $\{Y_k\}_{k\in \mathbf{Z}}$. The linear estimates of X_k over $\{Y_k\}_{k\in \mathbf{Z}}$ have the following form:

$$\hat{X}_k = \sum_{n=0}^{N-1} \theta_n Y_{k-n} \qquad [9.3]$$

They depend only on a finite number of observations $\{Y_k, k = 0..N\text{-}1\}$. It is more interesting in this case to adopt matrix notations. Thus, the N observations can be considered as a vector $Y = [Y_0, \ldots, Y_{N-1}]$, whose autocorrelation matrix is denoted by $\Gamma_{YY} = E[Y \cdot Y^T]$.

Let us also denote by $\Gamma_{XX} = E[X_l \cdot X_{l-k}]$ the autocorrelation function of X_k and by $\Gamma_{XY} = E[X_k \cdot Y]$ the cross-correlation vector between X_k and Y. Equation [9.3] can then be rewritten in the form:

$$\hat{X}_k = \Theta^T Y \text{, with } \Theta^T = \left[\theta_0, \cdots, \theta_{N-1}\right] \qquad [9.4]$$

The optimal linear estimate is therefore obtained as follows:

$$\Theta_{\text{opt}} = \arg \min_{\Theta\in R^N} E\left[\left(\Theta^T Y - X_k\right)^2\right] \qquad [9.5]$$

According to the orthogonal projection theorem, this estimate also meets the condition below:

$$E\left[(\hat{X}_k - X_k) Y\right] = 0 \qquad [9.6]$$

This results in the following expression of a discrete Wiener filter provided that Γ_{YY}^{-1} exists:

$$\Theta_{\text{opt}} = \Gamma_{YY}^{-1} \Gamma_{XY} \qquad [9.7]$$

Taking advantage of the orthogonal relationship between $X_k - \hat{X}_k$ and $\hat{X}_k$ the estimation error is thus simply obtained in the form:

$$\varepsilon_{\min}^2 = E\left[(\hat{X}_k - X_k)^2\right] = E\left[X_k(X_k - \hat{X}_k)\right] = \Gamma_{XX}(0) - \Theta_{\text{opt}}^T \Gamma_{XY} \qquad [9.8]$$

PROPERTIES

1. As the matrix Γ_{YY}^{-1} is positive defined, the following equation holds:

$$\Theta_{\text{opt}}^{T}\Gamma_{XY} = \Gamma_{XY}^{T}\Gamma_{YY}^{-1}\Gamma_{XY} \geq 0 \qquad [9.9]$$

and thus:

$$\varepsilon_{\min}^{2} \leq \Gamma_{XX}(0) \qquad [9.10]$$

If observation Y is not correlated with X, no information on X can be inferred from Y. In this case $\varepsilon_{\min}^{2} = \Gamma_{XX}(0)$. In other words, if only the first two moments of a random process are known, its minimal variance estimate is its mean value.

2. If Y is linearly related to X_k, the estimation error is zero. Obviously, this statement is valid only if the filter length is large enough to represent the true linear relationship between the two quantities.

3. The autocorrelation matrix Γ_{YY} and the cross-correlation vector Γ_{XY} has to be *a priori* known. Otherwise, they have to be estimated prior to the Wiener filter calculation. While the estimation of Γ_{YY} is straightforward because Y is the observed quantity, the estimation of Γ_{XY} is more difficult because the statistical characteristics of X may be unknown.

Kalman filter

Kalman proposed a filter structure based on an internal signal representation. According to this representation any random process is seen as the output of a linear filter driven by a white noise.

In fact, according to the Wiener-Khintchine theorem, the power spectral density of a linear filter output signal is the product between the input signal power spectral density and the square of the transfer function magnitude. If the input signal is a unitary white noise its power spectral density is constant and equal to 1. Consequently, the power spectral density $\gamma_{XX}(\nu)$ of the wide sense stationary process represented by the output signal $\{X_k\}_{k\in\mathbb{Z}}$ can be expressed as $\gamma_{XX}(\nu) = |G(\nu)|^2$, where $G(\nu)$ is the transfer function of a minimal phase linear filter (i.e., with all the poles and zeros inside the unit circle).

If the filter transfer function is rational, this relationship corresponds in the discrete-time domain to a difference equation with constant coefficients. If the filter is not homogenous these coefficients vary over time. This type of internal representation is especially preferred in control theory since some physical properties of the system can be integrated into the model. Furthermore, since the internal approach is recursive, it is particularly suitable for non-stationary systems.

These two aspects make the Kalman approach an interesting, sometimes indispensable, alternative to the Wiener filter.

The main features of a Kalman filter result from the internal signal representation and are listed below:

– state-space system model;

– recursive formulation of the mean-square optimal solution;

– appropriate integration of non-stationary variations.

The state-space system model is represented by the following state and measurement equations:

$$X_k = A_k X_{k-1} + W_{k-1} \qquad [9.11]$$

$$Y_k = C_k X_k + V_k \qquad [9.12]$$

and involves the elements indicated below:

– X_k: state vector; this fully describes the state of the system at moment k (example: position and speed of harmonic oscillator);

– A_k: transition matrix; this determines the free evolution of the system state (without excitation $W_k = 0$); the state-space model is stable if all the eigenvalues of A_k are inside the unit circle;

– W_k : zero-mean white noise; this represents the input of the state-space model and is sometimes called system or process noise;

– Q_k : covariance matrix of W_k ;

– Y_k : observation vector at moment k;

– C_k : observation or measurement matrix; this indicates how the state vector is observed;

– V_k : stands for the noise which corrupts the observation vector (measurement noise);

– R_k : covariance matrix of V_k ;

– ψ_k : the space spanned by observations $\{Y_1, \cdots, Y_k\}$.

If W_k is Gaussian, then X_k becomes Gaussian, because of the linearity of the state model. The two noises W_k and V_k are assumed to be uncorrelated.

The measurement equation shows that Y_k is just a partial noisy observation of the state vector X_k. The Kalman filter provides the best linear estimate of X_k, according to the LMS criterion, based on the observations measured up to moment k.

The optimal LMS estimate of X_k based on the observations measured until the moment m is denoted by $\hat{X}_{k|m}$. An estimation error $\tilde{X}_{k|m} = X_k - \hat{X}_{k|m}$ and its covariance matrix $P_{k|m} = E\left[\tilde{X}_{k|m}\tilde{X}_{k|m}^T\right]$ are thus associated with this estimate.

The Kalman filter provides the LMS optimal estimate in a recursive form including the following equations:

$$\begin{cases} P_{k|k-1} = A_K P_{k-1|k-1} A_k^T + Q_{k-1} & \text{(variance of the prediction error)} \\ K_k \ = P_{k|k-1} C_k^T (R_k + C_k P_{k|k-1} C_k^T)^T & \text{(Kalman gain)} \\ P_{k|k} \ = (I - K_k C_k) P_{k|k-1} & \text{(variance of the estimation error)} \\ \hat{X}_{k|k-1} = A_k \hat{X}_{k-1|k-1} & \text{(linear prediction)} \\ \hat{X}_{k|k} \ = X_{k|k-1} = K_k (Y_k - C_k \hat{X}_{k|k-1}) & \text{(state estimation)} \end{cases} \quad [9.13]$$

Deterministic approach

In the framework of the two previously presented stochastic approaches, the observation is considered an outcome of a random process. The 2nd order statistical properties of this process are assumed to be completely known. Consequently, the associated processing algorithms do not depend on a particular outcome of the observed process, they only depend on its statistical properties *a priori* known. In this way, the most probable outcomes are favored against the less probable ones.

The processing algorithm is, on the contrary, adapted to a particular outcome if a deterministic approach is considered. The random aspect of the process is secondary in this case. A data dependent algorithm is obtained which provides the best estimate for a particular outcome of the observed process.

Obviously, the two approaches converge in practice whenever the statistical properties of the process have to be estimated using its outcomes. Thus, the link between the two approaches is represented by the ergodicity property of the observed process.

9.1.3. *Least squares (LS) method*

The LS method was introduced by Gauss to estimate noisy parameters. Let us consider a series of k scalar measured values $\{y_n\}_{n=1..k}$ which linearly depend on an observable vector set $\{X_n\}_{n=1..k}$:

$$y_n = \theta_{opt}^T X_n + b_n \text{ for } 1 \leq n \leq k \qquad [9.14]$$

where b_n stands for the noise samples.

In practice, these measures may be issued from a set of N sensors. Let us consider the following notations:

– $\theta_{opt} = [\theta_1, \cdots, \theta_N]^T$ vector of the parameters to be estimated,

– $X_n = \left[x_{n_1}, \cdots, x_{n_N}\right]^T$ for $n = 0..k$ the observed vector at the moment n,

– $\Gamma = \left\{x_{ij}\right\}_{i=1..k,\ j=1..N}$ matrix of all observations,

– $Y = [y_1, \cdots, y_k]^T$.

θ_{opt} is the LS solution of the following matrix equation:

$$Y = \Gamma \cdot \theta_k \qquad [9.15]$$

The only interesting case is $k \geq N$ (overdetermined system). The LS solution of the system at moment k can be obtained as follows:

$$\theta_k = \arg \min_{\theta \in \mathbb{R}^N} \left\{(Y - \Gamma\theta)^T (Y - \Gamma\theta)\right\} \qquad [9.16]$$

This results in:

$$\theta_k = \left[\left(\Gamma^T \Gamma\right)^{-1} \Gamma^T\right] Y \qquad [9.17]$$

According to the orthogonal projection theorem, the optimal estimate is the orthogonal projection of Y on the observation space. In fact, $(\Gamma^T\Gamma)^{-1}\Gamma^T$ (known as the right pseudo-inverse of Γ) is the orthogonal projection operator on the image of Γ. If Γ is square ($k = N$) and invertible, the pseudo-inverse $(\Gamma^T\Gamma)^{-1}\Gamma^T$ becomes simply the inverse matrix; a Wiener solution is thus obtained as: $\theta = \Gamma^{-1}Y$.

The following recursive algorithm is finally derived using the Woodbury identity:

$$\begin{cases} K_k = P_{k-1}X_k / (1 + X_k^T P_{k-1} X_k) \\ \theta_k = \theta_{k-1} + K_k (y_k - \theta_{k-1}^T X_k) \\ P_k = P_{k-1} - K_k X_k^T P_{k-1} \end{cases} \qquad [9.18]$$

where:

$$P_k = \left(\sum_{n=1}^{k} X_n X_n^T \right)^{-1} \qquad [9.19]$$

9.1.4. *LS method with forgetting factor*

If vector θ is time-variant, the LS method cannot be directly applied. The algorithm should "forget" the old process "history" in this case because the corresponding information is not relevant for the current value of parameter θ. The basic idea is to observe only the data inside a sliding window. Its task is to weigh the data according to its credibility. The simplest and most used window is the exponential function. In this case the optimal LS estimate is given by:

$$\theta_* = \arg\min_{\theta \in R} \sum_{n=0}^{k} \lambda^{k-n} e_n^2 \text{ , with } e_n = y_n - \theta x_n \qquad [9.20]$$

Forgetting factor λ takes values between 0 and 1. The following function is thus minimized:

$$J(\theta) = e_k^2 + \lambda e_{k-1}^2 + \lambda^2 e_{k-2}^2 + \cdots + \lambda^k e_0^2 \qquad [9.21]$$

The closer to zero λ is, the larger the forgetting rate is. For λ =1 the particular case of the exact LS method is obtained. The more recent an observation is, the larger its weight is in estimating the current value of θ. The LS algorithm with forgetting factor (RLS, recursive least square) has the following form:

$$\begin{cases} K_k = \lambda^{-1} P_{k-1} X_k / (1 + \lambda^{-1} X_k^T P_{k-1} X_k) \\ \theta_k = \theta_{k-1} + K_k (y_k - \theta_{k-1}^T X_k) \\ P_k = \lambda^{-1} P_{k-1} - \lambda^{-1} K_k X_k^T P_{k-1} \end{cases} \qquad [9.22]$$

9.2. Solved exercises

EXERCISE 9.1.

A digital signal containing 20 binary elements is transmitted on a communication channel. This signal has the following characteristics:

– binary rate: 1000 symbols/s,

– sampling frequency: 10 kHz,

– symbol duration: 1 ms,

– code: unipolar without return to zero.

1. Using the function subplot plot the signal, the matched filter impulse response and its output signal. Define the decision threshold and the optimal decision moments.

```
Rb = 1000;  % binary rate
fs = 10000; % sampling frequency
Ts = 1/fs;  % sampling period
Tb = 1/Rb;  % binary element duration
sequence=[0 1 1 1 0 1 0 0 1 0 1 1 1 0 1 0 0 1 0 1];
no= length(sequence);  % number of bits
no_ech = no * Tb/Ts;   % number of samples
t = [0:(no_ech-1)]*Ts; % sampling moments
pulse =ones(1,fs/Rb);  % waveform (unipolar without return to zero)
x=(sequence'*pulse)';  % signal generation
x=x (:);
t_time = Ts * [0:no_ech];
figure
subplot(311);
plot(t_time,[x;0]);
axis([0 0.021 0 1.2]);grid
title('Bandbase signal')
g = (pulse);
h = g(length(g):-1:1); % matched filter impulse response
subplot(312);
plot([0:size(h,2)+1]*Ts,[h 1 0]);
axis([0 0.021 0 1.2]);grid
title('Matched filter impulse response')
y_fil = conv(h,x);      % matched filter output signal
y_fil = [y_fil( :)' 0];
subplot(313);
plot([0:size(y_fil,2)]*Ts,[0 y_fil]);
axis([0 0.021 0 1.2*max(y_fil)]);grid
xlabel('Time [s]');
title('Matched filter output signal')
```

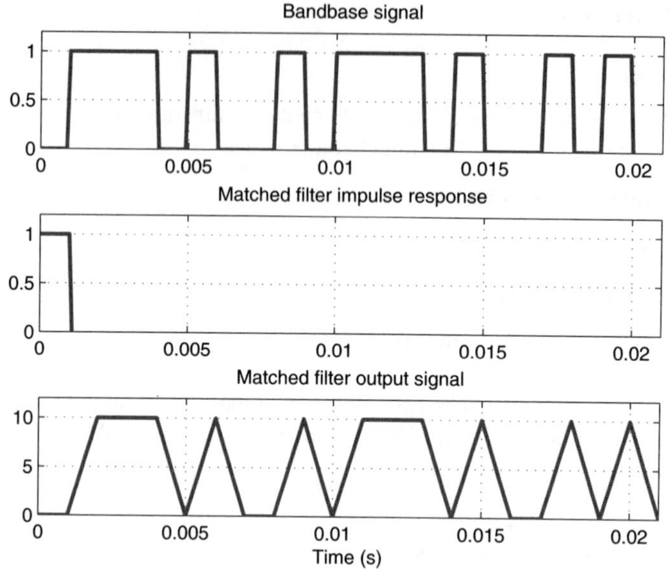

Figure 9.1. *Input signal, matched filter impulse response and its output signal*

As the impulse response width is 1 ms, the optimal decision moments should be multiples of this quantity.

For equiprobable symbols, the optimal decision threshold is equal to half of the symbol amplitude according to the maximum likelihood criterion.

The amplitude of the matched filter output signal is equal to the energy of the projection of the received signal on the matched filter impulse response.

2. Real communication channels are not flat; they have limited passband and suffer from additional noise.

Write a MATLAB code to simulate a channel which is characterized by a bandwidth of 4,900 Hz and a zero-mean white Gaussian additive noise with the variance 1. Filter the transmitted signal using an IIR 8th order Chebychev type I filter.

Plot on the same figure the ideal signal, the noisy signal and the matched filter output signal.

Decode the previous message using the noisy signal and the symbol length as time unit. Repeat this procedure using the matched filter output signal and the same detection threshold. Explain why the matched filter performs better.

```
ondul = 0.1;    % passband ripple
ord_filt = 8;   % filter order
fc = 4900;      % cutoff frequency
[Bc,Ac] = cheby1(ord_filt,ondul,fc/(fs/2));
y = filter(Bc,Ac,x);          % signal filtering
r = randn(size(y)); noise_power=1;
y = y + sqrt(noise_power)*r; % adding the zero-mean Gaussian white noise
figure ; subplot(311) ; plot(t_time, [y ;0]);
axis([0 0.021 -1.2*max(y) 1.2*max(y)]);
grid on; title('Baseband noisy signal')
subplot(312); plot([0:size(h,2)+1]*Ts, [h 1 0]);
axis([0 0.021 0 1.2]); grid on;
title('Matched filter impulse response')
y_fil = conv(h,y)*Ts;     % matched filter output signal
y_fil = [y_fil( :)' 0];
subplot(313); plot([0 :size(y_fil,2)]*Ts, [0 y_fil]);
axis([0 0.021 -1.2*max(y_fil) 1.2*max(y_fil)])
grid on; xlabel('Time [s]');
title('Matched filter output signal')
```

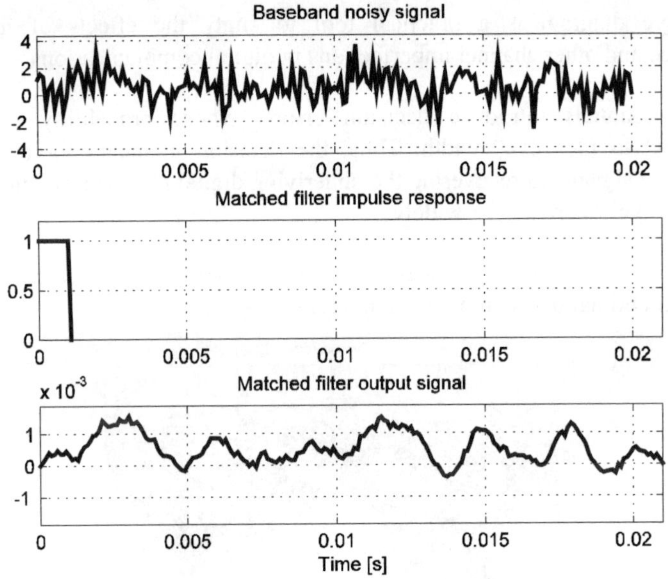

Figure 9.2. *Noisy input signal, matched filter response and output signal*

3. The choice of the decision threshold and of the decision moments is a very interesting point to investigate. This choice is made difficult by symbol interferences due to the symbol delay and the channel features. The eye diagram allows these quantities to be optimized.

Generate the eye diagram using a binary sequence of 1,000 symbols and the channel previously defined.

```
sequence=round(rand(1,1000));
x=(sequence'*pulse)';          % signal generation
x=x( :); y = filter(Bc,Ac,x); % signal filtering
r = randn(size(y)); noise_power=0.25;
y = y + sqrt(noise_power)*r;      % adding noise
no_eyes=2;                    % number of diagram eyes
no_block = fix(tot_ech/(no_eyes*no)); % number of samples blocks
index  = [1:(no_eyes*no_ech)];
t_time = Ts *[0:no_eyes*no_ech];
time   = Ts *[1:no_eyes*no_ech];
foo = zeros((no_eyes*no_ech), no_block-1);
foo(:) = y_fil((no_eyes *no_ech)+1: no_block* no_eyes *no_ech);
first_row = foo(no_eyes *no_ech,:);
first_row = [y_fil(maxindex) first_row(1:no_block-2)];
foo = [first_row; foo];
plot(t_time, foo,'b');
```

The eye diagram is a practical tool to study the effects of intersymbol interference and other channel imperfections in digital communications.

Noisy channels introduce decision errors, whose probability is inversely proportional to the "eye" width. Thus, the decision moments for demapping a demodulated signal and recovering the underlying digital message, should be chosen where the "eye" is open most widely.

The horizontal opening provides information about the influence of the distance between an optimal decision moment and real ones.

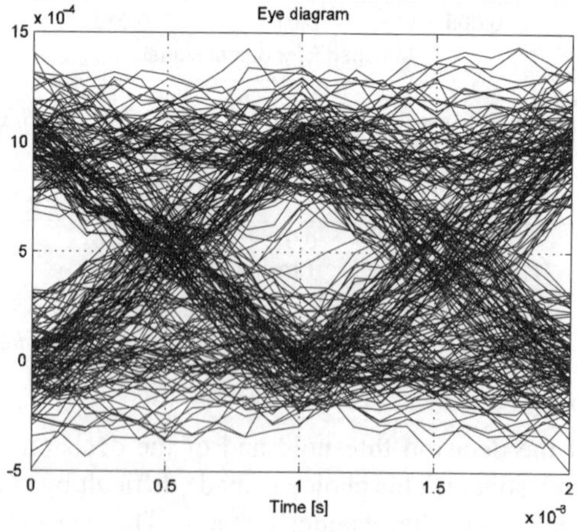

Figure 9.3. *Eye diagram*

EXERCISE 9.2.[1]

This exercise is aimed at comparing results provided by three types of 2D filters: Prewitt, Sobel and Roberts. An image containing a vertical outline is generated at this end and two quantities are estimated for each type of filter:

– the detection probability: percentage of well detected outline points;

– the false alarm probability: percentage of points taken from outside the vertical outline and considered as outline points.

In order to simplify the comparison, detection thresholds are set to obtain roughly the same false alarm probability level. The three filters are then compared using only the detection probability.

```
I=0.6*ones(64,64); I(:,1:32)=0.5*ones(64,32);
J=I+sqrt(.001)*randn(64,64);
fil_pre_h=[1 1 1 ;0 0 0 ;-1 -1 -1] ; fil_pre_v=-fil_pre_h';
fil_sob_h=[1 2 1;0 0 0;-1 -2 -1]; fil_sob_v=-fil_sob_h';
fil_rob_h=[1 0;0 -1]; fil_rob_v=[0 1;-1 0];
d_pre_h=filter2(fil_pre_h,J);
d_pre_v=filter2(fil_pre_v,J);
d_sob_h=filter2(fil_sob_h,J);
d_sob_v=filter2(fil_sob_v,J);
d_rob_h=filter2(fil_rob_h,J);
d_rob_v=filter2(fil_rob_v,J);
cont_pre=sqrt(d_pre_h.* d_pre_h+ d_pre_v.* d_pre_v);
cont_sob=sqrt(d_sob_h.* d_sob_h+ d_sob_v.* d_sob_v);
cont_rob=sqrt(d_rob_h.* d_rob_h+ d_rob_v.* d_rob_v);
% contour thresholding
cont_pre_s=( cont_pre>0.25);
cont_sob_s=( cont_sob>0.35);
cont_rob_s=( cont_rob>0.145);
% plotting the filtering results
subplot(221);
imagesc(J);
title('noisy image');
subplot(222);
imagesc(cont_pre_s);
title('contours obtained with the Prewitt mask');
subplot(223);
imagesc(cont_sob_s);
title('contours obtained with the Sobel mask');
subplot(224);
imagesc(cont_rob_s);
title('contours obtained with the Roberts mask');
% Detection and false alarm probability calculation
```

1 This exercise was proposed by Gilles Burel, from the University of West Brittany, Brest, France.

```
medium=cont_pre_s(2:63,2:63);
det=(medium(:,31)+medium(:,32))>0;
nbdet=sum(det);
pdet_prewitt=nbdet/62
medium(:,31:32)=zeros(62,2);
nbfa=sum(sum(medium));
pfa_prewitt=nbfa/(60*62)
medium=cont_sob_s(2:63,2:63);
det=(medium(:,31)+medium(:,32))>0;
nbdet=sum(det);
pdet_sobel=nbdet/62
medium(:,31:32)=zeros(62,2);
nbfa=sum(sum(medium));
pfa_sobel=nbfa/(60*62)
medium=cont_rob_s(2:63,2:63);
det=(medium(:,31)+medium(:,32))>0;
nbdet=sum(det);
pdet_roberts=nbdet/62
medium(:,31:32)=zeros(62,2);
nbfa=sum(sum(medium));
pfa_roberts=nbfa/(60*62)
```

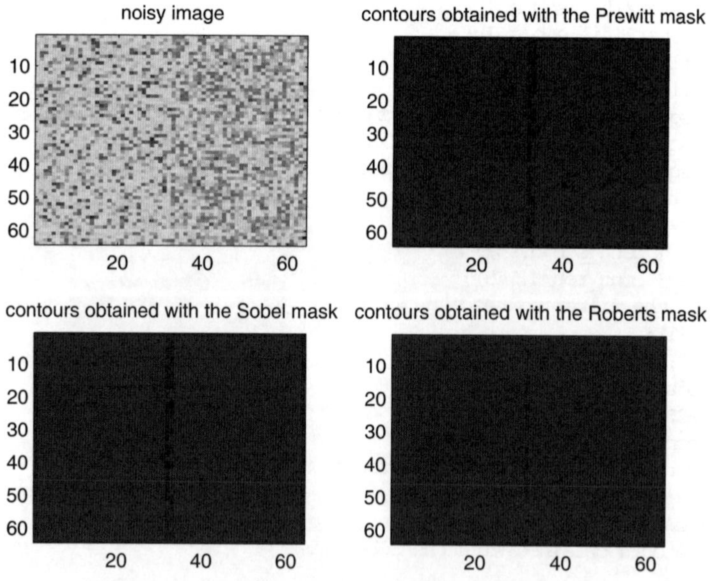

Figure 9.4. *Contour detection using different operators*

Prewitt's filters detect both horizontal and vertical outlines. That can be obtained by thresholding the "norm" of the two images issued from the filtering process.

Thus, the Prewitt filters do not perform very well for detecting some types of outlines, which are neither vertical nor horizontal, such as closed forms.

The same comments are also valid for Sobel's filters. Note that the threshold choice is essential. In fact, the first two images are quite similar before the thresholding, so the final results are also similar.

Roberts filters are quite useful for detecting diagonal contours, while the associated run-time is considerably reduced due to the mask dimension of 2. However, the threshold choice is more difficult. In fact, the obtained image contains many short separated straight lines, which make the contour detection more complex.

Prewitt's filter outperforms the two other filters in this example. In fact, it provides the best detection rate, for roughly the same level of the false alarm rate, while the outline is well detected and continuous.

EXERCISE 9.3.

Consider a system having the transfer function below, which is assumed unknown and has to be estimated:

$$H(z) = \frac{z^{-1} - 0.5z^{-2}}{1 - 0.8z^{-1}}$$

Write a MATLAB code in order to estimate the model parameters using a RLS algorithm and a binary pseudorandom sequence (BPRS) containing 1,023 values (this length is large enough with respect to the system frequency bandwidth).

Plot the time variation of the estimated coefficients and evaluate the autocorrelation function of the error signal defined as the difference between the system outputs corresponding to its actual and estimated coefficients respectively.

The problem can be formulated in an equivalent manner as a model identification problem for the following process:

$$y(t) = a_1 y(t-1) + b_1 u(t-1) + b_2 u(t-2) + v(t)$$

where $a_1 = 0.8$, $b_1 = 1$, $b_2 = -0.5$ and $v(t)$ is a zero-mean white additive Gaussian noise with a unit variance.

```
a1=0.8; b1=1; b2=-0.5;
para=[a1 b1 b2];
```

```
n=1000; np=3;
u=1:10;u=(1).^u;
for i=11:n
     u(i)=-u(i-7)*u(i-10); % generation of the BPRS
end
figure
plot([1:n],u);
axis([0 100 -1.1 1.1]);
xlabel('time');
ylabel('amplitude');
title('Binary pseudo-random sequence');

p=1e8*eye(np); teta=zeros(np,2);
v=randn(1,n); y=[0 0];
for i=3:n
     x(:,1)=[y(i-1) u(i-1) u(i-2)]';
     y(i)=para*x(:,1)+v(i);
    k=(p*x)/(1+x'*p*x);
     teta(:,i)=teta(:,i-1)+k*(y(i)-x'*teta(:,i-1));
    p=p-k*x'*p;
end
para_fin=teta(:,i);
figure
subplot(211)
for i=1:np
     plot(1:n,teta(i,:)); hold on
end
axis([0 1000 -1.5 1.5]);
title('parameters variation');
xlabel('time');grid on;
ys=[0 0];
for i=3:n
     ys(i)=[ys(i-1) u(i-1) u(i-2)]*para';
end
err=ys-y; average=mean(err);
var=std(err-average)^2;
err=err-average; cor_err=xcorr(err,err);
subplot(212)
plot([-length(err)+1:length(err)-1],cor_err);
grid; xlabel('time');
title('autocorrelation function of the error signal');
```

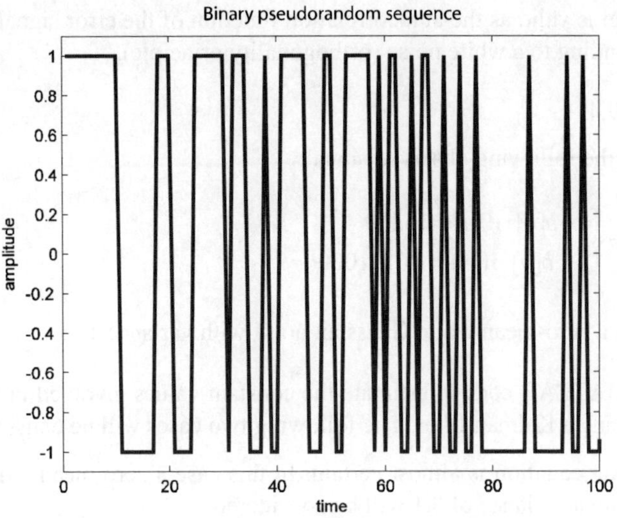

Figure 9.5. *Binary pseudorandom sequence*

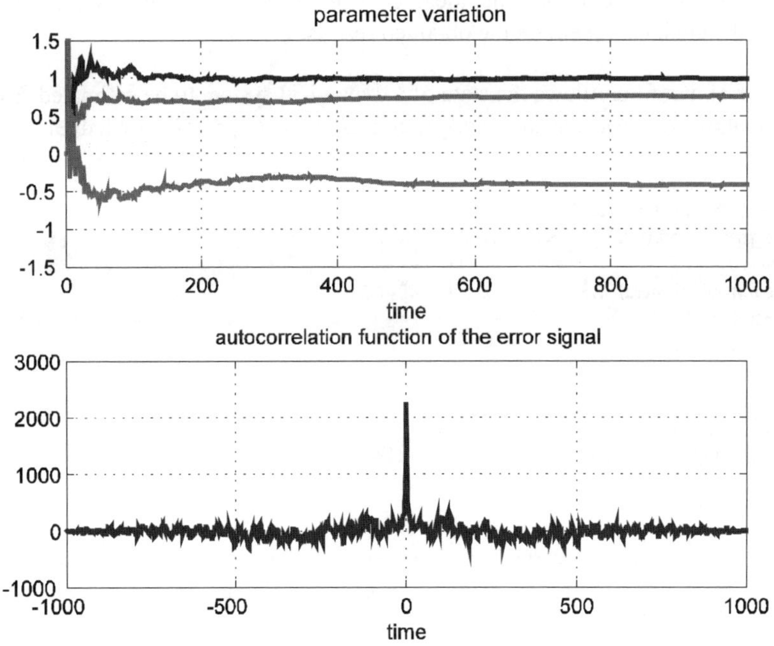

Figure 9.6. *Parameter estimation (top) and the autocorrelation function of the error signal (bottom)*

The model is valid as the autocorrelation function of the error signal is similar to that corresponding to a white noise (orthogonality principle).

EXERCISE 9.4.

Consider the following observed signal:

$$y[n] = \begin{cases} 1 + b[n] & \text{if } n = 1..200 \\ 6 + b[n] & \text{if } n = 201..1000 \end{cases}$$

where $b[n]$ is a zero-mean white Gaussian noise with variance 1.

Write a MATLAB code to estimate the constant values involved in the observed signal $y[n]$ using a Kalman filter. The following two cases will be considered:

1. The state equation is almost certain. In this case a zero-mean white Gaussian state noise with a variance of 0.1 will be considered.

2. The state model is assumed to be unknown. In this case a zero-mean white Gaussian state noise with a variance 0.64 will be considered.

Plot on the same figure the ideal constant signal, the noisy signal and the estimated signal. Comment on the obtained results.

The Kalman filter allows the state of a dynamical system to be estimated from its inputs-outputs (*a posteriori* knowledge) and its model (*a priori* knowledge).

```
sigmau=1; T=1000;
x=[ones(200,1);6*ones(T-200,1)];
y=x+sigmau*randn(T,1) ;t=(0 :T-1);
sigmaw=.01;
s_n=(sigmaw/sigmau)^2;
moy_x=zeros(T,1);
g(1)=sigmau^2;
moy_x(1)=y(1);
for k=2:T
    g(k)=(s_n+g(k-1))/(1+s_n+g(k-1));
moy_x(k)=moy_x(k-1)+g(k-1)*(y(k)-moy_x(k-1));
end
subplot(211);
plot(t,x,'-',t,y,' :',t,moy_x,'o');
title('sigmau=1, sigmaw=.01')
legend('ideal signal','noisy signal','estimated signal',0)
axis([0 500 -2 8]);
xlabel('time [s]');
ylabel('Amplitude [V]')
sigmau=1;
```

```
sigmaw=.8;
T=1000 ;t=(0 :T-1);
y=x+sigmau*randn(T,1);
s_n=(sigmaw/sigmau)^2;
moy_xx=zeros(T,1);
g(1)=sigmau^2;
moy_xx(1)=y(1);
for k=2 :T
    g(k)=(s_n+g(k-1))/(1+s_n+g(k-1));
    moy_xx(k)=moy_xx(k-1)+g(k-1)*(y(k)-moy_xx(k-1));
end
subplot(212)
plot(t,x,'-',t,y,' :',t,moy_xx,'o');
title('sigmau=1, sigmaw=.8')
legend('ideal signal','noisy signal','estimated signal',0)
axis([0 500 -2 8]);
xlabel('time [s]');
ylabel('Amplitude [V]')
```

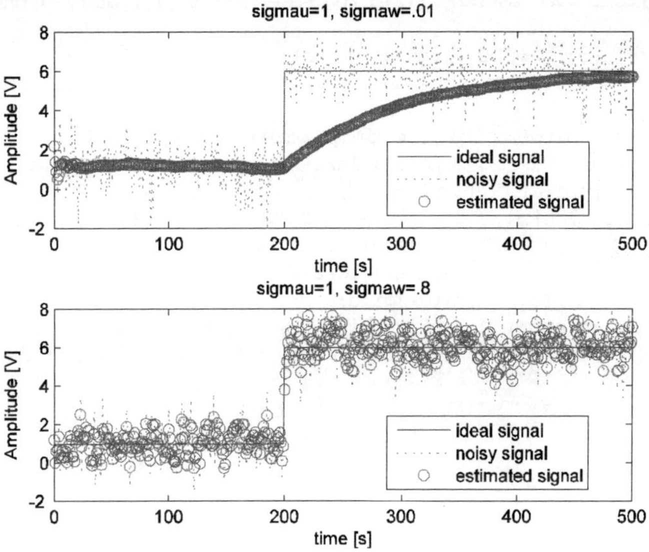

Figure 9.7. *Kalman filter performance*

The model and characteristics of the noise are assumed to be known in the case of the standard Kalman filtering algorithm. However, this is not always true in practice. Thus, in the first case, when the state equation is considered almost certain, the filter heavily follows the slow signal variations and completely ignores the fast ones. In the second case, a rapid variation of the signal to be estimated is expected. The filter then follows the signal variations very closely, but proposes an estimation with a significant variance.

EXERCISE 9.5.

Write a MATLAB code to illustrate the influence of the two basic signal parameters (frequency bandwidth B and pulse length T) on the matched filter performance. Consider a *chirp* signal sampled at 12 kHz in the following two cases:

– the frequency is linearly swept between 3,500 and 4,000 Hz (B = 500 Hz) and T = 25 ms,

– the frequency is linearly swept between 500 and 5,000 Hz (B = 4,500 Hz) and T = 50 ms.

Plot the two autocorrelation functions without normalization between -0.01 s and 0.01 s and explain the obtained results.

```
% Fs: sampling frequency in Hz
% pulselength: pulse length in seconds,
Fs=12000;; pulselength=0.025; t1=(0:1/Fs:pulselength); fmin=3500;
fmax=4000;chirp1=vco(sawtooth((2*pi/pulselength)*t1,1),[fmin/Fs,fmax/Fs]*Fs,F
s);
pulselength=0.05; t2=(0:1/Fs:pulselength);
fmin=500;fmax=5000;chirp2=vco(sawtooth((2*pi/pulselength)*t2,1),[fmin/Fs,fmax
/Fs]*Fs,Fs);
c_chirp1=xcorr(chirp1,chirp1); c_chirp2=xcorr(chirp2,chirp2);
lc=length(c_chirp1); t1=linspace((-lc/2+1)*1/Fs,(lc/2-1)*1/Fs,lc);
subplot(211); plot(t1,c_chirp1);grid;
title('Autocorrelation of a chirp with T=25 ms, B=500 Hz ')
axis([-0.01 0.01 -200 200]); xlabel('time [s]'); ylabel('amplitude')
lc=length(c_chirp2); t2=linspace((-lc/2+1)*1/Fs,(lc/2-1)*1/Fs,lc);
subplot(212); plot(t2,c_chirp2);grid;
title('Autocorrelation of a chirp with T=50 ms, B=4500 Hz')
axis([-0.01 0.01 -100 400]); xlabel('time [s]'); ylabel('amplitude')
```

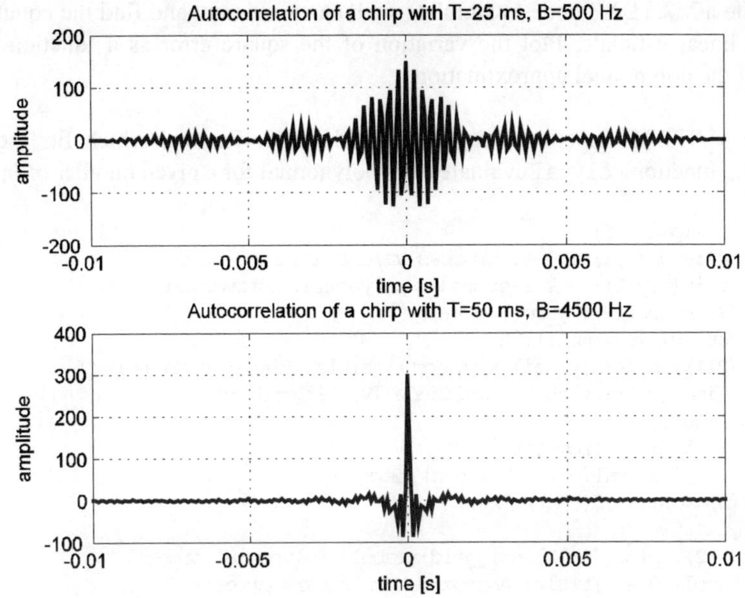

Figure 9.8. *Autocorrelation function of a chirp signal*

The width of the autocorrelation function of a *chirp* signal is about 1/B. In the first case, the autocorrelation function is large due to the narrow frequency bandwidth, while in the second case, the large frequency bandwidth leads to a an autocorrelation function with a very narrow mainlobe.

The amplitude of the autocorrelation function at origin is equal to the signal energy. Since the two signals have the same amplitude, the ratio of their energies is the ratio of their lengths and can be also retrieved from the graph above.

A high performance matched filter requires a signal with both a large time length and a broad frequency bandwidth. The first condition allows the detection probability to be maximized, while keeping the level of false alarm probability low. The second provides a high time resolution at the matched filter output.

However, the signal time length and frequency bandwidth are limited in practice by the transmitter power supply or the associated furtivity requirements and by the transmission channel constraints respectively.

EXERCISE 9.6.

Assume that the results issued from an experiment seem to be outcomes of a linear process (approximation by a 1st order polynomial function).

Write a MATLAB code to simulate such an experiment and find the equation of the LS linear estimate. Plot the variation of the square error as a function of the order of the polynomial approximation.

The MATLAB function polyfit provides a polynomial which fits the data, while the function polyval evaluates this polynomial for a given number of inputs.

```
x=[0:10]; sigma=3.5;
y=x+sigma*randn(1,11); % 11 process values are simulated
coef=polyfit(x,y,1);   % 1-st order polynomial estimation
m=coef(1); b=coef(2); y_meil=m*x+b;
er_quadrat=sum((y-y_meil).^2)
subplot(211); plot(x,y_meil,x,y,'xr'); title('Linear regression')
xlabel('Time [s]'); ylabel('Amplitude [V]'); grid on;
newx=(0:1:10);
for n=1:11 % order from 1 to 11
     f( :,n)=polyval(polyfit(x,y,n),newx)';
     er(n)=sum((y'-f(:,n)).^2);
end
subplot(212); plot([1:11],er);grid;xlabel('Polynomial order');
ylabel('Amplitude');title('Approximation square error')
```

When the polynomial order is equal to 10 or higher, the error is zero because this approximation includes all the available information (11 coefficients or more to be estimated from 11 recorded data values). The problem is well-conditioned and the solution is unique in this case.

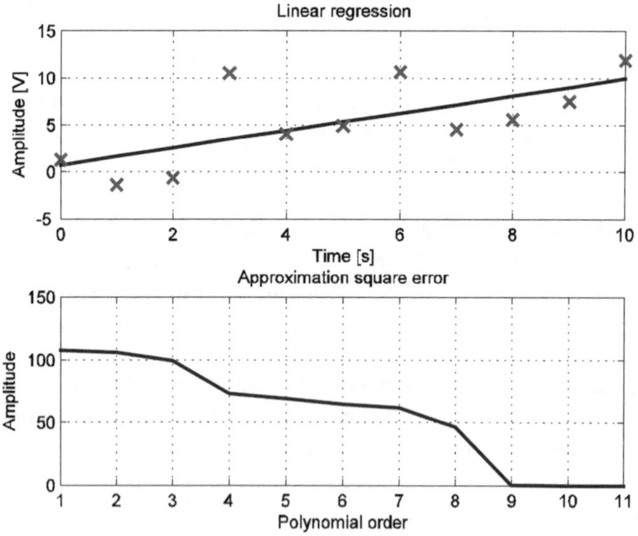

Figure 9.9. *Illustration of a linear regression and variation of the quadratic error as a function of the approximation order*

EXERCISE 9.7.

a. Generate the three signals defined below over 1,024 samples:

– a sum of two sinusoids with normalized frequencies 0.075 and 0.175,

– a 50 point length pulse,

– a signal obtained at the output of a bandpass elliptic filter driven by a zero-mean white Gaussian noise with the variance 1; the filter specifications are the following: passband frequency edges 0.205 and 0.245, stopband frequency edges 0.195 and 0.255, passband ripple 0.5 dB, stopband minimum attenuation 50 dB.

Add a zero-mean white Gaussian noise with a variance of 3 to each signal.

b. Filter the three noisy signals using the corresponding Wiener filters. Explain the filtering effect in each case.

a.

```
% Signal generation
nivbr=3;
s1=sin(2*pi*.075*[1:1024])+sin(2*pi*.175*[1:1024]);
sf1=abs(fft(s1));
s2=[ones(1,50) zeros(1,1024-50)];
sf2=abs(fft(s2));
[n,wn] = ellipord([0.205,0.245],[0.195,0.255],0.5,20)
[b0,a0] = ellip(n,0.5,50,wn);
[h0,w0]=freqz(b0,a0,512);
s3=filter(b0,a0,randn(1,1024)); sf3=abs(fft(s3));

% Noise adding
x1=s1+nivbr*randn(1,1024); x1f=abs(fft(x1));
x2=s2+nivbr*randn(1,1024); x2f=abs(fft(x2));
x3=s3+nivbr*randn(1,1024); x3f=abs(fft(x3));
```

b.

```
% Wiener filtering
nb=9; na=10; % filter numerator and denominator orders
hz=fft(xcorr(x1,s1))./fft(xcorr(x1));
[b,a]=invfreqz(hz,[1:2047]*2*pi/2047,nb,na);
[h1,w1]=freqz(b,a,512); sr1=filter(b,a,x1); srf1=abs(fft(sr1));
hz=fft(xcorr(x2,s2))./fft(xcorr(x2));
[b,a]=invfreqz(hz,[1:2047]*2*pi/2047,nb,na);
[h2,w2]=freqz(b,a,512); sr2=filter(b,a,x2); srf2=abs(fft(sr2));
hz=fft(xcorr(x3,s3))./fft(xcorr(x3));
[b,a]=invfreqz(hz,[1:2047]*2*pi/2047,nb,na);
[h3,w3]=freqz(b,a,512); sr3=filter(b,a,x3); srf3=abs(fft(sr3));

% Plotting the results
```

```
figure; subplot(311)
plot(sf1(1:512)/max(sf1)); hold on
plot(x1f(1:512)/max(x1f),'m:'); plot(abs(h1)/max(abs(h1)),'k—')
axis([50 250 0 1.2]); legend('initial','noisy','Wiener')
subplot(312)
plot(sf2(1:512)/max(sf2)); hold on
plot(x2f(1:512)/max(x2f),'m:'); plot(abs(h2)/max(abs(h2)),'k—')
axis([0 250 0 1.2]); legend('initial','noisy','Wiener')
subplot(313)
plot(abs(h0)/max(abs(h0))); hold on
plot(x3f(1:512)/max(x3f),'m:'); plot(abs(h3)/abs(h3),'k—')
axis([50 250 0 1.2]); legend('initial','noisy','Wiener')
```

In the first two cases, the useful signal spectrum is plotted as a solid line. The transfer function of the bandpass filter used to generate the third signal is plotted in the same way on the last image since it corresponds to its power spectral density.

In all three cases the noisy signal spectrum is plotted as a dotted line. The transfer function of the Wiener filter corresponding to each case is plotted as a dashed line. Notice that the Wiener filter is able to find the frequency bands which concentrate the signal energy despite of the noise high level. However, it is less accurate for the first signal because of the very narrow frequency support of the two mixed sinusoids.

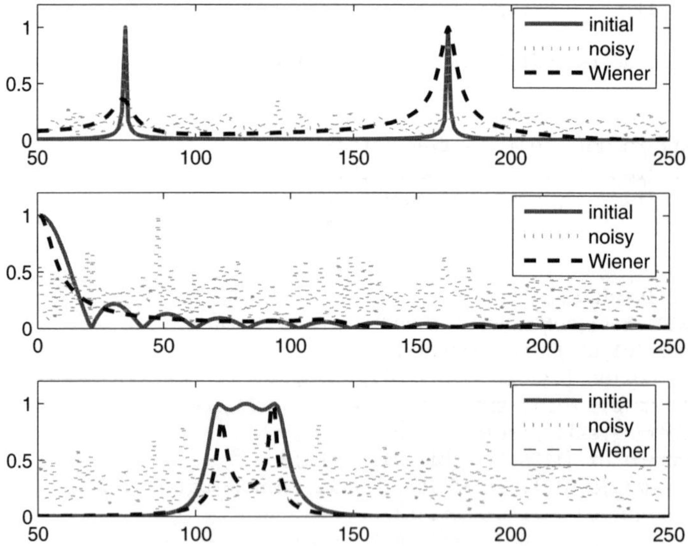

Figure 9.10. *Illustration of the Wiener filtering*

9.3. Exercises

EXERCISE 9.8.

Repeat exercise 9.1 using a different basic pulse for the transmitted waveform:

– bipolar code without return to zero;

– triangular pulse;

– Manchester code.

Compare the theoretical and experimental error probabilities obtained from a large number of simulations.

EXERCISE 9.9.

An active early warning radar transmits a pulse burst (1200 pulses/s) with an amplitude A, using a uniformly rotating antenna (15 rotations/min) with the mainlobe width of 1.5°.

1. How many pulses are backscattered by a scanned target?

2. Using an appropriate process, the N pulses backscattered by the scanned target form an observation vector having N components, corrupted by a zero-mean additive white Gaussian noise with a standard deviation σ.

The detection is performed using the Neyman-Pearson criterion, which fixes the false alarm probability level and maximizes the detection probability. Prove that this criterion is equivalent to comparing the sum of N components to a threshold which should be defined. Evaluate the results obtained for: $A = 1$, $\sigma = 0.6$, $N = 20$, $P_{FA} = 10^{-10}$.

Write a MATLAB code to predict the detection performance for different values of the involved parameters and to plot the ROC (receiver operating characteristics) curves. These curves illustrate the variation of the detection probability as a function of the false alarm probability, for different values of the signal-to-noise ratio.

EXERCISE 9.10.

Consider a random signal $s(t)$ corrupted by a white noise $b(t)$. $s(t)$ and $b(t)$ are uncorrelated and have the following power spectral densities: $\gamma_s(\nu) = \dfrac{3}{4 + 4\pi^2\nu^2}$ and $\gamma_b(\upsilon) = 2$ W/Hz..

Write a MATLAB code to estimate $s(t)$ from the observation $x(t)=s(t)+b(t)$, using a Wiener filter.

EXERCISE 9.11.

A measured signal $x(t)$ is the sum of a useful component $s(t)$ with a power spectral density $\gamma_s(\upsilon) = \dfrac{1.848}{0.48 + 4\pi^2\upsilon^2}$, and a zero-mean noise $n(t)$. The two components of $x(t)$ are assumed to be uncorrelated.

The noise power spectral density is $\gamma_b(\upsilon) = 2$ W/Hz between $-1/2T$ and $1/2T$ and is zero otherwise.

1. Write the state equations of the system.

2. Sample the signal $x(t)$ with a sampling period of 1 second to perform a discrete Kalman filtering on $x[k]$ and to estimate $s[k]$.

3. Convert the state equations of the analog system (differential equations) into discrete state equations (difference equations) using the function `c2d`.

4. Find the discrete state-space model.

5. Write a MATLAB code to implement the filter.

Chapter 10

Power Spectral Density Estimation

10.1. Theoretical background

Spectral analysis is one of the most important signal processing techniques and consists of identifying the spectral content of a time-varying quantity. The usual spectral analysis methods can be divided into two main classes:

– direct methods (selective filtering, periodogram);

– indirect methods (correlogram, parametric methods, etc.).

In an experimental framework, a signal spectral analysis can be performed using spectral analyzers, which represent an essential investigation tool in many applications. In real world applications, the observation period (continuous-time signals) or the number of samples (discrete-time signals) is finite. Thus, the spectral content will be estimated using the limited available information.

10.1.1. *Estimate properties*

Let us consider a stochastic process $x(n)$ depending on a parameter a, and an estimate of this parameter obtained from N outcomes of x:

$$\hat{a}= F\left[x(0) \ldots x(N\text{-}1)\right] \tag{10.1}$$

The quality of an estimate is evaluated using its bias and its variance defined below:

$$B = a - E\left[\hat{a}\right] \tag{10.2}$$

$$\operatorname{var}[\hat{a}]=E\left[\left(\hat{a}-E\left[\hat{a}\right]\right)^2\right] \tag{10.3}$$

A low variance indicates a low dispersion of the estimate values around its mean $E[\hat{a}]$. An estimate converges if its bias and variance will vanish when the number of observations N becomes infinite (estimate asymptotic characteristics).

Let us consider the example of an autocorrelation function, which is generally calculated using one of the two following estimates:

$$\text{a. } \hat{\Gamma}_0(k) = \frac{1}{N-|k|} \sum_{i=0}^{N-|k|-1} x_N(i) x_N(i+k) \quad , |k| < N-1 \tag{10.4}$$

It can easily be seen that this estimate is unbiased and it can be also shown that its variance is asymptotically zero.

$$\text{b. } \hat{\Gamma}_1(k) = \frac{1}{N} \sum_{i=0}^{N-|k|-1} x_N(i) x_N(i+k), \quad |k| < N-1 \tag{10.5}$$

which is the same as: $\hat{\Gamma}_1(k) = \frac{N-|k|}{N} \hat{\Gamma}_0(k)$ [10.6]

and thus: $E\left\{\hat{\Gamma}_1(k)\right\} = \left(1 - \frac{|k|}{N}\right)\Gamma(k)$ [10.7]

The second estimate is therefore asymptotically unbiased. It is equivalent to the first estimate if $|k| << N$ and becomes even better when $|k|$ comes close enough to N. In fact, the variance of the first estimate significantly increases in the second case.

10.1.2. *Power spectral density estimation*

Periodogram

Let us estimate the power spectral density (PSD) using the discrete Fourier transform of $\hat{\Gamma}_1(k)$, i.e.:

$$\hat{\gamma}(v) = \sum_{k=0}^{N-1} \hat{\Gamma}_1(k) e^{-j2\pi kv} \tag{10.8}$$

It can be shown that:

$$\hat{\gamma}(v) = \frac{1}{N} |X_N(v)|^2 \text{ with } X_N(v) = \sum_{k=0}^{N-1} x_N(k) e^{-j2\pi kv} \tag{10.9}$$

This estimate, called a periodogram, can thus be directly evaluated as the square of the Fourier transform of a recorded signal. Since the evaluation can be performed using FFT algorithms, this solution seems to be very attractive.

However, this estimate is not a good one because it is biased and its variance does not decrease when the number of samples increases. In fact, the estimate variance is proportional to the square of the actual spectrum magnitude irrespective of the value of N.

Averaged periodogram

Let us divide the length-N recorded signal into K length-M signals. The kth sample of the ith signal is defined by:

$$x_M^{(i)}(k) = x_M(iM+k), \quad i = 0..K-1, k = 0..M-1 \qquad [10.10]$$

The previously defined periodogram is straightforwardly obtained using each signal:

$$\hat{\gamma}^{(i)}(v) = \frac{1}{M}\left|X_M^{(i)}(v)\right|^2, \text{ with } X_M^{(i)}(v) = \sum x_M^{(i)}(k)e^{-j2\pi kv} \qquad [10.11]$$

Finally, the partial PSD estimates obtained in this way are averaged to obtain the final estimate:

$$\hat{\gamma}_m(v) = \frac{1}{K}\sum_{i=0}^{K-1}\hat{\gamma}^{(i)}(v) \qquad [10.12]$$

Suppose that the successive signals $x_M^{(i)}(k)$ are not correlated (i.e. M is higher than the process memory). The bias of the averaged periodogram is then related to the triangular window that weights the autocorrelation function, while its variance is divided by K.

However, the spectral resolution is $K = N/M$, which is lower than that obtained in the case of the periodogram. Therefore, a trade-off has to be found between the spectral resolution and the estimate variance for a given value of N.

Modified periodogram (Welch's method)

The modified periodogram, also known as Welch's method, is calculated using a similar approach as before, but multiplying each length-M signal with a weighting window. It is different from the rectangular window, which is considered by default in the case of the averaged periodogram in order to truncate the signal. The weighted samples and the new PSD estimate are expressed as follows:

$$x_M^{(i)}(k).w_M(k)=x_w^{(i)}(k) \quad [10.13]$$

$$\hat{\gamma}_w(v)=\frac{1}{K}\sum_{i=0}^{K-1}\frac{\left|X_w^{(i)}(v)\right|^2}{M} \quad [10.14]$$

Compared to the averaged periodogram, the new estimate does not reduce either the bias or the variance, and it even decreases the spectral resolution. Nevertheless, it highly improves the sidelobes' attenuation and it minimizes the risk to identify false spectral components.

A correction factor has to be added to the estimate defined above because the weighting window also modifies the signal mean power. The final expression of the modified periodogram becomes:

$$\hat{\gamma}_w(v)=\frac{M}{\sum\limits_{k=0}^{M-1} w_M^2(k)}\frac{1}{K}\sum_{i=1}^{K}\frac{\left|X_w^{(i)}(v)\right|}{M} \quad [10.15]$$

Notice that just as for the averaged periodogram, a trade-off between the spectral resolution and the estimate variance is required by the Welch's method. A useful idea to obtain a minimum variance for a given spectral resolution is to consider partially superposed signals instead of separate signals. However, the reduction of the variance is less important in this case because the signals become more and more correlated. A signal recovery of 50% is generally considered a good trade-off between the variance reduction and the additional required calculations.

Correlogram

This method estimates the autocorrelation function on the interval $[-M,+M]$ and then calculates its Fourier transform:

$$\hat{\Gamma}_M[k]=\frac{1}{N}\sum_{i=0}^{N-1-k}x[i]x[i+k], \quad k=0..M \quad [10.16]$$

$$\hat{\gamma}_M(v)=\sum_{k=-M}^{M}\hat{\Gamma}_M[k]e^{-j2\pi kv} \quad [10.17]$$

If the DFT is used instead of the Fourier transform, then:

$$\hat{\gamma}_M[n] = \sum_{k=-M}^{M} \hat{\Gamma}_M[k] e^{-j2\pi \frac{k}{2M+1} n} \qquad [10.18]$$

where $M << N$ (typically $M < N/10$).

Consequently, only the points which are well enough averaged are considered for the autocorrelation function estimate $\hat{\Gamma}_M[k]$.

The spectral resolution of $\hat{\gamma}_M(v)$ is thus reduced because of the truncation on the interval [-*M*, *M*], while the variance is improved compared to the periodogram. A weighting window is generally used to multiply $\hat{\Gamma}_M[k]$ in order to avoid sidelobe effects and to insure the positiveness of $\hat{\gamma}_M(v)$.

10.1.3. *Parametric spectral analysis*

The central concept of the previously presented methods is the Wiener-Khintchine theorem, which states that the power spectral density of a signal is the Fourier transform of its autocorrelation function. Since the calculations are always performed using a finite observation time or number of samples the spectral resolution becomes a problem, especially for short signals.

The parametric spectral analysis makes use of PSD models depending on a set of parameters instead of the Fourier transform. Estimating the signal PSD is then equivalent to calculating these parameters.

ARMA model

Let us consider a stable LTI system, characterized by the following rational transfer function:

$$A(z) = \frac{\alpha_0 + \alpha_1 z^{-1} + \ldots + \alpha_m z^{-m}}{1 + \beta_1 z^{-1} \ldots + \beta_n z^{-n}} \qquad [10.19]$$

Suppose that it is driven by a discrete-time white noise with the variance σ^2 (see Figure 10.1).

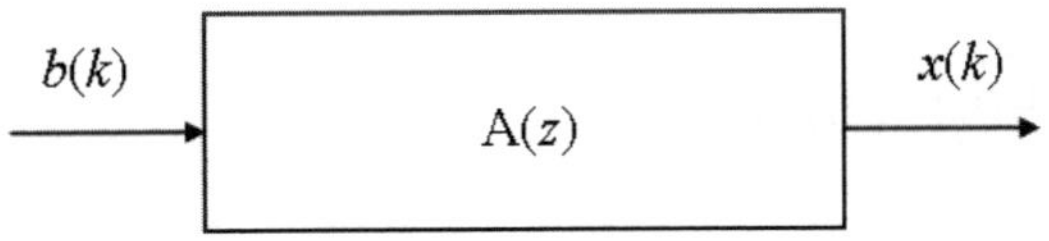

Figure 10.1. *LTI system driven by a white noise*

The input-output relationship can thus be written in the form:

$$\gamma_x(z) = A(z)A(z^{-1})\sigma^2 \qquad [10.20]$$

which results in:

$$\gamma_x(v) = \left|\gamma_x(z)\right|_{z=e^{j2\pi v}} = \left| \frac{\alpha_0 + \alpha_1 e^{-j2\pi v} ... + \alpha_m e^{-j2\pi m v}}{1+\beta_1 e^{-j2\pi v} ... + \beta_n e^{-j2\pi n v}} \right|^2 \sigma^2 \qquad [10.21]$$

The filtering can be seen in this context as a model for generating random signals, whose power spectral density is rational. The output signal can also be described using the following difference equation, also known as the *ARMA* model:

$$x(k) = -\beta_1 x(k-1) .... - \beta_n x(k-n) + \alpha_0 b(k) ... + \alpha_m b(k-m) \qquad [10.22]$$

A simplified model, called *MA* (*moving average*) or all-zeros method, is obtained under the condition below, and corresponds to a FIR filtering of a white noise:

$$\beta_i = 0 \;\; \forall \;\; i, \text{ then } x(k) = \alpha_0 b(k) ... + \alpha_m b(k-m) \qquad [10.23]$$

Another simplified model, called *AR* (*auto regressive*) or all-poles method, is obtained under the following condition:

$$\alpha_i = 0 \;\; \forall \;\; i > 0 \text{ then } x(k) = -\beta_1 x(k-1) ... - \beta_n x(k-n) + \alpha_0 b(k) \qquad [10.24]$$

The AR model is widely used in practice. It considers a linear relationship between the value of a signal at moment k and its previous values in addition with an uncorrelated new value, often called innovation.

The Yule-Walker equations can then be easily derived from equation [10.24]:

$$\Gamma_x(p) = -\sum_{i=1}^{n} \beta_i \Gamma_x(p-i), \text{ for } p > 0 \qquad [10.25]$$

For $p = 0$ this results in:

$$\Gamma_x(0) = -\sum_{i=1}^{n} \beta_i \Gamma_x(-i) + \sigma^2 \qquad [10.26]$$

AR model parameter estimation

Equation [10.24], which defines an AR model, can be rewritten in the following form:

$$x(k) = \sum_{i=1}^{n} -\beta_i \, x(k-i) + b(k) \qquad [10.27]$$

where $b(k)$ is a white noise with the variance σ^2.

The Yule-Walker equations can be also put into the following matrix form:

$$\begin{bmatrix} \Gamma(0) & \Gamma(-1) & \cdots & \Gamma(-n) \\ \Gamma(1) & \Gamma(0) & \dots & \Gamma(-(n-1)) \\ \vdots & \vdots & \ddots & \vdots \\ \Gamma(n) & \dots & \dots & \Gamma(0) \end{bmatrix} \begin{bmatrix} 1 \\ -\beta_1 \\ \vdots \\ -\beta_n \end{bmatrix} = \begin{bmatrix} \sigma^2 \\ 0 \\ \vdots \\ 0 \end{bmatrix} \qquad [10.28]$$

or: $$\left[\underline{\underline{\Gamma}}\right] \cdot \begin{bmatrix} 1 \\ \underline{\beta} \end{bmatrix} = \begin{bmatrix} \sigma^2 \\ 0 \end{bmatrix} \qquad [10.29]$$

Matrix $\left[\underline{\underline{\Gamma}}\right]$ is symmetric and has a Toeplitz structure.

If the $n+1$ values of Γ involved in this matrix are supposed known, the estimation of the AR model parameters (i.e. β_i and σ^2) is a straightforward problem.

The Levinson-Durbin algorithm is an effective recursive solution to this problem. In addition to its effectiveness, this algorithm allows the model order to be estimated, which is very useful when it is *a priori* unknown.

Note that identifying a 10th order AR model requires knowledge of the following 11 successive values of the autocorrelation function: $\Gamma_x(0)$, $\Gamma_x(1)$, ..., $\Gamma_x(10)$.

The AR model identified in this way provides a PSD estimate which is not related to the trade-off variance-spectral resolution, while the PSD estimate obtained by the DFT of the sequence $\{\Gamma_x(0), \Gamma_x(1), \cdots, \Gamma_x(10)\}$ would lead to a much lower spectral resolution.

Actually, values $\Gamma_x(i)$ are not *a priori* known and they also have to be estimated, which requires the acquisition of much more data than the AR model order.

A common choice for estimating the model order is Akaike's FPE (*final prediction error*) criterion:

$$FPE(p) = \frac{N+p}{N-p}\sigma^2(p) = \sigma^2(p) + \frac{2p}{N-p} \quad [10.30]$$

Another usual criterion is AIC (*Akaike information criterion*):

$$AIC(p) = NLn\sigma^2(p) + 2p \quad [10.31]$$

These two criteria tend to be equivalent for a large number of samples as the following relationship holds:

$$AIC(p) = N \cdot Ln\left(FPE(p)\right) \quad [10.32]$$

If the observed random signal is not an AR process, it can still be represented by an AR model of infinite order. In the same way, the model identification can also be performed using an estimate of the autocorrelation function.

Since the number of samples is finite, a trade-off bias-variance is necessary. In fact, the larger the model order is, the lower the error power is and the closer the predicted values are to the observed signal (the bias decreases).

However, increasing the model order also increases the estimate variance because the autocorrelation function values corresponding to larger time-delays are required.

MA model

The all-zeros model for the signal and its power spectral density is given below:

$$x(k) = \sum_{j=0}^{M} b(j)e(k-j) \quad [10.33]$$

$$\gamma_x(z) = \sigma^2 B(z)B(z^{-1}) \quad [10.34]$$

where $e(k)$ is a zero-mean discrete-time white noise with the variance σ^2 and:

$$B(z) = \sum_{j=0}^{M} b(j)z^{-j} \quad [10.35]$$

Coefficients $b(j)$ are related to the autocorrelation function by a non-linear convolution relationship, which cannot be easily solved:

$$\hat{\Gamma}_x(k) = E\left[x(i)x(k+i)\right] \quad [10.36]$$

$$\hat{\Gamma}_x(k) = \begin{cases} \sum_{j=0}^{M-k} b(j)b(k+j), & k = 0..M \\ 0, & k > M \end{cases} \quad [10.37]$$

Therefore, the calculation of these coefficients requires an estimation of $\Gamma_x(k)$ for $k = 0..M$. Nevertheless, if we are only interested in estimating the signal PSD, it is not useful to perform such an estimation. In fact, once the autocorrelation function of the observed signal is estimated, its PSD can be obtained directly by taking its Fourier transform.

A possible solution to calculate coefficients $b(j)$ without estimating the signal autocorrelation function is to find a high order AR model and evaluate an inverse filter (a MA model) which minimizes the square error.

In fact, an MA model can be seen as an infinite order AR model, since a zero of the PSD can always be approximated by a series of poles. This statement can be illustrated as follows:

$$a(m) + \sum_{k=1}^{\infty} b(i)a(m-i) = \delta(m) \quad [10.38]$$

Since the number of coefficients $b(i)$ is finite, error $e(m)$ associated with this model is given by the following equation:

$$a_p(m) + \sum_{i=1}^{M} b(i)a_p(m-i) = e(m) \quad [10.39]$$

In this case, the problem becomes the minimization of $E[e^2(m)]$, which is equivalent to solving the Yule-Walker equations by exchanging $x(i)$ with $a(i)$. In order to find coefficients $b(i)$, the Levinson algorithm is applied twice: first on the data $x(i)$, then on coefficients $a(i)$ from the first modeling.

10.1.4. *Super-resolution spectral analysis methods*

These methods consider a signal model in the form of a mixture of noisy complex exponentials:

$$x(n)=\sum_{k=1}^{N} c_k \cdot \exp(j \cdot 2\pi \cdot f_k \cdot t_n)+w(n)=s(n)+w(n)\ ,\ n=0..N_s-1 \quad [10.40]$$

where $x(n)$ and $w(n)$ stand for N_s samples of the observed noisy signal, $\{f_k\}_{k=1..N}$ designate the frequencies of the complex exponentials, and $t_n = nT_s$ are the sampling instants.

Super-resolution methods aim to separate the observation space in a *signal subspace*, containing only useful information, and its orthogonal complement, called *noise subspace*. This decomposition makes the spectral analysis more robust and highly improves the spectral resolution.

The above equations can be expressed in the following matrix form:

$$\mathbf{x}=\sum_{k=1}^{N} c_k \cdot \mathbf{e_k}+\mathbf{w}=E \cdot \mathbf{c}+\mathbf{w} \quad [10.41]$$

where: $\mathbf{e_k}=\begin{bmatrix}1 & \exp(j \cdot 2\pi \cdot T_s \cdot f_k) & \cdots & \exp(j \cdot 2\pi \cdot (N_e-1) \cdot T_s \cdot f_k)\end{bmatrix}^T$

$$\mathbf{w}=\begin{bmatrix}w(0) & w(1) & \cdots & w(N_s-1)\end{bmatrix}^T$$

$$E=\begin{bmatrix}\mathbf{e_1} & \mathbf{e_2} & \cdots & \mathbf{e_N}\end{bmatrix},\quad \mathbf{c}=\begin{bmatrix}c_1 & c_2 & \cdots & c_N\end{bmatrix}^T$$

Autocorrelation matrix estimation

The first step for implementing super-resolution methods is the eigenanalysis of the data autocorrelation matrix:

$$R_x=E\left[\mathbf{x} \cdot \mathbf{x}^H\right] \quad [10.42]$$

Since R_x is not generally known, it is estimated from the acquired data samples. A widely used approach consists of averaging the N_{obs} data vectors:

$$\hat{R}_x=\frac{1}{N_{obs}}\sum_{i=1}^{N_{obs}} \mathbf{x}_i \cdot \mathbf{x}_i^H \quad [10.43]$$

The $\hat{R}_x$ matrix must be of full rank in order to properly separate the two subspaces as well as the spectral components. The spatial smoothing method can be used to insure the full rank property, even though only one data vector is available. According to this method, the autocorrelation matrix of order $p > N$ is estimated as follows:

$$R_p = D_p \cdot D_p^H \qquad [10.44]$$

where D_p is the data matrix defined below:

$$D_p = \begin{bmatrix} x(0) & x(1) & \cdots & x(N_e - p) \\ x(1) & x(2) & \cdots & x(N_e - p + 1) \\ \vdots & \vdots & \ddots & \vdots \\ x(p-1) & x(p) & \cdots & x(N_e - 1) \end{bmatrix} \qquad [10.45]$$

This matrix is obtained by shifting a p-length window along the data snapshot. The $N_e - p + 1$ columns of the D_p matrix can be considered as the new observation vectors. They form the observation space which has the dimension p. Consequently, the dimension of signal and noise vectors will be also limited to p.

MUSIC algorithm

The MUSIC (*MUltiple Signal Classification*) algorithm is the most well known super-resolution method and can be seen as a generalization of Pisarenko's method. Equation [10.41] results in:

$$R_p = E \cdot R_c \cdot E^H + \rho_w \cdot I_p = S_p + W_p \qquad [10.46]$$

where R_c is the autocorrelation matrix of vector $\mathbf{c}$, ρ_w represents the noise power and I_p stands for the p-order identity matrix.

Autocorrelation matrices S_p and W_p correspond to the signal and the noise respectively. Since the "signal" vectors $\mathbf{e_k}$ are linearly independent, the rank of matrix E is equal to N. Matrix R_c has the same rank, so that the rank of matrix S_p should also be equal to N.

Let us denote by μ_i and $\mathbf{v_i}$ the eigenvalues and eigenvectors of matrix S_p. This matrix can thus be decomposed as follows:

$$S_p = \sum_{i=1}^{p} \mu_i \cdot \mathbf{v_i} \cdot \mathbf{v_i^H} = \sum_{i=1}^{N} \mu_i \cdot \mathbf{v_i} \cdot \mathbf{v_i^H} \qquad [10.47]$$

In the equation above $\mu_i = 0$ for $i = N+1..p$ due to the rank of S_p. On the other hand, because all the eigenvalues of the identity matrix are equal to 1 and any othonormal vector can be its eigenvector, the following relationship holds:

$$I_p = \sum_{i=1}^{p} \mathbf{v_i} \cdot \mathbf{v_i^H} \qquad [10.48]$$

Finally, combining equations [10.47], [10.48] and [10.46] yields:

$$R_p = \sum_{i=1}^{N} \mu_i \mathbf{v_i} \mathbf{v_i^H} + \rho_w \sum_{i=1}^{p} \mathbf{v_i} \mathbf{v_i^H} = \sum_{i=1}^{N} (\mu_i + \rho_w) \mathbf{v_i} \mathbf{v_i^H} + \sum_{i=N+1}^{p} \rho_w \mathbf{v_i} \mathbf{v_i^H} \qquad [10.49]$$

Thus, matrix R_p has the same eigenvectors as S_p, and the eigenvalues $(\mu_i + \rho_w)$ up to the order N and ρ_w for higher orders.

Consequently, the autocorrelation matrix eigenvectors corresponding to the largest N eigenvalues, known as main eigenvectors, span the same subspace as the signal vectors, while the other eigenvectors span the noise subspace. Since all the eigenvectors of an autocorrelation matrix are orthogonal, the two subspaces are also orthogonal.

Let us define the following mode vector:

$$\mathbf{a_i} = [1 \quad \exp(j2\pi T_s \nu_i) \quad \cdots \quad \exp(j2\pi(p-1)T_s \nu_i)]^T \qquad [10.50]$$

For different values of ν_i, the mode vector sweeps the whole observation space. It will be in the signal subspace for $\nu_i = f_k$ (detection of a spectral component). Therefore, it becomes orthogonal to any linear combination of the eigenvectors spanning the noise subspace. This is the main idea of the MUSIC estimate which is defined below using the projection operator on the noise subspace $\Pi_n^{\perp} = V_n V_n^H$, where V_n is the column matrix of the noise subspace eigenvectors:

$$PSD_{MUSIC}(\nu_i) = \frac{1}{\sum_{l=N+1}^{p+1} \left| \mathbf{a_i^H} \cdot \mathbf{v_l} \right|^2} = \frac{1}{\mathbf{a_i^H} \cdot \left(\sum_{l=N+1}^{p+1} \mathbf{v_l} \cdot \mathbf{v_l^H} \right) \cdot \mathbf{a_i}} = \frac{1}{\mathbf{a_i^H} \cdot \Pi_n^{\perp} \cdot \mathbf{a_i}} \qquad [10.51]$$

According to equation [10.51] the estimate value tends theoretically towards infinity whenever it is evaluated at a frequency corresponding to a signal spectral component. Actually, the estimate values are always finite because of the limited calculation accuracy and estimation errors.

The spectral resolution is highly improved since very sharp peaks are detected, but their amplitudes lose any physical significance. Due to this particular behavior, the MUSIC algorithm is often considered a frequency estimate rather than a PSD estimate. However, the spectral amplitude information can be recovered, if necessary, using an additional LS optimization procedure.

Root-MUSIC algorithm

The main advantage of the Root-MUSIC algorithm is the possibility of making a direct calculation of the signal spectral components, instead of the maxima search performed by the MUSIC algorithm. The basic idea is to consider the polynomials:

$$P_l(z) = \mathbf{v}_l^H \cdot \mathbf{q}(z), \quad l = N+1..p \qquad [10.52]$$

with $\mathbf{q}(z) = \begin{bmatrix} 1 & z & \cdots & z^p \end{bmatrix}^T$.

It can easily be seen that the N zeros of each polynomial defined above are represented by $z_k = \exp\left(j2\pi T_s f_k\right), k = 1..N$. In fact, in this case vector $\mathbf{q}(z)$ becomes a signal vector, i.e. orthogonal to any eigenvector $\mathbf{v}_l$.

The problem is now to calculate the zeros of the polynomials $P_l(z)$. Bearing in mind that the useful zeros should be located on the unit circle, the roots of the polynomial below are calculated instead, in order to move away the other zeros and minimize the error probability:

$$P(z) = \mathbf{q}(z)^H \cdot V_n \cdot V_n^H \cdot \mathbf{q}(z) \qquad [10.53]$$

Due to the Hermitian operator, this polynomial depends on both z and z^*. It is transformed in the *root-MUSIC polynomial*, which only depends on z:

$$P_{RM}(z) = z^p \cdot \mathbf{q}(z^{-1})^T \cdot V_n \cdot V_n^H \cdot \mathbf{q}(z) \qquad [10.54]$$

The frequencies corresponding to different signal spectral components are then obtained in the form:

$$f_k = \frac{1}{2\pi \cdot T_s} \arg\left(z_k\right) \quad k = 1..N \qquad [10.55]$$

ESPRIT algorithm

The principle of the ESPRIT (*estimation of signal parameters via rotational invariance techniques*) algorithm can be easily understood from the following relationships:

$$R_p = E \cdot R_c \cdot E^H + \rho_w \cdot I = V_s \cdot \Lambda_s \cdot V_s^H + \rho_w \cdot V_n \cdot V_n^H \qquad [10.56]$$

$$I = V_s \cdot V_s^H + V_n \cdot V_n^H \qquad [10.57]$$

where Λ_s stands for the diagonal matrix of the eigenvalues corresponding to the signal subspace.

The above equations result in:

$$E \cdot R_c \cdot E^H + \rho_w \cdot \left(V_s \cdot V_s^H + V_n \cdot V_n^H\right) = V_s \cdot \Lambda_s \cdot V_s^H + \rho_w \cdot V_n \cdot V_n^H$$

$$\Rightarrow E \cdot R_c \cdot E^H + \rho_w \cdot V_s \cdot V_s^H = V_s \cdot \Lambda_s \cdot V_s^H \Rightarrow E \cdot R_c \cdot E^H \cdot V_s + \rho_w \cdot V_s = V_s \cdot \Lambda_s$$

$$\Rightarrow \begin{cases} V_s = E \cdot T \\ T = R_c \cdot E^H \cdot V_s \cdot (\Lambda_s - \rho_w \cdot I)^{-1} \end{cases} \qquad [10.58]$$

Let us define matrices E_1 and E_2 obtained from the E matrix by taking off its last and first row respectively. It is then easy to show that the following relationship holds:

$$E_2 = E_1 \cdot \Phi \qquad [10.59]$$

where:

$$\Phi = \mathrm{diag}\left[\exp(j2\pi T_s f_1) \quad \exp(j2\pi T_s f_2) \quad \cdots \quad \exp(j2\pi T_s f_N)\right]^T$$

Using equation [10.58] the following two matrices can be defined:

$$\begin{cases} V_{s_1} = E_1 \cdot T \\ V_{s_2} = E_2 \cdot T \end{cases} \qquad [10.60]$$

Also taking into account equation [10.59], we obtain:

$$\begin{cases} V_{s_2} = E_1 \cdot \Phi \cdot T \\ E_1 = V_{s_1} \cdot T^{-1} \end{cases} \Rightarrow V_{s_2} = V_{s_1} \cdot T^{-1} \cdot \Phi \cdot T \qquad [10.61]$$

Finally, with the following notation:

$$\Psi = T^{-1} \cdot \Phi \cdot T \qquad [10.62]$$

equation [10.61] becomes:

$$V_{s_2} = V_{s_1} \cdot \Psi \qquad [10.63]$$

The eigenvalues of matrix Φ have the form $\eta_k = \exp(j2\pi T_s f_k)$, $k = 1..N$, because:

$$\det[\Phi - \eta \cdot I] = 0 \Rightarrow \prod_{k=1}^{N} (\eta_k - \eta) = 0 \quad [10.64]$$

Matrix Φ and Ψ are related by a similarity transformation and therefore their eigenvalues are the same. Consequently, the solution is obtained from the eigenvalues of the Ψ matrix, determined by solving equation [10.63]. A direct solution can be developed using the Moore-Penrose pseudo-inverse of matrix V_{s_1}, denoted by $V_{s_1}{}^{\#}$:

$$\Psi = V_{s_1}{}^{\#} \cdot V_{s_2} \quad [10.65]$$

Estimation of the signal subspace dimension

The previously introduced super-resolution methods need accurate estimation of the signal subspace dimension, i.e. of the number of signal spectral components.

It is well known that the variation of the autocorrelation matrix eigenvalues $\{\lambda_k\}_{k=1..p}$ is directly related to the signal subspace dimension. Thus, in the noise-free case the number of non-zero eigenvalues equals N.

When the signal is noisy, the N most important eigenvalues are still associated with the eigenvectors which span the signal subspace, but it is no longer possible to make a robust decision using only their simple variation. Nevertheless, they can still be used, in a different form, in order to obtain a robust estimate of N.

The most well known technique for estimating the signal subspace dimension is the Akaike information criterion (AIC). The number of signal spectral components is estimated to perform the best trade-off between the model and the observation data. Analytically, this condition is expressed in the form:

$$\hat{N} = \arg\left\{ \min_{n=1,p} [C(n)] \right\} \quad [10.66]$$

where $C(n)$ is a cost function, which has the following general expression:

$$C(n) = -L(n) + N_{PL} \quad [10.67]$$

with:

$$L(n) = (N_e - p - 2)\cdot(p-n)\cdot\log\left[\frac{MA(n)}{MG(n)}\right] \quad [10.68]$$

$L(n)$ is the maximal value of the log-likelihood ratio with respect to the model parameters for $N = n$, and:

$$MA(n) = \frac{1}{p-n}\sum_{i=n+1}^{p}\mu_i \text{ and } MG(n) = \left(\prod_{i=n+1}^{p}\mu_i\right)^{\frac{1}{p-n}} \quad [10.69]$$

stand for the arithmetic and geometric means respectively of the $p-n$ last eigenvalues.

The AIC is obtained when N_{PL} is equal to the number of free parameters of the model, i.e.:

$$N_{PL} = n[2p-n] \quad [10.70]$$

The main drawback of Akaike's criterion is that it yields an inconsistent estimate that tends, asymptotically, to overestimate the number of signal spectral components.

To overcome this problem Rissanen proposed the MDL (*minimum description length*) criterion. It only changes the expression of N_{PL} as indicated below in order to obtain an asymptotical overestimation probability equal to zero:

$$N_{PL} = 0.5\cdot n\cdot[2\cdot p - n]\cdot\log(N_e - p - 2) \quad [10.71]$$

Although the estimate of N is consistent now, the signal subspace dimension tends to be underestimated when the number of samples is small.

10.1.5. *Other spectral analysis methods*

In the literature, we can find many other spectral analysis methods beside the methods already presented in this chapter. Some of these methods assume, just like the previously described algorithms, a particular signal model (mixture of noisy sinusoids).

Thus, Prony's method is an extension of this idea to the non-stationary case, since it considers as signal model the mixture of noisy damped sinusoids.

Capon's method, sometimes inappropriately called the maximum likelihood method, is at the border between parametric and nonparametric approaches. Based on the signal selective filtering, it can be decomposed into three simple steps: Fourier transform, square detection and averaging.

Its spectral resolution depends on the S/N ratio: for low S/N ratios, it is similar to the periodogram, while for high S/N ratios it is close to the MUSIC algorithm.

Generally, there is no optimal spectral estimation method. The choice of the most suitable spectral analysis algorithm highly depends on both the specific considered application and on the *a priori* information about the signal features.

10.2. Solved exercises

EXERCISE 10.1.

Consider a zero-mean white Gaussian process P with the variance $\sigma_p^2 = 0.25$, defined on 2,048 samples. Calculate its periodogram and find out the bias and the variance of this PSD estimate. The process is assumed ergodic.

Repeat the same procedure for 512, 1,024 and 4,096 samples.

Periodogram

```
% This function allows the signal periodogram to be calculated
function [PSD_periodogram,frequency] = simple_periodogram (x,N)
K = N;
sequence = x;
Y = fft(sequence,K);
PSD=Y.*conj(Y);
PSD_periodogram = PSD/K;
PSD_periodogram = PSD_periodogram(1:1+K/2);
frequency=(0:N/2)/N;       % 0<V<0.5

% The function defined above is now applied to the signal
N = 2048;       % change then this value with 512, 1024, and 4096
x = randn(1,N)*sqrt(0.25);
[PSD_periodogram,frequency] = simple_periodogram(x,N);
semilogy(frequency, PSD_periodogram,'b');
xlabel('normalized frequency');
ylabel('Amplitude');
title('Periodogram of a zero-mean white Gaussian process');
```

The bias and the variance of this estimate can be calculated as indicated below:

```
bias = mean(PSD_periodogram)-0.25          % estimate bias
```

```
variance = std(PSD_periodogram)^2           % estimate variance
```

Since the process is white, its power spectral density is constant and equal to its variance. It can be thus verified that the periodogram is a biased estimate and that its variance does not decrease when the observation time gets longer; thus, it is an inconsistent estimate.

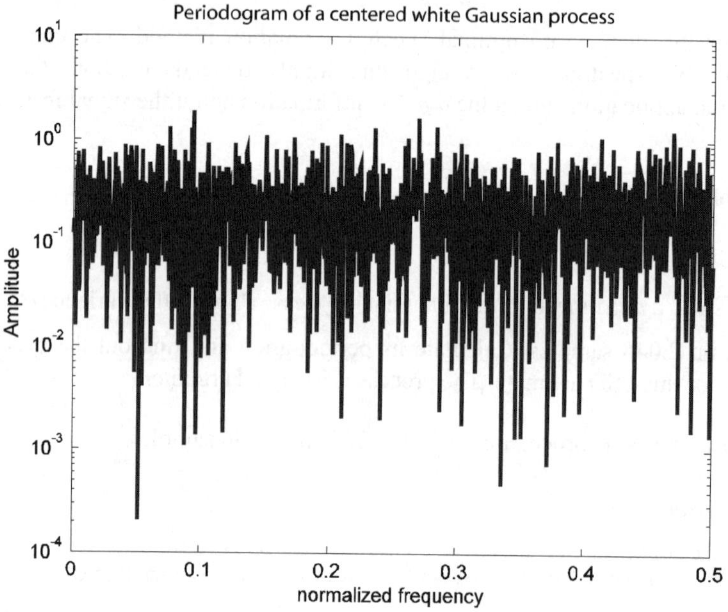

Figure 10.2. *Estimation of the white noise PSD using the periodogram*

EXERCISE 10.2.

Consider again the Gaussian white process previously generated. Calculate and plot its averaged and modified periodogram. Find out their bias and variance. Change the number of averaged sequences and study the effect of different windows on the modified periodogram using the MATLAB functions `boxcar`, `triang`, `hamming`, `hanning`, and `blackman`.

Averaged periodogram

The averaged periodogram in calculated with the MATLAB code below, without sequence recovery.

```
% This function allow calculating the signal averaged periodogram
function [PSD_averaged_periodogram,frequency]=averaged_periodogram(x,K)
N=length(x); L=N/K; origin=0; PSD=0;
for index_loop=1:L
    sequence=x(origin+1:origin+K); origin = origin + K;
    Y=fft(sequence,K); PSD = PSD+Y.*conj(Y);
end
PSD_averaged_periodogram=PSD/(L*K);
PSD_averaged_periodogram=PSD_averaged_periodogram(1:1+K/2);
frequency = (0:K/2)/K; % 0<V<0.5

% The function defined above is now applied to the signal
K=256;
[PSD_averaged_periodogram,frequency]=averaged_periodogram(x,K);
semilogy(frequency,PSD_averaged_periodogram,'b:');
xlabel('normalized frequency');
ylabel('Amplitude'); hold on;

K=32;
[PSD_averaged_periodogram,frequency]=averaged_periodogram(x,K);
semilogy(frequency,PSD_averaged_periodogram,'-r');
title('Averaged periodogram of a zero-mean white Gaussian process');
legend('averaging on 8 sequences','averaging on 64 sequences');
```

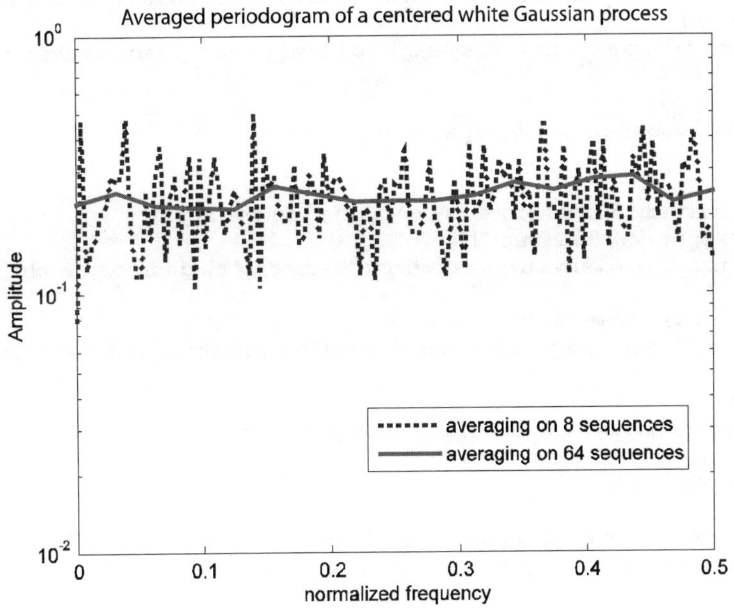

Figure 10.3. *Results obtained with the averaged periodogram*

It can be seen that this estimate is still biased, but its variance decreases when the number of averaged sequences increases (see Table 10.1). Note also that the spectral resolution gets lower when the length of the averaged sequences decreases.

Length of the averaged sequences	Normalized spectral resolution $\Delta\nu/\nu_s$	Variance
512	0.00195	0.0157
256	0.0039	0.0081
128	0.0078	0.0032
64	0.0156	0.0025

Table 10.1. *Variation of the spectral resolution and variance with the length of the averaged sequences*

Modified periodogram

The same algorithm as before is used in this case, except that each sequence is multiplied by a weighting window before the FFT calculation.

```
% This function allows the signal modified periodogram function to be calculated
[PSD_modified_periodogram,frequency]=modified_periodogram(x,K,weighting_window)
...
sequence=sequence.* weighting_window;
...

% The function defined above is now applied to the signal
weighting_window=[blackman(K)];
weighting_window=weighting_window*sqrt(K/sum(weighting_window.*weighting_window));
weighting_window=weighting_window.';
[PSD_modified_periodogram,frequency]=modified_periodogram(x,K,weighting_window);
figure
semilogy(frequency, PSD_modified_periodogram,'b');
xlabel('normalized frequency');
ylabel('Amplitude');
hold on;
% different other weighting windows may be then tried
title('Modified periodogram using different weighting windows');
```

The results provided by different types of weighting windows can thus be compared for a given signal.

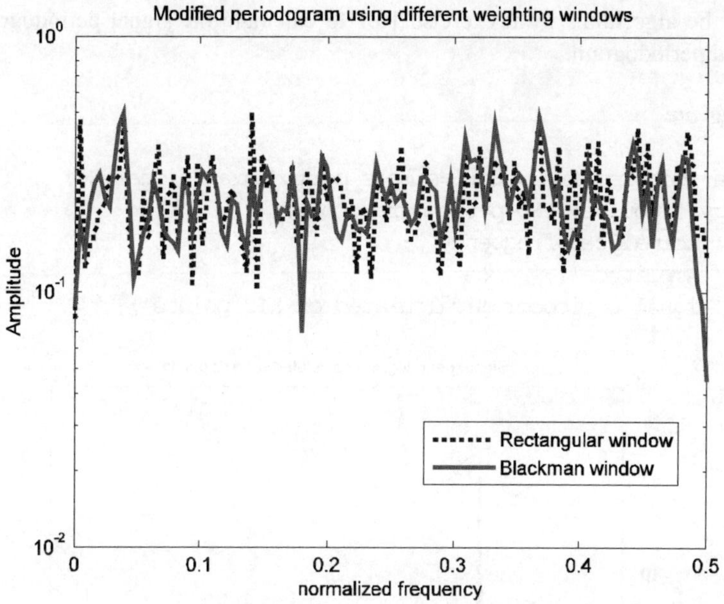

Figure 10.4. *Results obtained with the modified periodogram*

EXERCISE 10.3.

Perform the spectral analysis of the mixture of two noisy sinusoids having the parameters indicated below:

1st sinusoid:	2nd sinusoid:	Gaussian noise:
frequency = 25 Hz	frequency = 50 Hz	mean value = 0
amplitude = 1	amplitude = 0.01	standard deviation = 0.031

The sampling frequency is 200 Hz. Consider the following signal lengths: K = {512, 1,024, 2,048, 4,096} and calculate its periodogram in each case.

Then test the averaged periodogram in the following cases: {K=512, L=2}, {K=1,024, L=4}, {K=2,048, L=8}, {K=4,096, L=16} (L is the number of sequences).

```
N=512;    % change then this value with 1024, 2048 and 4096
t=0:1/200:(N/200)-1/200;
y1=sin(2*pi*25*t); y2=0.01*sin(2*pi*50*t);
noise=randn(1,N);
signal=y1 + y2 + noise * 0.031;
```

Use the algorithms from exercise 10.1 to calculate the signal periodogram and averaged periodogram.

Periodogram

```
[PSD_periodogram,frequency]=simple_periodogram(signal,N);
semilogy(frequency,PSD_periodogram,'b');
xlabel('normalized frequency');
ylabel('Amplitude');
title('Signal periodogram calculated on 512 points');
```

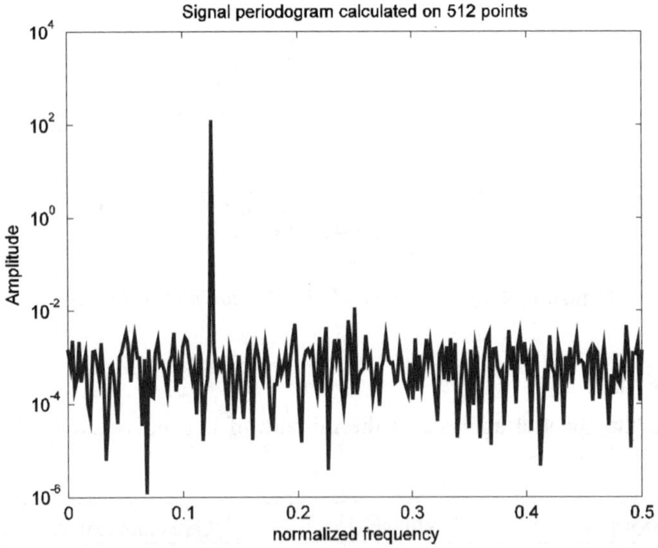

Figure 10.5. *Signal spectral analysis using the periodogram*

The peak corresponding to the high amplitude signal spectral component is clearly detected by the periodogram. However, as this estimate is not consistent the identification of the weak signal spectral component is much more difficult.

Therefore, the periodogram is unable to detect weak sinusoids in white noise due to its large variance.

Averaged periodogram

```
K=128;
[PSD_averaged_periodogram,frequency]=averaged_periodogram(signal,K);
semilogy(frequency,PSD_averaged_periodogram,'b');
```

```
xlabel('normalized frequency');
ylabel('Amplitude');
title('Signal averaged periodogram calculated for N=512 and L=4');
```

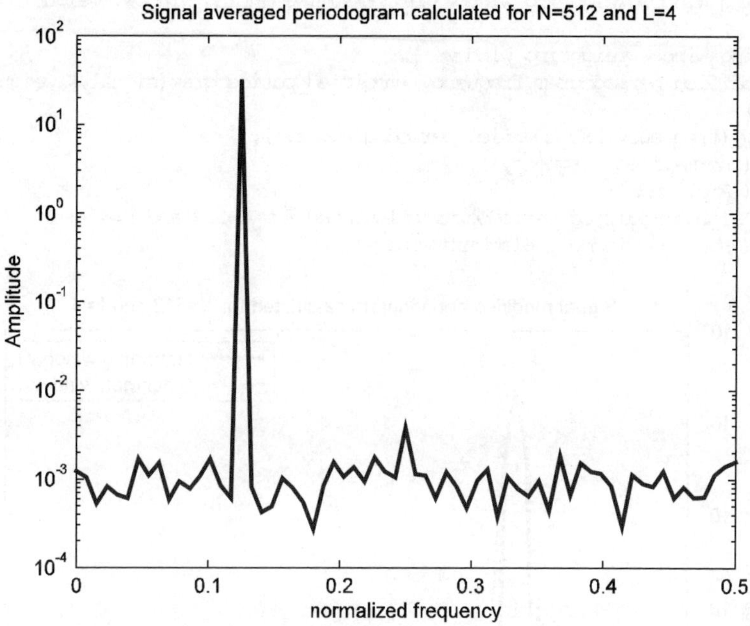

Figure 10.6. *Signal spectral analysis using the averaged periodogram*

In the case of the averaged periodogram, for the same spectral resolution, (K/L constant), increasing the number of averaged sequences L allows the noise variance to be reduced and thus the two sinusoids to be detected.

EXERCISE 10.4.

Repeat exercise 10.3 using the modified periodogram to perform the signal spectral analysis. Use a Hamming and then a Blackman window, with $K = 512$ and $L = 4$. Compare this result to those obtained previously. Use the algorithm from exercise 10.2 to calculate the signal modified periodogram.

```
K=128;
weighting_window=[hamming(K)];
weighting_window=weighting_window*sqrt(K/sum(weighting_window.*weighting_wind
ow));
weighting_window=weighting_window.';
[PSD_modified_periodogram,frequency]=modified_periodogram(signal,K,weighting_
window);
```

```
figure;
semilogy(frequency,PSD_modified_periodogram,'b');
hold on
weighting_window=[blackman(K)];
weighting_window=weighting_window*sqrt(K/sum(weighting_window.*weighting_wind
ow));
weighting_window=weighting_window.';
[PSD_modified_periodogram,frequency]=modified_periodogram(signal,K,weighting_
window);
semilogy(frequency,PSD_modified_periodogram,'r:');
xlabel('normalized frequency');
ylabel('Amplitude')
title('Signal modified periodogram calculated for N=512 and L=4');
legend('Hamming window','Blackman window')
```

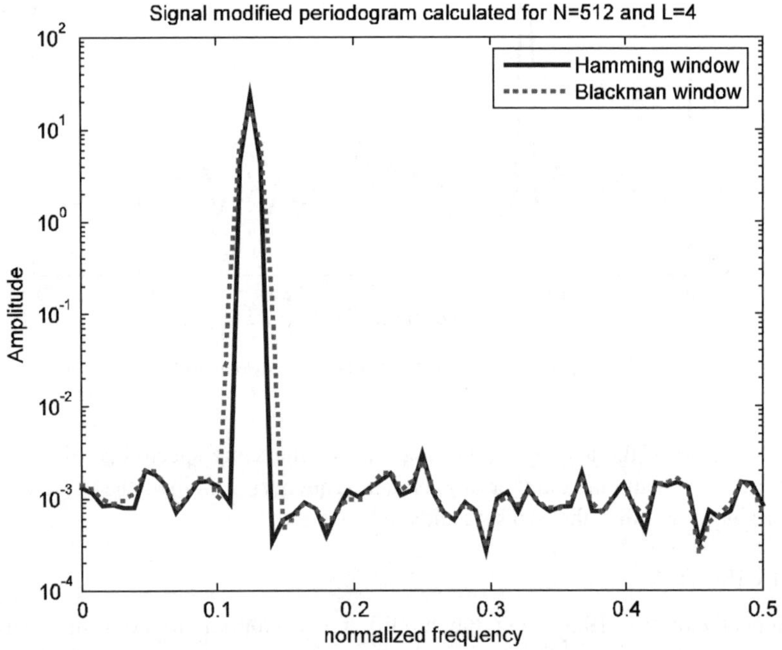

Figure 10.7. *Signal spectral analysis using the modified periodogram*

The two peaks corresponding to the two spectral components are detected again. However, compared to the averaged periodogram they are larger in this case due to the mainlobe width of the weighting window. This can be a shortcoming for the frequency estimation of the detected peaks, but it can be also useful when these frequencies are not harmonic (exercise 10.8).

EXERCISE 10.5.

Consider the filter having the transfer function given below and driven by a zero-mean white Gaussian noise with the variance $\sigma^2 = 0.25$:

$$H(z) = \frac{1}{1+0.5z^{-1}+0.75z^{-2}}.$$

The goal of this exercise is to identify the AR process and to retrieve parameters σ^2, a_1 and a_2 corresponding to the data. Suppose that the model order is known ($p = 2$). Prove that the signal spectrum is proportional to the filter transfer function.

```
function [coefficients,sigma2]=parameter_AR(signal,order);
length_signal=length(signal);
r=zeros(order+1,1);
autocor=xcorr(signal,'unbiased');
[maximum,position]=max(autocor);
for index=1:order+1
    r(index,1)=autocor(position+index-1);
end;
Matrice_toeplitz=toeplitz(r);
parameters=inv(Matrice_toeplitz)*[1;zeros(order,1)];
coefficients=parameters/parameters(1);
sigma2=1/parameters(1);
zplane(1,coefficients.');

% Signal generation and filtering
noise=randn(1,2048)*0.5;
b=1; a=[1 0.5 0.75];
y=filter(b,a,noise); y=y.';

% The function defined above is now applied to estimate the AR
% parameters
[coef,sigma2]=parameter_AR(y,2);
title('Poles and zeros of a 2^n^d order AR process');
```

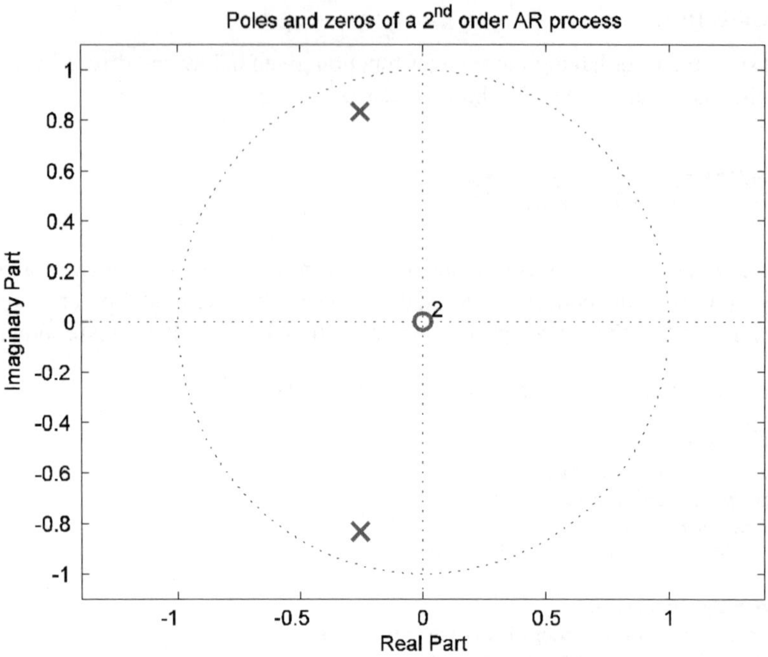

Figure 10.8. *Poles and zeros of a 2nd order AR process*

The following MATLAB code estimates the power spectral density of the filtered signal:

```
order=2;PSD=0;
for nu=0:200
     denominator=1;
     for ind=1:order
          coefm=coef(ind+1,1)*exp(-2*pi*j*ind*nu/200);
          denominator=denominator+coefm ;
     end;
     PSD=[PSD sigma2/(abs(denominator).^2)];
end;
nu=0:1/200:0.5;
plot(nu,PSD(1,1:length(nu)));
xlabel('normalized frequency');
ylabel('Amplitude');
```

It is clear from the figure below that the estimated PSD corresponds to the transfer function of a filter having two complex poles i.e. a bandpass filter.

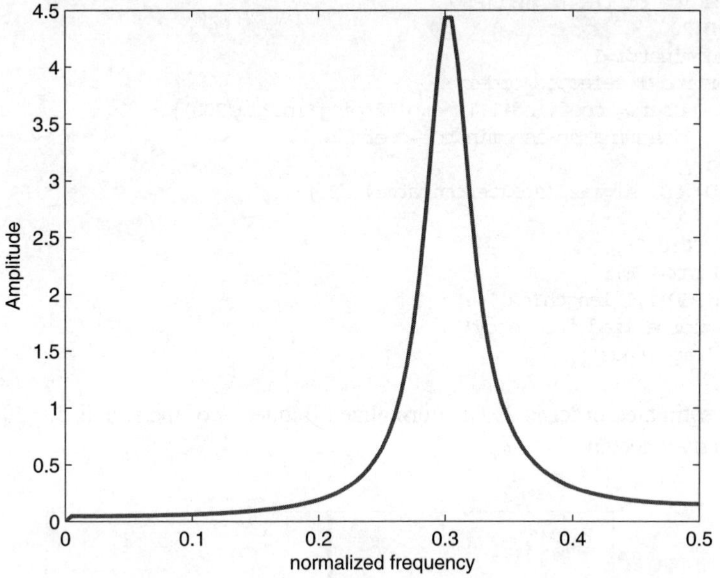

Figure 10.9. *Estimated signal PSD using a 2nd order AR model*

EXERCISE 10.6.

Generate a pure sinusoid with the frequency 50 Hz, sampled at 200 Hz (defined on 4,096 samples). Plot its power spectral density estimated using an AR model. Find out the order of this model using the AIC. Use the function `parameter_AR` previously defined.

```
% Signal generation
sig_length=4096;
t=0:1/200:(sig_length /200)-1/200;
y=sin(2*pi*50*t);y=y.';

% Akaike's criterion
PSD=[];
FPE=zeros(1,10);
for order=1:10
     [coef,sigma2]=parameter_AR(y,order);
     FPE(1,order)=((sig_length+order)/(sig_length-order))*sigma2;
end
FPE(FPE<1e-10)=0;
[mini,position]=min(FPE);
selected_order =position;
% Estimation of the parameters of the AR model
[coef,sigma2]=parameter_AR(y,selected_order);
```

```
% Estimation of the signal PSD
for nu=0:200
     denominator=1;
     for ind=1:selected_order
          coefm= coef(ind+1,1)*exp(-2*pi*j*ind*nu/200);
          denominator=denominator+coefm ;
     end;
     PSD=[PSD sigma2/(abs(denominator).^2)];
end;
nu=0:1/200:0.5;
PSD=PSD/max(PSD);
plot(nu,PSD(1,1:length(nu)));
xlabel('normalized frequency');
ylabel('Amplitude');
```

The estimated order is 2. The normalized frequency of the signal $50/200=0.25$ is perfectly detected.

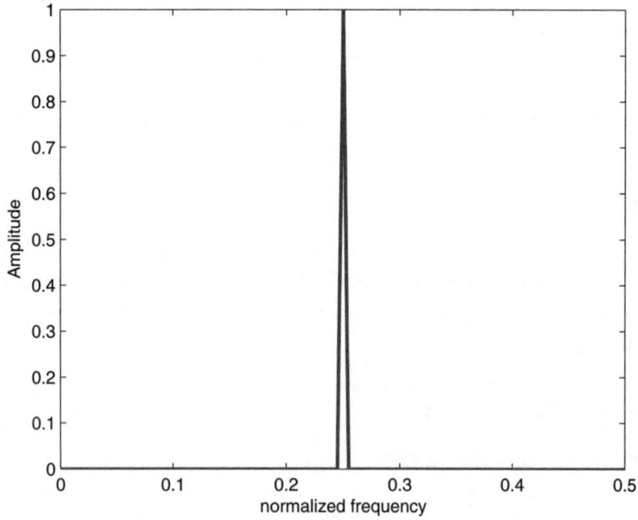

Figure 10.10. *Signal PSD estimated using an AR model*

EXERCISE 10.7.

a. Generate a mixture of 6 complex exponentials having unit amplitudes and the frequencies 100 Hz, 150 Hz, 175 Hz, 200 Hz, 375 Hz and 400 Hz during 15 ms. Add a DC component of 1 V and a white Gaussian noise. Sample this signal at 1 KHz. Calculate the number of samples and the spectral resolution corresponding to the considered observation time.

b. Perform the spectral analysis of the generated signal using the Fourier transform and the MUSIC algorithm. Use a number of calculation points 10 times higher than the number of samples and a large SNR (100 dB for example). For the Fourier transform use successively the rectangular, Bartlett and Blackman windows.

c. Consider the same signal, but containing only 5 complex exponentials having the frequencies 100 Hz, 150 Hz, 200 Hz, 350 Hz and 400 Hz. Perform again the spectral analysis of this signal using the Fourier transform and the algorithms MUSIC, root-MUSIC and ESPRIT for two SNR: 100 dB and 30 dB.

d. Perform a statistical study on the estimation of one spectral component with the frequency 250 Hz be means of the four methods previously used. Consider that the SNR sweeps the range from 0 dB to 30 dB and plot the variance variation of each obtained estimate over 1,000 outcomes.

a.

The number of samples and the spectral resolution corresponding to the given observation time are obtained as follows:

$$N_S = 1 + [T_{obs} \cdot F_S] = 1 + \left[0.015 \cdot 10^3\right] = 16$$

$$\Delta \nu = \frac{1}{T_{obs}} = \frac{1}{0.015} = 66.66 \text{ Hz}$$

Consequently, most frequency gaps between the signal spectral components are less than $\Delta \nu$.

```
Fs=1e3; time_obs=1.5e-2; comp_cont=1; snr=100;
freq_vect=[100 150 175 200 375 400]; amp_vect=[1 1 1 1 1 1];
freq_res=1/time_obs
N=1+round(time_obs*Fs); p=round(2*N/3); Np=100*N;
time_vect=linspace(0,time_obs,N); sig_util=zeros(N,1);
for k=1:length(amp_vect)
sig_util=sig_util+amp_vect(k)*exp(j*2*pi*freq_vect(k)*time_vect');
end
sig_util=sig_util+comp_cont; noise=randn(N,1)+j*randn(N,1);
pb=norm(abs(noise))^2/N; ec_type=(pb^-0.5)*10^(-snr/20);
sig=sig_util+ec_type*noise;
```

b.

The following MATLAB code performs the spectral analysis of the generated signal using the Fourier and MUSIC algorithms:

```
% Fourier transform based spectral analysis
```

```
sigrf=fftshift(abs(fft(sig.*hamming(N),Np)));
sigrf=sigrf-min(sigrf); sigrf=sigrf/max(sigrf);
% Estimation of the autocorrelation matrix
Rxx=zeros(p); J=fliplr(eye(p)); L=(N+1-p);
for k=1:N+1-p
    xv=sig(k:k+p-1); Rk=xv*xv'; Rxx=Rxx+Rk+J*conj(Rk)*J;
end
Rxx=Rxx/L;
% Autocorrelation matrix eigenanalysis
[vctp,valp]=eig(Rxx); valpv=abs(diag(valp));
[valpvs,idxvp]=sort(valpv); ld=flipud(valpvs);
idxvp=flipud(idxvp); vctps=vctp(:,idxvp);
% Estimation of the signal subspace dimension
ris=zeros(1,p-1); aic=zeros(1,p-1);
for m=1:p-1
    S=(sum(ld(m+1:p,1))/(p-m)).^(p-m); P=prod(ld(m+1:p,1));
    if P<1e-6; P=1e-6; end
    aic(m)=(N-p-2)*log(S/P)+m*(2*p-m);
    ris(m)=(N-p-2)*log(S/P)+.5*m*(2*p-m)*log(N-p-2);
end
[vmin,da]=min(aic);
[vmin,dr]=min(ris)
figure;plot(aic,'b--');
hold on;plot(ris,'r-')
xlabel('n');ylabel('Cost function value')
legend('AIC','MDL');grid
% Calculation of the projection operator on the noise subspace
vctpb=vctps(:,dr+1:p);
mb=vctpb*vctpb';
% Spectral analysis by means of the MUSIC algorithm
sigrm=zeros(Np,1);
xv=[0:Np-1]/Np; yv=[0:p-1];
am=xv'*yv;
am=exp(-j*2*pi*am);
matr_prod=am*mb*am';
vect_prod=diag(matr_prod);
sigrm=abs(1./vect_prod);
sigrm=fftshift(log10(abs(sigrm)));
sigrm=sigrm-min(sigrm); sigrm=sigrm/max(sigrm);
% Plotting the results
freqv=linspace(-Fs/2,Fs/2,Np+1);
figure
hold on
plot(freqv(1:Np),sigrf,'b--');
plot(freqv(1:Np),sigrm,'r-')
axis([-Fs/2 Fs/2 0 1.1]); grid
xlabel('Frequency [Hz]');
ylabel('Normalized amplitude')
legend('Fourier','MUSIC')
```

The Fourier transform based spectral analysis coded above makes use of a rectangular weighting window. To obtain the other required results on the first line "boxcar" must be replaced by "barlett" and then by "Blackman".

The spectral analysis results using the three types of weighting windows are shown in the figure below together with the estimation of the signal subspace dimension by means of the Akaike and Rissanen criteria. Note that the MUSIC algorithm is able to separate spectral components much closer than the resolution given by the inverse of the observation time.

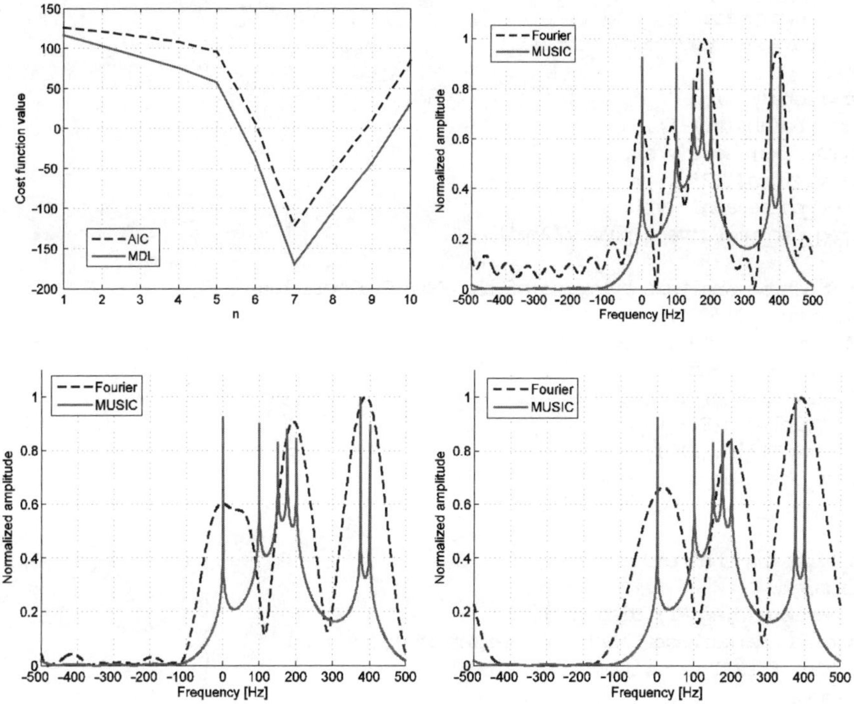

Figure 10.11. *MUSIC algorithm and Fourier transform based spectral analysis results*

c.

The signal generation and its spectral analysis using the Fourier transform and the MUSIC algorithm are performed with the previous MATLAB codes. A Hamming window is considered here for the Fourier transform based PSD estimate, due to the good trade-off between the associated spectral and dynamic resolutions.

The MATLAB codes corresponding to the root-MUSIC and ESPRIT algorithms are provided below. Only the $2N$ roots closest to the unit circle of the root-MUSIC polynomial are plotted for the first of the two algorithms. In the same way, the N eigenvalues of the Ψ matrix corresponding to the signal spectral components are plotted for the second algorithm.

```
% Spectral analysis by means of the root-MUSIC algorithm
pol=zeros(1,2*p-1);
for k=dr+1:p
    vect=vctps(:,k);
    vect1=fliplr(vect');
    vect2=vect.';
    pol=pol+conv(vect1,vect2);
end
vr=roots(pol);
vra=abs(abs(vr)-1);
[vrs,idxs]=sort(vra);
vf=vr(idxs(1:2*dr));
vect_phase=angle(vf);
freq_est_rm=Fs*vect_phase/(2*pi)

% Spectral analysis by means of the ESPRIT algorithm
Vs=vctps(:,1:dr);
Vs_sup=Vs(1:end-1,:);
Vs_inf=Vs(2:end,:);
F=pinv(Vs_inf)*Vs_sup;
[T,Tvlp]=eig(F);
Mf=inv(T)*inv(F)*T;
vect_phase=angle(diag(Mf));
freq_est_ep=Fs*vect_phase/(2*pi);

% Plotting the results
figure
hh=zplane(diag(Mf),diag(Mf));
set(hh,'MarkerEdgeColor','r','MarkerSize',[10])
title('ESPRIT algorithm eigenvalues')
figure
hh=zplane(vf,vf);
set(hh,'MarkerEdgeColor','r','MarkerSize',[10])
title('Useful roots of the root-MUSIC polynomial')
```

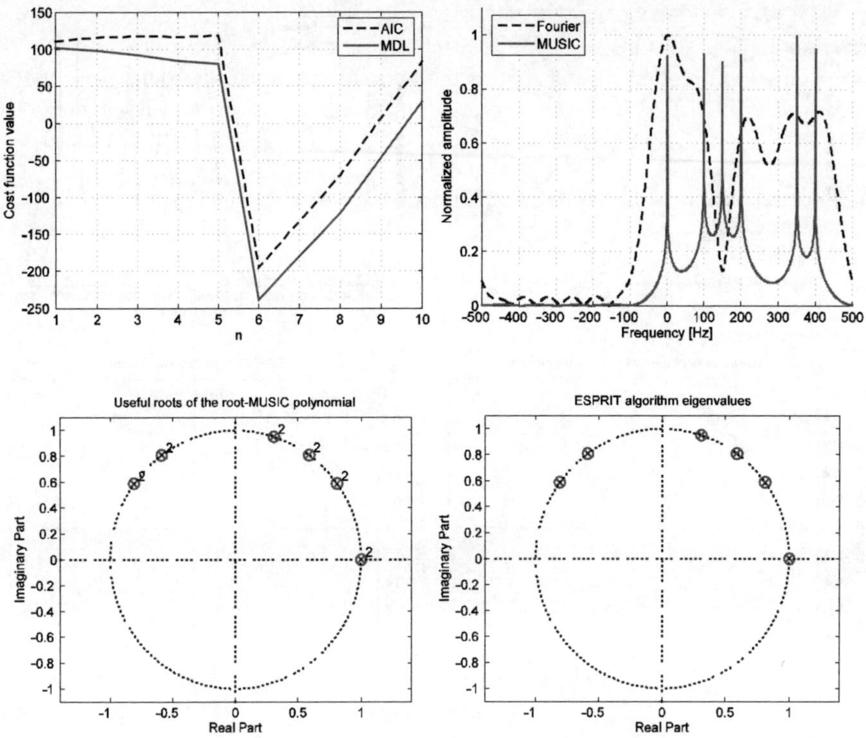

Figure 10.12. *Comparison of Fourier transform based and super-resolution spectral analysis methods*

For a SNR of 30 dB, the results are plotted on Figure 10.13. Compare these results to those previously obtained and notice the modifications which occur in this case.

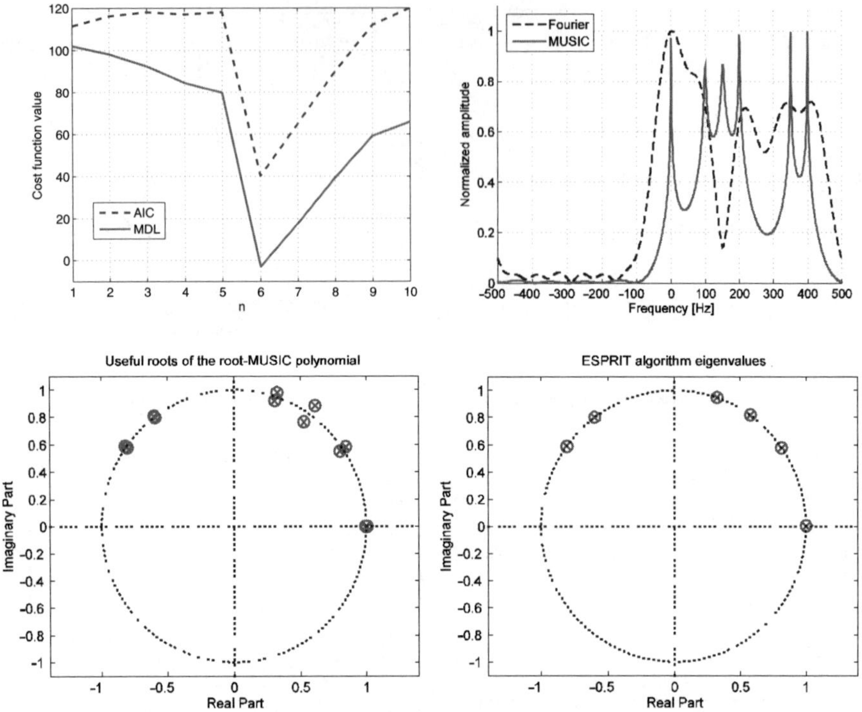

Figure 10.13. *Comparison of Fourier transform based and super-resolution spectral analysis methods for a SNR value of 30 dB*

d.

The code which allows this statistical study to be performed is provided below:

```
snr_dbv=[0:30]; Nr=1000;
Fs=1e3; time_obs=1.5e-2;
comp_cont=0; freq0=250;
N=1+round(time_obs*Fs);
p=round(2*N/3); Np=100*N;
freqv=linspace(-Fs/2,Fs/2,Np+1);
matr_fqf=zeros(Nr,length(snr_dbv));
matr_fqm=zeros(Nr,length(snr_dbv));
matr_fqe=zeros(Nr,length(snr_dbv));
matr_fqr=zeros(Nr,length(snr_dbv));
for k1=1:length(snr_dbv)
    snr=snr_dbv(k1)
    for k2=1:Nr
        % Signal generation
        time_vect=linspace(0,time_obs,N);
```

```
sig_util=exp(j*2*pi*freq0*time_vect');
noise=randn(N,1)+j*randn(N,1);
pb=norm(abs(noise))^2/N;
ec_type=(pb^-0.5)*10^(-snr/20);
sig=sig_util+ec_type*noise;
% Estimation of the autocorrelation matrix
Rxx=zeros(p); J=fliplr(eye(p)); L=(N+1-p);
for k=1:N+1-p
    xv=sig(k:k+p-1);
    Rk=xv*xv';
    Rxx=Rxx+Rk+J*conj(Rk)*J;
end
Rxx=Rxx/L;
% Autocorrelation matrix eigenanalysis
[vctp,valp]=eig(Rxx); valpv=abs(diag(valp));
[valpvs,idxvp]=sort(valpv); ld=flipud(valpvs);
idxvp=flipud(idxvp); vctps=vctp(:,idxvp); dr=1;
% Calculation of the projection operator
vctpb=vctps(:,dr+1:p); mb=vctpb*vctpb';
% Fourier transform based spectral analysis
sigrf=fftshift(abs(fft(sig.*hamming(N),Np)));
sigrf=sigrf-min(sigrf); sigrf=sigrf/max(sigrf);
% Spectral analysis by means of the MUSIC algorithm
sigrm=zeros(Np,1); xv=[0:Np-1]/Np; yv=[0:p-1];
am=xv'*yv; am=exp(-j*2*pi*am);
matr_prod=am*mb*am';
vect_prod=diag(matr_prod);
sigrm=abs(1./vect_prod);
sigrm=fftshift(log10(abs(sigrm)));
sigrm=sigrm-min(sigrm); sigrm=sigrm/max(sigrm);
% Spectral analysis by means of the root-MUSIC algorithm
pol=zeros(1,2*p-1);
for k=dr+1:p
    vect=vctps(:,k);
    vect1=fliplr(vect'); vect2=vect.';
    pol=pol+conv(vect1,vect2);
end
vr=roots(pol); vra=abs(abs(vr)-1);
[vrs,idxs]=sort(vra);
vf=vr(idxs(1:dr)); vect_phase=angle(vf);
 freq_est_rm=Fs*vect_phase/(2*pi);
% Spectral analysis by means of the ESPRIT algorithm
Vs=vctps(:,1:dr); Vs_sup=Vs(1:end-1,:); Vs_inf=Vs(2:end,:);
F=pinv(Vs_inf)*Vs_sup; [T,Tvlp]=eig(F);
Mf=inv(T)*inv(F)*T; vect_phase=angle(diag(Mf));
freq_est_ep=Fs*vect_phase/(2*pi);
% Estimation of the signal frequency
[valmx,idxmx]=max(sigrm); matr_fqm(k2,k1)=freqv(idxmx);
[valmx,idxmx]=max(sigrf); matr_fqf(k2,k1)=freqv(idxmx);
matr_fqe(k2,k1)=freq_est_ep;
```

```
        matr_fqr(k2,k1)=freq_est_rm;
    end
end
var_vectf=var(matr_fqf);
var_vectm=var(matr_fqm);
var_vectr=var(matr_fqr);
var_vecte=var(matr_fqe);
% Plotting the results
figure
plot(snr_dbv,moy_vectf,'-rx',snr_dbv,moy_vectm,'-bo',snr_dbv,moy_vectr,'-
k<',snr_dbv,moy_vecte,'-gs')
xlabel('SNR [dB]'); ylabel('Frequency [Hz]');
legend('Fourier','MUSIC','root-MUSIC','ESPRIT'); grid
figure
plot(snr_dbv,10*log10(var_vectf),'-rx',snr_dbv,10*log10(var_vectm),'-
bo',snr_dbv,10*log10(var_vectr),'-k<',snr_dbv,10*log10(var_vecte),'-gs')
xlabel('SNR [dB]'); ylabel('Variance [dB]');
legend('Fourier','MUSIC','root-MUSIC','ESPRIT'); grid
```

The results of this statistical study are shown in the figures below. The obtained curves demonstrate that the statistical behavior of the four estimates is similar for SNR larger than 2 dB.

The variances of the root-MUSIC and ESPRIT estimates are slightly lower than the ones for the two other estimates due to the exact calculation of the signal frequency, while the accuracy of the Fourier and MUSIC methods is limited by the number of considered calculation points. This phenomenon is much less visible for low signal-to-noise ratios, where the difference introduced by the calculation accuracy is small compared to the variance due to the noise.

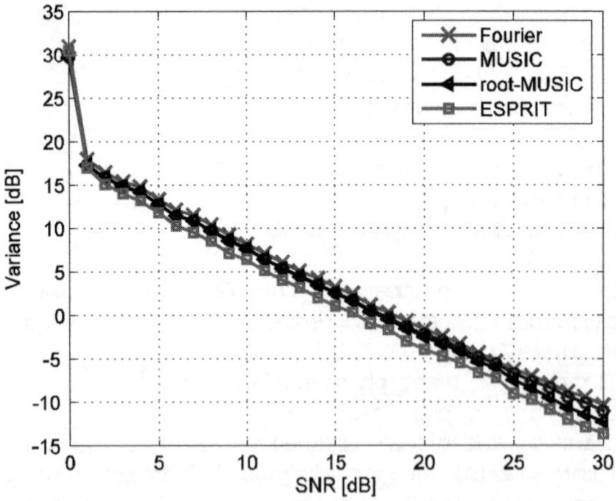

Figure 10.14. *Statistical behavior of some spectral analysis methods*

10.3. Exercises

EXERCISE 10.8.

Consider again the signal from exercise 10.3; calculate its modified periodogram (using the algorithm already proposed) for different windows and conclude.

EXERCISE 10.9.

Generate the mixture of the two sinusoids below, whose frequencies are not harmonic. Add a zero-mean white Gaussian noise with the variance 0.031.

1st sinusoid:	2nd sinusoid:
frequency = 25.4 Hz	frequency = 51.3 Hz
amplitude = 1	amplitude = 0.01

Calculate the periodogram of this signal for the following signal lengths: K = 256, 512, 1,024, and identify the harmonic frequencies as a function of K ?

Then test the averaged periodogram for $\{K=2{,}048,\ L=8\}$, $\{K=4{,}096,\ L=16\}$ and $\{K=4{,}096,\ L=8\}$. Which are, in each case, the harmonic frequencies? Do they depend on K? Comment on these results and highlight the spectral leakage phenomenon. Finally, apply the modified periodogram for different windows and conclude.

EXERCISE 10.10.

Consider again the signal previously generated. Plot its power spectral density using a parametric spectral analysis (assume an AR model for the signal and find out its order – see exercise 10.5). Compare the estimated PSD with that obtained using the averaged and the modified periodogram (see exercise 10.2).

EXERCISE 10.11.

Consider an ARMA(2,2) model having the following transfer function:

$$H(z)=\frac{1+0.9025z^{-2}}{1-0.5562z^{-1}+0.81z^{-2}}$$

driven by a zero-mean white Gaussian noise with the variance 1. Calculate and plot the corresponding power spectral density (dB scale).

EXERCISE 10.12.

Consider a white noise filtered with a 1st order recursive filter, characterized by the following difference equation:

$$y(k)-0.9y(k-1)=x(k)$$

The output signal is filtered again with a similar filter having the coefficient 0.8. Find out the power spectral density of the second filter output signal. Compare the simulation result to the theoretical calculation.

What is the variance of the white Gaussian noise mean, with $\sigma_n^2=1$, estimated from $K=100$ samples?

(answer: $\sigma_{\hat{m}_n}^2=0.01$)

The true value belongs to the interval $[\hat{m}_b-0.3, \hat{m}_b+0.3]$ with the probability 0.9973. Write a MATLAB code to retrieve this result from a large number of simulations.

Chapter 11

Time-Frequency Analysis

11.1. Theoretical background

11.1.1. *Fourier transform shortcomings: interpretation difficulties*

The Fourier transform is involved in a wide range of signal processing applications. However, despite its rigorous mathematical definition it may lead to some physical interpretation difficulties. It is, for example, clear from its definition that the evaluation of a spectrum value $X(\nu)$ requires the knowledge of all signal history, from $-\infty$ to $+\infty$. In the same way, the signal value at any time t is expressed by the inverse Fourier transform as a superposition of an infinite number of complex exponentials, i.e. eternal waves perfectly delocalized in the time domain. If this mathematical point of view is able to reveal interesting signal properties in many cases, it may also be sometimes rather inappropriate.

This is typically the case with transient signals, which are known to be bounded in time. The Fourier analysis can provide this image of the signal, but in a somehow artificial manner, i.e. as a sum of an infinite number of virtual sinusoids which cancel each other. Consequently, we obtain a "dynamic" zero on a time interval where the signal is vanishing, while from a physical point of view this should be a "static" zero, since the signal does not exist on this interval. Furthermore, the Fourier analysis expresses a finite energy signal as a linear superposition of infinite energy basic signals.

These physical interpretation difficulties suggest that the Fourier transform should be replaced in these cases by other signal analysis tools, such as those presented in this chapter. The latest are able to take into account both the signal spectral description and its time localization.

11.1.2. *Spectrogram*

The main idea of the spectrogram is to analyze the signal using a sliding window. The window length is chosen so that the signal may be considered to be almost stationary inside. In this case, its spectral energy density can be evaluated using the Fourier transform over each time interval obtained by shifting the sliding window. The spectrogram is thus defined in the following form:

$$S(k,v) = \left| \sum_{n=0}^{N-1} h(n)x(k+n)e^{-j2\pi vn} \right|^2 \quad [11.1]$$

It is obviously calculated using the FFT for the frequencies $v_i = i/N$.

The short-time Fourier transform (STFT) or its square magnitude (spectrogram) consider therefore a non-stationary signal as a concatenation of stationary signals within the sliding window *h*(*u*). Thus, the time resolution of this analysis is given by the window size, while the spectral resolution is proportional to its inverse. This means that it is not possible to increase the two resolutions simultaneously.

For highly non-stationary signals a fine time resolution is required; thus, window *h*(*u*) should be short in this case and the spectral resolution will be low. If a fine spectral resolution is required, window *h*(*u*) should be large, which reduces the time resolution.

In order to obtain the best trade-off between the spectral resolution and the time resolution, Gabor proposed the following transform:

$$G(t,\upsilon) = \int x(\theta)h(\theta - t)\exp(-j2\pi v\theta)d\theta \quad [11.2]$$

with: $h(\theta) = \exp\left(-\lambda\theta^2\right) \quad (\lambda > 0)$ [11.3]

Note that the time-shifted function $h(\theta - t)$ is a sliding window. The term $h(\theta - t)\exp(-j2\pi v\theta)$ can be seen as the impulse response of a frequency selective filter at v. Therefore, Gabor's transform can be considered as a bank of similar filters, shifted in the frequency domain.

According to the Heisenberg-Gabor uncertainty principle, the product of time and frequency resolutions $\Delta\theta \cdot \Delta v$ is minimal if function $h(\theta)$ is Gaussian. In this case, minimum surface cells are obtained in the time-frequency plane. However, the two resolutions are still limited and do not adapt to the signal variation.

11.1.3. *Time-scale analysis – wavelet transform*

The wavelet transform (WT) is an important evolution of the STFT as it makes use of a sliding window, whose length $\overline{\Delta t}$ is adapted to the analyzed spectral region, so that the ratio $\overline{\Delta \nu} / \nu$ is kept constant. A wavelet is the shortest vibration in a given frequency range, so it is very concentrated in time or in frequency.

The basic idea is to decompose a signal on a set of functions, obtained by shifting and scaling a single function $\psi(\theta)$, called the mother wavelet. The set of functions are therefore expressed in the form:

$$\frac{1}{\sqrt{a}}\psi\left(\frac{\theta - t}{a}\right) \qquad [11.4]$$

The time-scale transform is defined as follows:

$$WT(t,a) = \int x(\theta)\frac{1}{\sqrt{a}}\psi\left(\frac{\theta - t}{a}\right)d\theta \qquad [11.5]$$

It may also be seen as a bank of filters, whose selectivity depends on the parameter *a*, so that the product $\overline{\Delta \nu} \cdot \overline{\Delta t}$ has a minimum but constant value.

The time-scale analysis (TSA) can be linked to the time-frequency analysis (TFA) by the following relationship:

$$TFA(t,\nu) = TSA(t,\nu / a) \qquad [11.6]$$

Note:

a. It is possible to reconstruct the signal as the sum of its projections provided that the mother wavelet meets the admissibility constraint:

$$\int \psi(t)dt = 0 \qquad [11.7]$$

b. $\psi(t)$ should be selected so that its Fourier transform $TF\{\psi(t)\} = \hat{\psi}(\nu)$ meets the following constraints: $\int_0^{+\infty} \frac{|\hat{\psi}(\nu)|^2}{\nu} d\upsilon < \infty$ and $\hat{\psi}(\nu)$ is small enough in the neighborhood of $\nu = 0$;

c. the Gabor functions define a wavelet family, but the Gabor transform is very different from the wavelet analysis;

d. the wavelet transform is a “multiresolution” analysis in time and frequency.

Although it does not make use of the Fourier transform, the time and spectral resolutions also have inverse variations. However, compared to the STFT the two resolutions are not the same over the entire time-frequency plane:

– in the case of transient or localized structures, the wavelet transform will be concentrated essentially in the small scales domain which are able to highlight signal details; however, since these small scales correspond to an analyzing wavelet of reduced time support, the associated frequency support is large and the spectral resolution is low;

– in the other sense, a spectral resolution can be increased by only applying a long analyzing wavelet, i.e. large observation scales; a spectral resolution gain is then obtained, but it is paid for by the reduction of the time resolution.

Similar to the spectrogram, a filter bank interpretation of the wavelet transform is also possible. The signal is analyzed using a set of non-uniform filters, each of them being obtained by a homothetic transformation of a single bandpass filter.

This is a constant-Q filtering, which means that all filters have the same relative passband, defined as the ratio between the filter passband and its central frequency. Thus, the larger the frequencies corresponding to the analyzed scale, the lower the spectral resolution.

In order to insure a parsimonious signal representation and to be able to perfectly reconstruct the signal from its decomposition, Mallat and Meyer proposed in 1989 an effective and flexible tool, called *multiresolution analysis*, which has led to many applications so far.

It aims at approximating a signal $f(t)$ by its projections on a set of $L^2(R)$ subspaces, denoted by $\{V_m\}_{m \in Z}$, so that:

a. $f^L(t) = \mathrm{Proj}_{V_L}[f(t)] = P_L[f(t)]$ is obtained from $f(t)$ after a successive removal of its details $\{Q_m[f(t)]\}_{m=1..L}$;

b. the original signal $f(t)$ can be completely recovered from its approximation $P_L[f(t)]$ and all the details, which have been successively removed.

More precisely, the multiresolution analysis performs the signal decomposition using a set of approximation and detail coefficients, corresponding to the different scales. These coefficients are obtained by means of two mirror filters (lowpass and highpass), denoted by $h(n)$ and $g(n)$ respectively, according to the decomposition tree shown in Figure 11.1.

At each step, approximation and detail coefficients are obtained by filtering the signal approximation from the previous level with the two filters. The signal is therefore analyzed with different resolutions, which explains the name of the method.

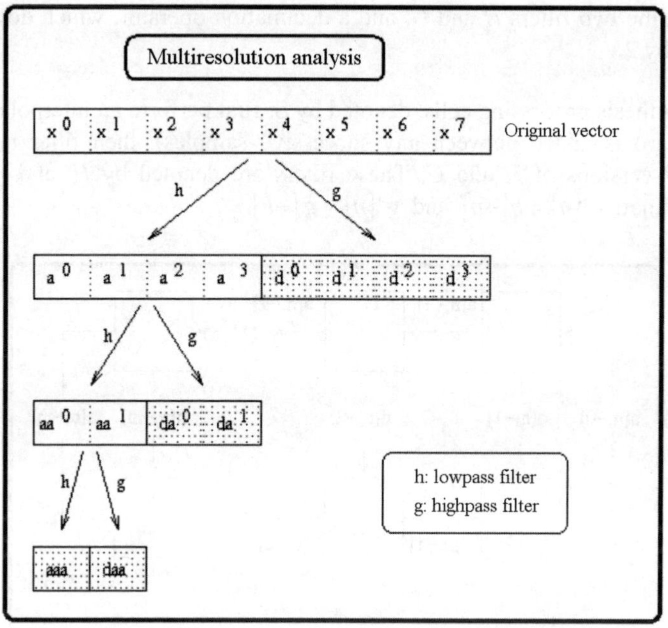

Figure 11.1. *Multiresolution analysis decomposition*

The signal is analyzed at low resolution with large wavelets and a few coefficients, in order to obtain a rough approximation. It is analyzed at higher resolutions, using many narrow wavelets in order to extract its details. This is why the wavelets may also be seen as a "mathematical microscope". In fact, when the wavelets are compressed the magnification of this microscope gets higher and it is able to reveal finer and finer signal details.

The analysis processing flow is reversible and leads to a dual synthesis algorithm. It allows the signal approximation corresponding to a given resolution level to be obtained from the approximation and the details associated with the previous resolution level.

Compared to the analysis algorithm, which consists of successive filtering-decimation steps, the synthesis algorithm performs an interpolation followed by the signal filtering at each step. The flowchart of the analysis and synthesis algorithms is illustrated in Figure 11.2.

It can be seen that the two algorithms are implemented as a series connection of a similar processing cell. In the case of the analysis algorithm these cells, denoted by A, contain the two filters *H* and *G*, and a decimation operator, which downsamples the signal by 2.

The synthesis processing cells, denoted by S, first perform an interpolation of the input (a zero is added between any successive samples), then filter it using the transposed versions of *H* and *G*. These filters are denoted by *H'* and *G'* and are defined so that: $h'[n]=h[-n]$ and $g'[n]=g[-n]$.

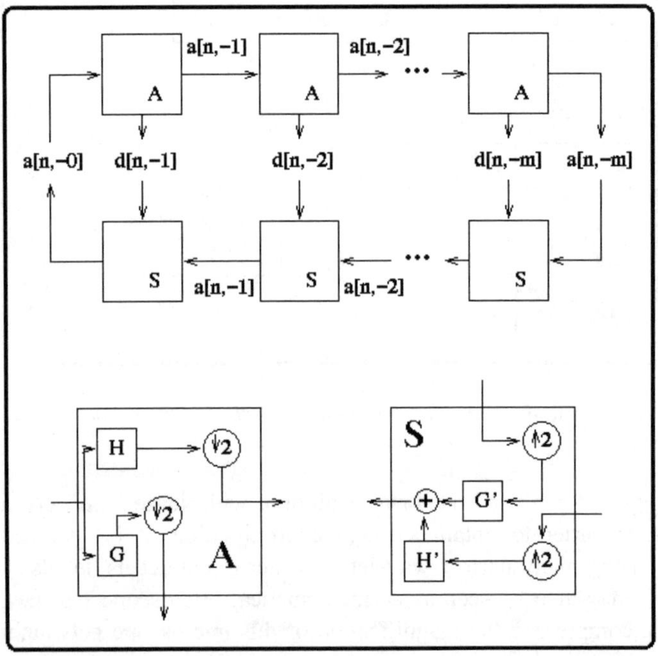

Figure 11.2. *Multiresolution analysis and synthesis flowchart*

11.1.4. *Wigner-Ville distribution*

This non-stationary signal analysis tool was introduced in 1948 by Ville. Actually, it represents an energy distribution and should be compared to the spectrogram or the scalogram rather than to the STFT or the wavelet transform. The

Wigner-Ville distribution (WVD) belongs to a more general distribution class, known as Cohen's class.

The main idea of this transformation is to measure the local symmetry of a signal around any point of the time-frequency plane. It does not suppose any *a priori* signal local stationarity.

The Wigner-Ville distribution is defined as follows, for the finite energy continuous-time signals:

$$\rho(t,\nu) = \int_{-\infty}^{\infty} x\left(t+\frac{\tau}{2}\right)x^*\left(t-\frac{\tau}{2}\right)e^{-j2\pi\nu\tau}d\tau \qquad [11.8]$$

while for the discrete-time signals it takes the form:

$$\rho(k,\nu) = 2\sum_{n=-\infty}^{+\infty} x(k+n)x^*(k-n)e^{-j4\pi\nu n} \qquad [11.9]$$

It is clear from the above equations that the WVD is a non-linear transformation.

The term $x(k+n)x^*(k-n)$ is an instantaneous correlation corresponding to the time delay $2n$ and measures the signal symmetry around instant k. Its Fourier transform is thus considered (abusively, as it will be shown later) as a time-frequency energy distribution.

Properties

a. The WVD is real since $x\left(t+\frac{\tau}{2}\right)x^*\left(t-\frac{\tau}{2}\right)$ is an even function of τ.

b. Energy conservation:

$$\int_{-\infty}^{\infty}\int_{-\infty}^{\infty} \rho(t,\nu)dtd\nu = \int_{-\infty}^{\infty} |x(t)|^2\, dt \qquad [11.10]$$

$$\sum_{k=-\infty}^{\infty}\int_{-1/4}^{1/4} \rho(k,\nu)d\nu = \sum_{k=-\infty}^{\infty} |x(k)|^2 \qquad [11.11]$$

The integration limits $(-1/4,+1/4)$ result from the downsampling by 2 performed when calculating the correlation term $x(k+n)x^*(k-n)$. This calculation supposes that there is no spectral aliasing, i.e. that the spectrum $X(\nu)$ of $x(k)$ is confined within the normalized frequency band $(-1/4,+1/4)$.

c. The following equalities are valid for all $\nu \in \boldsymbol{R}$:

$$\int \rho(t,\nu)dt = |X(\nu)|^2 \ , \ \sum_k \rho(k,\nu) = |X(\nu)|^2 \qquad [11.12]$$

$$\int \rho(t,\nu)d\nu = |x(t)|^2 \ , \ \int_{-1/4}^{+1/4} \rho(k,\nu)d\nu = |x(k)|^2 \qquad [11.13]$$

d. The WVD preserves signal time and frequency supports.

e. The WVD of an analytic signal $z(t)$ associated with $x(t)$ is expressed in the form:

$$\rho_z(t,\nu) = \int z\left(t+\frac{\tau}{2}\right)z^*\left(t-\frac{\tau}{2}\right)e^{-j2\pi\nu\tau}d\tau \qquad [11.14]$$

This allows the signal instantaneous frequency $f(t)$ to be defined as indicated below:

$$f(t) = \frac{\int \nu\rho_z(t,\nu)d\nu}{\int \rho_z(t,\nu)d\nu} \qquad [11.15]$$

f. The WVD is a reversible transform.

Nevertheless, the WVD also leads to some interpretation difficulties. In fact, it may have negative values, which are not compatible with an energy density. It produces interference terms between signal components, due to its non-linearity.

Thus, a point (t_1,ν_1) belonging to the WVD of the signal component x_1 interacts with a point (t_2,ν_2) belonging to the WVD of the signal component x_2 and creates an interference located in $\{(t_1+t_2)/2,(\nu_1+\nu_2)/2\}$. The interference amplitude $I(t,\nu)$ meets the following constraint: $I(t,\nu) \le 2\sqrt{\sup\left[\rho_{x_1}(t,\nu)\right]\sup\left[\rho_{x_2}(t,\nu)\right]}$.

By construction, the interferences have an oscillating nature. The oscillations occur across the axis linking the centers of $\rho_{x_1}(t,\nu)$ and $\rho_{x_2}(t,\nu)$. Their period is equal to the inverse of the distance between the two centers.

11.1.5. *Smoothed WVD (SWVD)*

The two main problems of the WVD are therefore the negative values and the interference terms. An interesting solution for facilitating its interpretation is to apply a smoothing operator in the time-frequency plane. In fact, this operation has little effect on signal components, but it highly attenuates the interference terms because of their oscillating nature.

The time-frequency smoothing leads to a much more readable distribution, but this gain is paid for by the failure of some theoretical properties of the WVD (marginal densities and instantaneous frequency).

In practice, the use of a separable time-frequency smoothing window is preferred. In the continuous-time case this results in:

$$F(t,\nu) = g(t)\, H(\nu) \text{ with } \iint F(t,\nu)dtd\nu = 1 \qquad [11.16]$$

The SWVD is defined as follows:

$$\begin{aligned} SWVD_x(t,\nu) &= \int_{-\infty}^{\infty}\int_{-\infty}^{\infty} \rho(\theta,w)F(t-\theta,\nu-w)dwd\theta \\ &= \int_{-\infty}^{\infty}\int_{-\infty}^{\infty}\int_{-\infty}^{\infty} x\left(\theta+\frac{\tau}{2}\right)x^*\left(\theta-\frac{\tau}{2}\right)g(t-\theta)H(\nu-w)e^{-j2\pi w\tau}dwd\theta d\tau \end{aligned} \qquad [11.17]$$

Finally, we obtain:

$$SWVD_x(t,\nu) = \int_{-\infty}^{\infty}\int_{-\infty}^{\infty} x\left(\theta+\frac{\tau}{2}\right)x^*\left(\theta-\frac{\tau}{2}\right)g(t-\theta)h(\tau)e^{-j2\pi\nu\tau}d\theta d\tau \qquad [11.18]$$

It can be readily understood from the above equation that the WVD of a signal *x*(*t*) is separately smoothed in time and frequency. The smoothing characteristics can thus be independently controlled along the two axes, time and frequency.

The frequency domain smoothing, obtained using sliding window h, is useful for removing the interferences which occur along the time axis. The time domain smoothing, performed by the filter with the impulse response *g*, is effective against the interferences which occur along the frequency axis.

Many other time-frequency distributions are available in the Cohen's class. Compared to the WVD they differ by the form of the kernel defining the smoothing operators in the time-frequency domain.

11.2. Solved exercises

EXERCISE 11.1.

A chirp signal, whose instantaneous frequency linearly sweeps the band between f_1 and f_2, is expressed as follows:

$$s(t) = \sin\left[(2\pi f_1 + 2\pi\beta t)t + \Phi_0\right]$$

with $\beta = (f_2 - f_1)/(2PulseLength)$, *Pulselength* being the signal duration.

a. Generate this signal using a MATLAB code for $\Phi_0 = 0$.

b. Plot the generated signal and its power spectral density.

c. It is difficult to obtain a complete image about the signal structure from these partial representations. In fact, they are not able to clearly indicate the modulation parameters or the time evolution of the signal spectral content. This information can be easily retrieved in the time-frequency plane. Illustrate this capability of the time-frequency distributions using the spectrogram for example.

a.

```
% Generation of a linear frequency modulated signal
f1=2000; f2=8000;
pulselength=0.025;
Fs=20000; % Sampling frequency
% Warning: Fs should verify the Nyquist constraint: Fs>2*max(f1,f2)
t=(0:1/Fs:pulselength);
beta=(f2-f1)/(2*pulselength);
chirp1=sin(2*pi*(f1+beta*t).*t);
% Another way to generate the chirp signal
chirp2 = vco(sawtooth((2*pi/pulselength)*t,1),[f1/Fs,f2/Fs]*Fs,Fs);
% chirp1 and chirp2 are similar up to a phase term
```

b.

```
figure; clf;
subplot(211)
plot(t,chirp1);
xlabel('Time [s]');
ylabel('Amplitude');
title('Time variation of a chirp signal')
C=fftshift(abs(fft(chirp1)).^2);
lc=length(chirp1); mc=lc/2;
freq=(-mc:1:mc-1)*Fs/lc;
subplot(212)
plot(freq,C);
xlabel('Frequency [Hz]');
ylabel('Power spectral density')
```

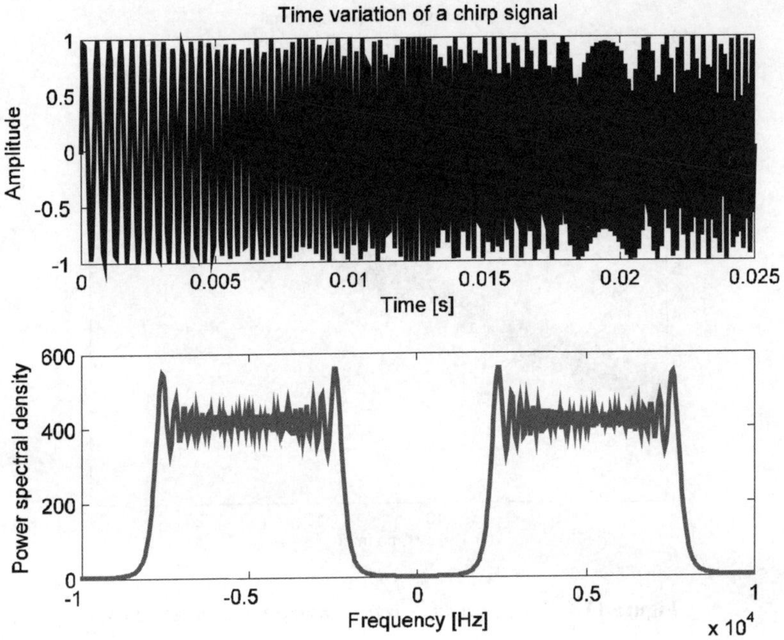

Figure 11.3. *Time and frequency representations of a chirp signal*

c.

```
Wsize = 32;
N_recover = 16;
NFFT = 1024;
[Cspec,F,T] = spectrogram(chirp1,hamming(Wsize),N_recover,NFFT,Fs);
figure;clf
imagesc(1000*T,F/1000,abs(Cspec).^2);
colormap(flipud(hot))
axis xy
xlabel('Time [ms]');
ylabel('Frequency [kHz]');
title('Spectrogram of a chirp signal')
```

Notice that the linear variation of the signal frequency in time is clearly indicated in the figure below.

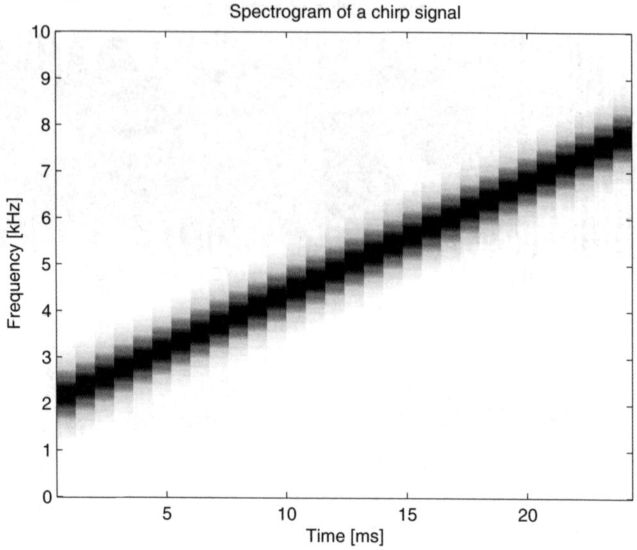

Figure 11.4. *Time-frequency representation of a chirp signal*

EXERCISE 11.2.

The spectrogram is the most traditional time-frequency analysis tool. However, it requires a trade-off between the time and frequency resolutions. Illustrate this spectrogram drawback using the procedure indicated below:

a. Generate a mixture of a Dirac pulse, which occurs at 26 ms, and a sinusoid with the frequency of 1 kHz, which occurs between 5 and 16 ms.

b. Plot the spectrogram of this signal for different lengths of the analysis window and comment on the results obtained.

a.

```
% Signal generation
Fs=10000; % Sampling frequency
f1=1000; Td=0.016;
t=[0.005:1/Fs:Td];
delta=0.005*Fs; % Time delay of 5ms between any couple of signal components
sig=[zeros(1,delta),    sin(2*pi*f1*t),    zeros(1,2*delta),    5*ones(1,1),
zeros(1,delta)];
```

b.

```
% Time-frequency analysis
figure; subplot(311);
```

```
plot(0:1e3*(1/Fs):1e3*(length(sig)/Fs-1/Fs),sig);
axis([0 31 -2 6]);
title('Analyzed signal');
xlabel('Time [ms]'); ylabel('Amplitude');
[S,F,T] = spectrogram(sig,hamming(64),0,128,Fs);
subplot(312);
imagesc(T*1000,F/1000,abs(S).^2);
colormap(flipud(hot));
axis xy; axis([0 31 0 5])
xlabel('Time [ms]');
ylabel('Frequency [kHz]');
title('Spectrogram with a window length of 64 points'),
[S,F,T] = spectrogram(sig,hamming(16),0,128,Fs);
subplot(313);
imagesc(T*1000,F/1000,abs(S).^2);
colormap(flipud(hot));
axis xy; axis([0 31 0 5])
xlabel('Time [ms]'); ylabel('Frequency [kHz]');
title('Spectrogram with a window length of 16 points')
```

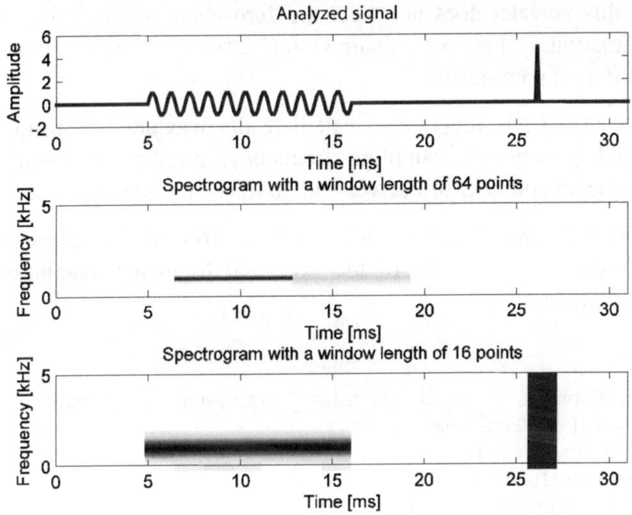

Figure 11.5. *Signal spectrogram calculated using two window lengths*

The above figure clearly illustrates the influence of the analysis window length. In fact, it can be seen that if the window is long, the time resolution is low. It is thus difficult to verify that the sinusoid occurs between 5 and 16 ms, and the Dirac pulse at 26 ms. However, the spectral resolution is high in this case and allows an accurate identification of the sinusoid frequency.

A short analysis window increases the time resolution, but significantly reduces the spectral resolution. Thus, a trade-off between the two resolutions is always required in the case of the spectrogram.

EXERCISE 11.3.

The scalogram is also an energy distribution obtained by taking the square of the continuous wavelet transform. Similarly to the spectrogram, it is also submitted to the Heisenberg-Gabor inequality involving the time and frequency resolutions.

a. Write a MATLAB function to generate a Morlet wavelet, having as input parameters its length, the central frequency and the sampling frequency.

b. The Morlet wavelet is defined by the following modulated Gaussian function:

$$h(t) = \left(\pi t_0^2\right)^{-1/4} \exp\left(-\frac{1}{2}\left(\frac{t}{t_0}\right)^2\right) \exp(-i2\pi \nu_0 t)$$

Although this wavelet does not meet the zero-mean admissibility constraint, a good approximation is however obtained for $2\pi \nu_0 t_0 = 5.4285$. Plot the Morlet wavelet for different parameters.

c. Write a MATLAB function to calculate the wavelet transform, having the following input parameters: sampling frequency, number of calculated points, maximum frequency, number of octaves and number of voices per octave.

d. Generate the sum of a Dirac pulse and two truncated sinusoids and plot its scalogram. Discuss the variation of the time and frequency resolutions over the time-frequency plane.

a.

```
function [waveform,ts] = ond_mor(wavelet_length,analyzed_freq,FE)
W0=5.4285; ts = [-n2:1:n2]/FE;
t0=W0/(2*pi*analyzed_freq);
n2 = (wavelet_length-1)/2;
wave = exp(-i*2 *pi*ts*analyzed_freq);
EnvGauss = exp(-0.5*(ts/t0).^2);
waveform=(pi*t0^2)^(-1/4).* wave.* EnvGauss; % Morlet wavelet
```

b.

```
[waveform,ts] = ond_mor(129,10,100);
figure; clf; subplot(211); plot(ts,real(waveform));
hold on; plot(ts,abs(waveform),'r');
grid; title('Morlet wavelet')
```

c.

```
% FE: sampling frequency
% NB_TIME: number of calculated points for the wavelet transform
% FMAX: maximum frequency
% NB_OCT: number of octaves
% NB_VOICES: number of voices per octave

function [TOnd_SIG, analyzed_freqs] = morlet(SIG, FE, NB_TIME, FMAX, NB_OCT,
NB_VOICES )
a0 = 2; W0=5.4285;
Sigma0 = W0/(2*pi*FMAX/FE );
NB = length(SIG);
total = NB_OCT*NB_VOICES-1;
TOnd_SIG = zeros(NB_OCT*NB_VOICES,NB_TIME );
% Calculation of the associated analytic signal
% Adding zeros allows shifting the signal instead of shifting the wavelet
function
length_max = fix(7*Sigma0*a0^(total/NB_VOICES)/2)+1;
ze = zeros(1,length_max);
ze = ze + ze.*i;
s = [ze hilbert(SIG) ze];
analyzed_freqs=[];
PT = length_max + 1;
for compt = 0:total
   a = a0^(compt/NB_VOICES );
   wavelet_length = fix(7*a*Sigma0 );
   if rem(wavelet_length,2 )==0
      wavelet_length = wavelet_length + 1;
   end;
   analyzed_freq = FMAX/a;
   analyzed_freqs = [analyzed_freqs,analyzed_freq];
   [Wavelet, vectime] = ond_mor(wavelet_length,analyzed_freq,FE);
   for index = 1:NB_TIME,
      signal=s(PT-(wavelet_length-1)/2:PT + (wavelet_length-1)/2);
      val = sum( signal .* Wavelet)/sqrt(a);
      TOnd_SIG(compt + 1,index) = val;
      PT = PT + NB/NB_TIME;
   end
   PT = length_max + 1;
end
```

d.

```
% Signal generation
Fs=64000;
f1=500; f2=8000; T=0.01;
t=[0:1/Fs:T];
delta=0.005*Fs;
sig1=[zeros(1,delta), sin(2*pi*f1*t), zeros(1,2*delta)];
sig2=[zeros(1,delta), sin(2*pi*f2*t), zeros(1,2*delta)];
```

```
sig=sig1 + sig2;
sig(4*delta+1:4*delta+5) = 10*ones(1,5);
subplot(212); plot(sig);
title('Analyzed signal: sum of 2 sinusoids and a Dirac pulse')
```

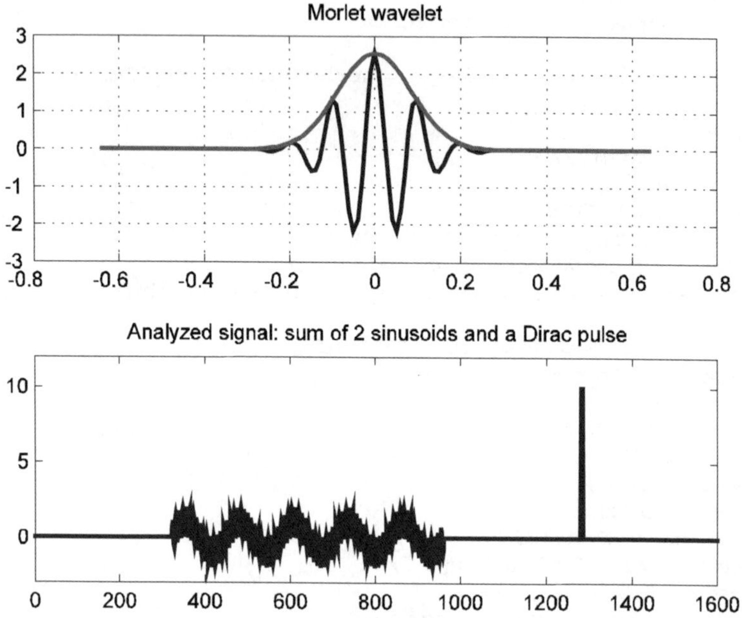

Figure 11.6. *Morlet wavelet (top) and analyzed signal (bottom)*

```
NB_OCT=8; NB_VOICES=1;
[TOnd_SIG,freqs]=morlet(sig,Fs,length(sig),Fs/2,NB_OCT,NB_VOICES);
vec_time=[0:1/Fs:length(sig)/Fs]*1000; vec_freqs=[0:1:7];

figure; clf; colormap(gray);
imagesc(vec_time,vec_freqs,2*log10(abs(TOnd_SIG)));
axis xy; set(gca,'YTickLabels',250*2.^([0:7]));
title('Scalogram of a Dirac pulse and 2 sinusoids');
xlabel('Time [ms]'); ylabel('Frequency [Hz]')
```

The two sinusoids seem to be detected with the same resolution. However, this is not true since the vertical axis is logarithmic. Thus, the rectangle corresponding to the sinusoid at 8 kHz covers a frequency band four times larger than the sinusoid at 500 Hz. As a general rule, the higher the frequency, the lower the spectral resolution.

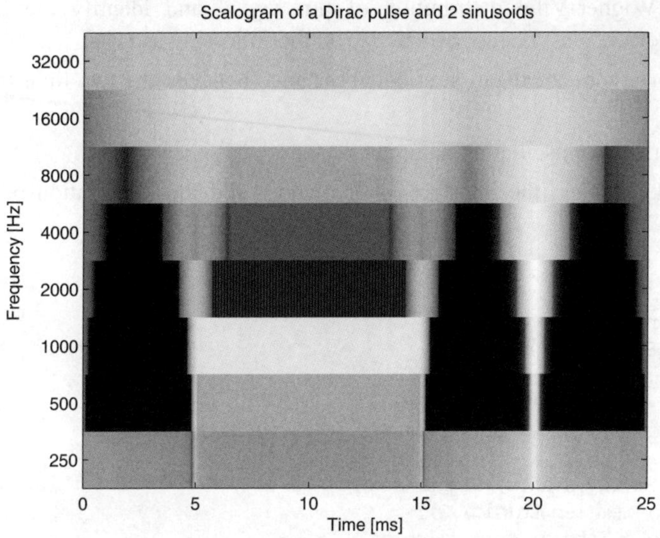

Figure 11.7. *Scalogram of a Dirac pulse and two sinusoids*

The wavelet transform of the Dirac pulse also illustrates the variation of the time resolution with the frequency. Indeed, as can be seen in the above figure, the lower the frequency, the better the time resolution. The wavelet transform is therefore well suited for detecting transient signals, which exhibit a large frequency band during a short period of time.

EXERCISE 11.4.

a. Calculate the Wigner-Ville distribution of a linear frequency-modulated signal, and verify the following properties:

– the WVD is real,

– the WVD contains negative terms,

– the WVD conserves the signal time and frequency supports,

– it is possible to obtain 1D energy distributions (spectral density and instantaneous power) as marginal distributions of the WVD.

b. Illustrate the main drawback of this distribution, related to the interference terms, which occur if the signal contains two or more atoms. Generate a mixture of three truncated sinusoids, two of them starting at time t_1, and having the frequencies f_1 and f_2, and the third one starting at time t_2, and having the frequency f_1.

Plot the Wigner-Ville distribution of this signal and identify the time and frequency interferences. Focus particularly on the following aspects:

– the interference localization at equal distance between the two time-frequency atoms;

– the interference oscillating nature;

– the link between the interference frequency and the localization of the two atoms.

a.

```
%% Signal generation
f1=2000; f2=8000; T=0.008;
Fs=20000; t=(0:1/Fs:T);
beta=(f2-f1)/(2*T); delta=0.001*Fs;
x=[zeros(1,delta), sin(2*pi*(f1+beta*t).*t), zeros(1,delta)];

function [W,Wr,fwv,x_axis]=wigner(sig,Fs)
N=length(sig)+rem(length(sig),2);
length_FFT=N; % Take an even value of N
if N~=length(sig); sig=[sig 0];end
length_time=length(sig);posi=length_time;
%% Generation of the associated analytic signal and adding zeros to allow
translating the sliding window, even on the signal borders
s=[zeros(1,length_time) hilbert(sig) zeros(1,length_time)];
s_conj=conj(s);
W=zeros(length_FFT,length_time);
tau1=linspace(0,N/2,N/2+1);
tau2=linspace(2,N/2,N/2-1);
%% Calculation of the Wigner-Ville distribution
for t=1:length_time,
   R(tau1+1,t)=(s(posi+tau1).*s_conj(posi-tau1)).';
   R(N+2-tau2,t)=conj(R(tau2,t)); posi=posi+1;
end
W=fft(R,length_FFT)/(2*length_FFT); Wr=real(W);
%%Plotting the results
x_axis=1e3*linspace(0,N*1/Fs,N);
fwv=linspace(0,Fs/2,N)/1e3;
f=linspace(0,Fs/2,N/2)/1e3;
Sig=abs(fft(sig));
figure; subplot(232)
plot(x_axis,sig); title('Time domain');
xlabel('time [ms]');
subplot(233); plot(f,Sig(1:N/2));
title('Spectral domain'); xlabel('frequency [kHz]');
subplot(234); plot(x_axis,Wr(N/2,:));
title('WVD frequency cross section'); xlabel('time [ms]');
subplot(231)
imagesc(x_axis,fwv,Wr); axis xy; colormap(flipud(hot))
title('Wigner-Ville distribution ');
```

```
xlabel('time [ms]'); ylabel('frequency [kHz]');
subplot(235); plot(x_axis,sum(W));
title('Marginal in time'); xlabel('time [ms]');
subplot(236); plot(fwv,(sum(W.')));
title('Marginal in frequency');
xlabel('frequency [kHz]');

% The function defined above is now applied to the signal
[W,Wr,x_axis]=wigner(x,Fs);
```

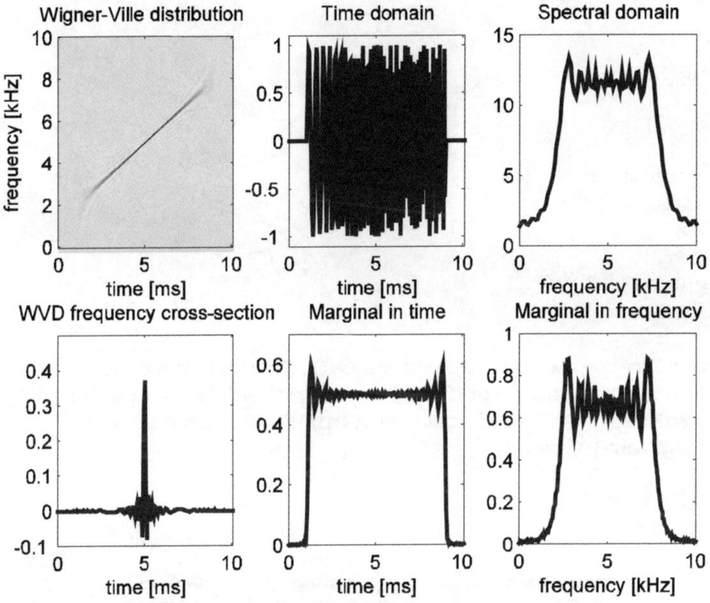

Figure 11.8. *Wigner-Ville distribution results*

Using the commands below it can be seen that the imaginary part of the WVD is zero (actually it is equal to the calculation accuracy).

```
max(max(real(W)))

ans = 0.3959

max(max(imag(W)))

ans = 6.5943e-015
```

The signal modulation type can be readily identified from its signal time-frequency representation. It is also clear from this image that the time support (1-9 ms) and the frequency support (2-8 kHz) are conserved.

The frequency cross-section highlights the negative values of the WVD. The last property can be easily verified by comparing the time and frequency marginals and the initial signal time and frequency representations.

Finally, the signal energy can be retrieved from its Wigner-Ville distribution as indicated below:

```
sum(sum(W))

ans = 79.9730 - 0.0000i

sum(x.^2)

ans = 79.9883
```

b.

```
% Signal generation
f1=1000; deltaf=3000;
f2=f1+deltaf;
Fs=round(3*f2);
T=0.01;t=[0:1/Fs:T];
deltat=0.01*Fs;
marge=20;
sig1=[zeros(1,marge),sin(2*pi*f1*t),zeros(1,T*Fs+deltat+marge)];
sig2=[zeros(1,marge),sin(2*pi*f2*t),zeros(1,T*Fs+deltat+marge)];
sig3=[zeros(1,marge + T*Fs+deltat),sin(2*pi*f1*t),zeros(1,marge)];
sig=sig1+sig2+sig3;

% WVD calculation
[W,Wr,fwv,x_axis]=wigner(sig,Fs); [N1,N2]=size(W);
```

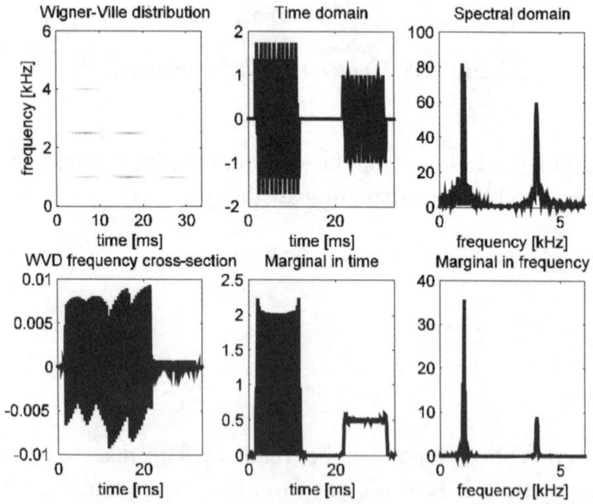

Figure 11.9. *Wigner-Ville distribution of the sum of three truncated sinusoids*

```
figure
subplot(311)
middlef = round(((f1+f2)/2)/(Fs/2)*N1);
plot(x_axis,Wr(middlef,:))
xlabel('time [ms]');
title('Interference illustration')
subplot(312)
middlet = round(marge + T*Fs + deltat/2);
plot(fwv,Wr(:,middlet));
xlabel('frequency [kHz]');
subplot(313)
index = round(marge + T/2*Fs);
plot(fwv,Wr(:,index));
xlabel('frequency [kHz]')
```

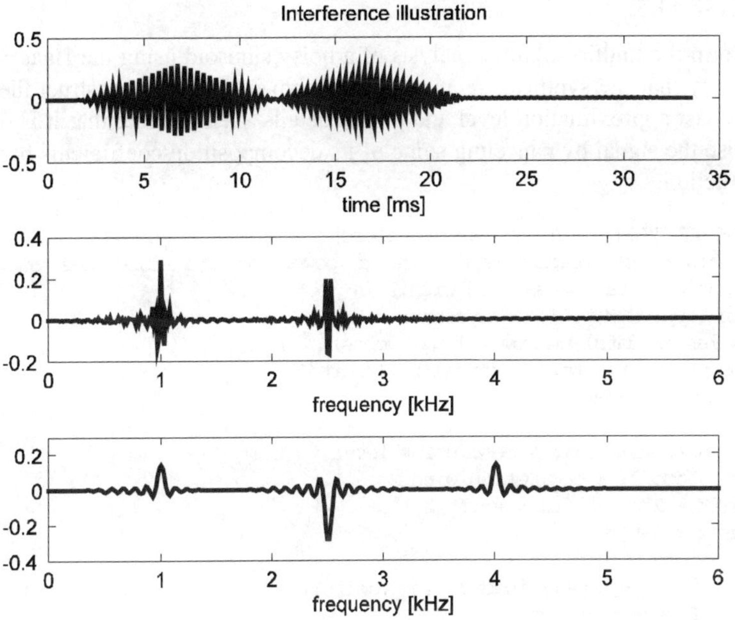

Figure 11.10. *Interferences related to the Wigner-Ville distribution*

Let us first have a look to the interferences which occur along the horizontal axis, around 2,500 Hz. This is the mean of the frequencies corresponding to two of the signal components (f_1 = 1,000 Hz, f_2 = 4,000 Hz). A cross-section along this axis (first curve) shows the oscillating nature of these interferences. The oscillating period is 0.33 ms, which is the inverse of the distance separating these two first atoms (3 kHz).

The two signal components having the same frequency (f_1), but delayed in time with 20 ms, lead to time interferences. Their period is the inverse of this time-delay (0.05 kHz) and can be read on the cross-section along the frequency axis (second curve).

The third interference occurs between the two sinusoids with different localizations both in time and frequency.

Finally, note that the sinusoids are not represented as straight lines in the time-frequency plane. In fact, they are bordered by several evanescent parallel lines. The third cross-section indicates a "sinc"-type variation of their amplitudes, which is the consequence of the sinusoid truncation.

EXERCISE 11.5.

Perform the multiresolution analysis of a noisy sinusoid using the Haar wavelet. Then verify that the synthesis reciprocal algorithm is able to reconstruct the signal from the last approximation level and all the details. Demonstrate that it is possible to denoise the signal by canceling some of its decomposition coefficients before the reconstruction.

```
% MULTIRESOLUTION ANALYSIS (Haar wavelets)
% The number of points should be a power of 2, since the successive
approximations are downsampled with 2
function [approx,detail]=analyzehaar(data)
numpts= length(data);numrows= log2(numpts);
top = numpts; ctr= numrows;normalize=1/sqrt(2);
apx(numrows+1,:)=data;
while ctr >= 1
    n = numrows-ctr+1; % resolution level
    top = top/2; % decimation with 2
    shift = 2^n; shift1 = shift/2;
    for k = 1:top,
        jump = shift*(k-1);
        index1=jump+1; index2= jump+shift1+1;
        firsttrm=apx(ctr+1,index1);
        secndtrm=apx(ctr+1,index2);
        apx(ctr,index1) = normalize*(firsttrm+secndtrm);
        d(ctr,index1) = normalize*(firsttrm-secndtrm);
        for j = 0:(shift-1),
            if j < shift1,
                detail(ctr,index1+j) = d(ctr,index1);
            else
                detail(ctr,index1+j) = -d(ctr,index1);
            end
            approx(ctr,index1+j) = apx(ctr,index1);
        end
    end
```

```
    ctr = ctr-1;
end

%%% SYNTHESIS ALGORITHM
function sig=synthesehaar(approx,detail)
% The reconstruction is performed from the last approximation level and all
the successively removed details
[numrows,numpts]=size(detail);normalize=1/sqrt(2);
newsig=zeros(size(approx));
newsig(1,:)=approx(1,:);
for ctr=1:numrows
   n= numrows-ctr+1;top = numpts/(2^n);shift  = 2^n;
   shift1 = shift/2;
  for k=1:top
    jump = shift*(k-1);index1=jump+1;index2= jump+shift1+1;
    firsttrm=newsig(ctr,index1);secndtrm=detail(ctr,index1);
    newsig(ctr+1,index1)=normalize*(firsttrm+secndtrm);
    newsig(ctr+1,index2)=normalize*(firsttrm-secndtrm);
  end
end
sig=newsig(size(newsig,1),:);

%%% Signal generation
clear; Fs=5000; wavelet_length=1024;
time=[1:1:wavelet_length]/Fs;
numpts=length(time);
data=sin(2*pi*25*time)+0.25*randn(1,wavelet_length);

%%% Analysis
[approx,detail]=analyzehaar(data);
%%% Reconstruction
newsig=synthesehaar(approx,detail);
%%% Signal denoising
selection=detail; selection(8:10,:)=zeros(3,numpts);
denoised_sig =synthesehaar(approx,selection);

%%% Plotting the results%%%%%%
figure; clf; zoom on
for plt = 1:size(detail,1),
    s=detail(plt,1:numpts);
    subplot(10,2,2*(plt-1)+1),plot(time,detail(plt,1:numpts));grid;
    axis([0 max(time) -1.2*max(abs(s)) 1.2*max(abs(s))])
    if (plt == 1),
       title('Detail');
    end;
    if (plt == 10);
       xlabel('time [s]');
    end
 end
for plt = 1:size(approx,1),
    s=approx(plt,1:numpts);
```

```
    subplot(10,2,2*plt),plot(time,approx(plt,1:numpts));grid
    axis([0 max(time) -1.2*max(abs(s)) 1.2*max(abs(s))])
    if (plt == 1),
        title('Approximation');
    end;
    if (plt == 10);
        xlabel('time [s]');
    end
end
```

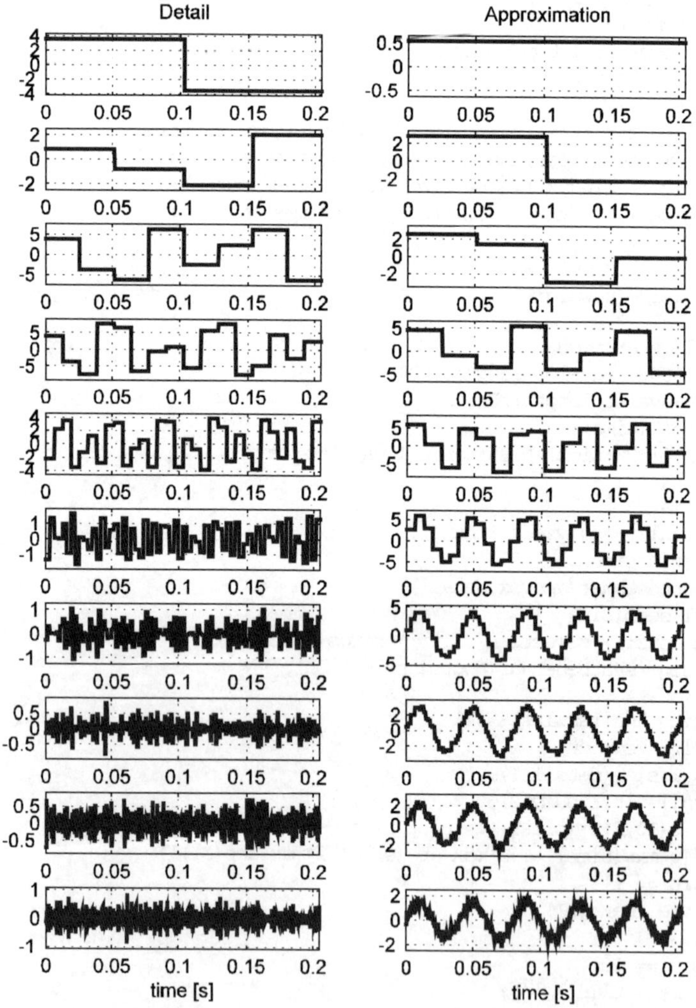

Figure 11.11. *Multiresolution analysis of a noisy sinusoid*

The approximation coefficients are related to the low frequencies and especially to the sinusoid, unlike the detail coefficients, which are related to the signal high frequency content. The last approximation level is a constant related to the signal DC component. It can be verified that this constant is equal to `mean(data)*sqrt(2)^10`, which results from the sum of the lowpass Haar filter coefficients ($\sqrt{2}$).

```
figure;clf;zoom on;
subplot(311);plot(time,data);title('Original signal');
axis([0 max(time) -1.2*max(data) 1.2*max(data)])
subplot(312);plot(time,newsig);title('Reconstructed signal');
axis([0 max(time) -1.2*max(newsig) 1.2*max(newsig)])
subplot(313);plot(time,denoised_sig);
axis([0 max(time) -1.2*max(denoised_sig) 1.2*max(denoised_sig)])
title('Denoised signal');xlabel('time [s]')
```

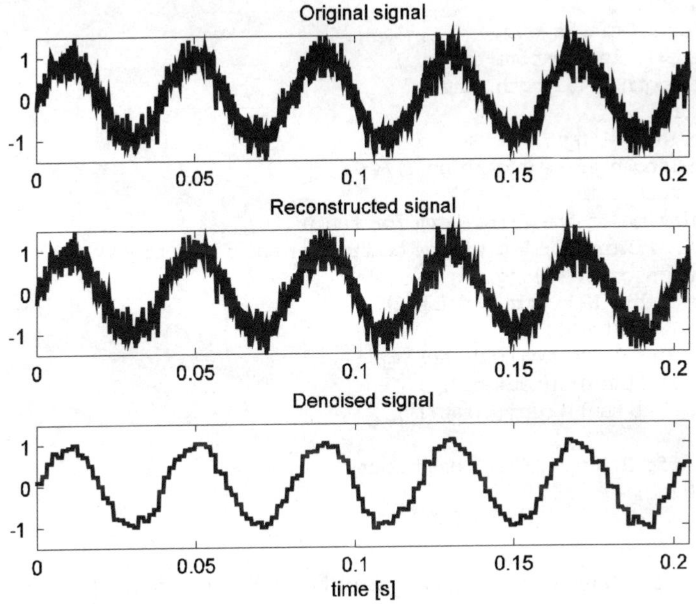

Figure 11.12. *Signal denoising by multiresolution analysis*

The signal reconstructed from the last level approximation and all the successively removed details is the same as the original signal. The denoised signal is not smoothed enough because the Haar wavelet is not appropriate for the continuous signal analysis, unlike some more complex wavelets (Daubechies wavelets for example). However, the Haar wavelet remains interesting for the analysis of some pulse signals, such as those used in digital communications.

11.3. Exercises

EXERCISE 11.6.

Consider again the signal generated in exercise 11.4, whose WVD contains interference terms. Write a MATLAB function to calculate the SWVD of this signal and choose the appropriate length for windows h and g. Verify the effect of these windows on the time and frequency resolutions. Note also the failure of the WVD theoretical properties, such as the time and frequency support conservation.

Note: modify the loop of the function `wigner` (see exercise 11.4) using the code lines below to obtain the new time-frequency distribution:

```
% Defining the windows
length_FFT=N;nh=N;NG=16;
g=hamming(2*NG+1)';
h=hamming(nh);
% Variables initialization
ECH=t_s;inc=1; length_time=t_s/inc;
WX=zeros(length_FFT,length_time);
A=zeros(1,ng);
X=linspace(0,ng-1,ng);
coef_norm=length_FFT*sum(g)*h(nh/2)/4;
for t=1:length_time,
  % Calculation of the first term for tau=0
  ind=X+ECH-fix(ng/2); % g window centred on the first signal value
  A=s(ind).*s_conj(ind).*g;
  R=zeros(1,nh); R(1)=sum(A)*h(nh/2)
  for tau=1:fix(nh/2);
    A=s(ind+tau).*s_conj(ind-tau).*g;
    R(tau+1)=sum(A)*h(tau+nh/2) ;
    R(fix(nh+1-tau))=conj(R(tau+1)) ;
  end
  WX(:,t)=fft(R,length_FFT)'/coef_norm;
  ECH=ECH+inc;
end
```

Other weighting windows may also be used (functions `boxcar`, `triang`, `hanning`, etc.). Remember that the multiplication by window h leads to a spectral smoothing and removes the frequency interferences if its length is smaller than the time distance between the time-frequency atoms. In the same way, window g performs a time domain smoothing and removes the time interferences if its length is smaller than the inverse of the frequency distance between the time-frequency atoms.

EXERCISE 11.7.

Consider two *chirps* $s_1(t)$ and $s_2(t)$, and denote by $f_1(t)$ and $f_2(t)$ the associated instantaneous frequencies. It can be demonstrated that the sum signal $s(t) = s_1(t) + s_2(t)$ has the instantaneous frequency $f(t)=(f_1(t)+f_2(t))/2$. Thus, unlike a time-frequency representation, the instantaneous frequency is not suitable to identify the two signal components.

Illustrate this limitation of the instantaneous frequency using such a signal, sum of two time-frequency atoms. Plot its Wigner-Ville distribution and show that it allows the two linear frequency modulations to be identified properly. Verify that the instantaneous frequency of $s(t)$ can be obtained as the associated analytical signal distribution mean.

EXERCISE 11.8.

The first step of the function `wigner`, proposed in exercise 11.4, consists of calculating the analytical signal associated with the considered real signal. In this way the spectral support is divided by 2 and the aliasing phenomenon is avoided. Illustrate the advantage of this procedure by comparing it with the direct calculation of the Wigner-Ville distribution (without the use of the analytical signal). Consider a real signal, containing several time-frequency atoms, sampled at the Nyquist frequency.

EXERCISE 11.9.

Verify that it is possible to obtain the spectrogram from the smoothed Wigner-Ville distribution. In fact, the equivalence between the two representations can be obtained using a separable time-frequency smoothing (see exercise 11.6), with Gaussian windows h and g.

EXERCISE 11.10.

Write a MATLAB code to calculate the time-frequency distributions associated with other kernels (Choï-Williams, Rihaczek, etc.) and verify their theoretical properties.

Chapter 12

Parametrical Time-Frequency Methods

12.1. Theoretical background

In many classification problems, salient features may be searched and extracted from the time-frequency plane. This process is optimized if the time-frequency representation (TFR) is matched to the processed signal.

For simple signals, the matched TFR may be one of the Cohen's class representations. For more complicated signals the construction of the matched TFR is a difficult problem, since the standard kernels do not preserve some properties required by the optimization procedure.

This chapter deals with some recently proposed approaches for constructing the matched TFR corresponding to a given signal. The optimization procedure aims to cancel the interference terms and to obtain the best time-frequency resolution (in an equivalent manner to the chirp representation with the Wigner-Ville distribution).

12.1.1. *Fractional Fourier transform*

The a-order fractional Fourier transform (FRFT) is defined, for $0 < |a| < 2$, by means of its kernel:

$$\begin{aligned}
&\left\{\Im^a f\right\}(t_a) = \int_{-\infty}^{\infty} K_a(t_a,t)\, f(t)\, dt \\
&K_a(t_a,t) = K_\phi \exp\left[j\pi\left(t_a^2 \cot\phi - 2t_a t\right)\csc\phi + t^2 \cot\phi \right]
\end{aligned} \qquad [12.1]$$

where $\phi = a\pi/2$ and $K_\phi = \exp\left[-j\left(\pi\,\mathrm{sgn}(\phi)/4 - \phi/2\right)\right] \Big/ \left|\sin(\phi)\right|^{0.5}$.

For $a = 0$ and $a = \pm 2$ the kernel $K_a(t_a, t)$ is defined by $K_0(t_a, t) = \delta(t_a - t)$ and $K_{\pm 2}(t_a, t) = \delta(t_a + t)$. This definition can be extended to orders higher than 2 or lower than -2, using the following FRFT periodicity property:

$$\left\{\Im^{4l+a} f\right\}(t_a) = \left\{\Im^{a} f\right\}(t_a) \qquad [12.2]$$

for any integer l.

The FRFT physical interpretation as the counter-clockwise rotation with the angle ϕ of the time-frequency coordinate system, is shown in Figure 12.1.

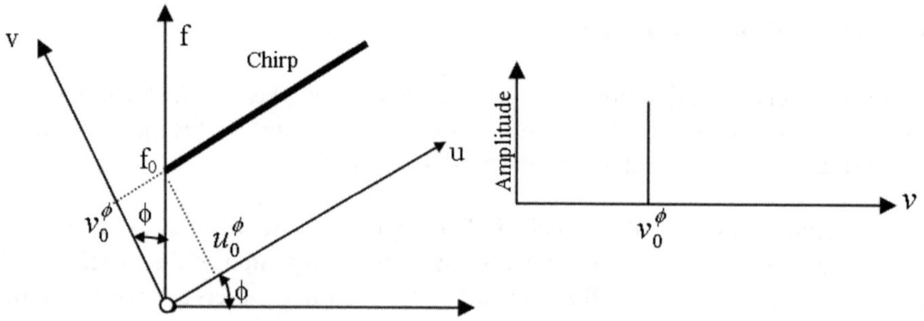

Figure 12.1. *Fractional Fourier transform*

The FRFT verifies the following properties:

a. Unitarity

This property insures the exact and univocal inverse FRFT transform:

$$\left(\Im^{a}\right)^{-1} = \left(\Im^{-a}\right) = \left(\Im^{a}\right)^{*} \qquad [12.3]$$

where "*" denotes here the Hermitian operator.

b. Additivity

This property is expressed as follows:

$$\Im^{a_1}\Im^{a_2} = \Im^{a_2}\Im^{a_1} = \Im^{a_1 + a_2} \qquad [12.4]$$

12.1.2. *Phase polynomial analysis concept*

The signals having non-linear time-frequency content, such as non-linear frequency modulations, can be characterized using a polynomial modeling[1] of their phase function, as indicated below:

$$y(t) = A\exp\left[j\phi(t)\right] \cong A\exp\left[j\sum_{k=0}^{K} a_k t^k\right] \qquad [12.5]$$

where $\{a_k\}$ stand for the polynomial coefficients and K is the approximation order of the signal phase $\phi(t)$.

The polynomial modeling of a signal consists of estimating the associated $\{a_k\}$ coefficients, in order to obtain the best approximation of the non-linear phase function. Compared to the deformation based time-frequency representations, which have to adapt to each type of non-linearity, the polynomial modeling tools are more adequate, since they can be similarly used for almost any signal of interest.

The currently used approach to estimate the parameters of a polynomial phase signal (PPS) is based on the high order ambiguity function (HAF) and has the following shortcomings:

1. the masking effects for low signal-to-noise ratios,

2. the presence of interferences in the case of multi-component PPS (mc-PPS).

Different methods have been recently proposed to overcome these limitations. The main idea consists of using the *multi-lags* concept to calculate the HAF.

High order ambiguity function

The high order ambiguity function has been initially proposed to estimate the parameters of a polynomial phase signal expressed according to equation [12.5]. $y(t)$ corresponds to a sinusoid for $K = 1$, to a linear *chirp* for $K = 2$ or to a cubic frequency modulation for $K = 3$.

If integer q counts how many times the conjugate operator is applied then:

$$y^{(*q)}(t) = \begin{cases} y(t), & \text{if } q \text{ is even} \\ y^*(t), & \text{if } q \text{ is odd} \end{cases} \qquad [12.6]$$

1 This polynomial modeling is based on the Weierstrass theorem, which states that any non-linear function can be decomposed as a polynomial series.

The Kth order *high order instantaneous moment* (HIM) is defined by the following relationship:

$$M_K\left[y(t);\tau\right]=\prod_{q=0}^{K-1}\left[y^{(*q)}(t-q\tau)\right]^{C_{K-1}^{q}} \qquad [12.7]$$

where:

$$C_{K-1}^{q}\equiv\binom{K-1}{q}=\frac{(K-1)!}{q!(K-1-q)!} \qquad [12.8]$$

are the binomial coefficients.

The main HIM properties are given below:

1. the HIM can be iteratively calculated using:

$$M_K\left[y(t);\tau\right]=M_2\left[M_{K-1}\left[y(t);\tau\right];\tau\right] \qquad [12.9]$$

2. the HIM calculation is equivalent to differentiating the PPS phase;
3. the 2nd order HIM reduces a Kth order PPS to a $(K\text{-}1)^{th}$ order PPS;
4. for $y(t)$ defined by equation [12.5] the following relationship holds:

$$M_k\left[y(t);\tau\right]=A^{2^{K-1}}\exp\left[j\left(\tilde{\omega}_k\cdot t+\tilde{\phi}_k\right)\right] \qquad [12.10]$$

where:

$$\tilde{\omega}_k=k!\tau^{k-1}a_k \qquad [12.11]$$

$$\tilde{\phi}_k=(k-1)!\tau^{k-1}a_{k-1}-0.5k!(k-1)\tau^k a_k \qquad [12.12]$$

The above equation means that the Kth order HIM applied to a Kth order PPS is a complex exponential having a constant amplitude $A^{2^{K-2}}$, a frequency $\tilde{\omega}$ and a phase $\tilde{\phi}$. The Kth order ambiguity function of signal $y(t)$ is defined by the Fourier transform of $M_K[y(t);\tau]$:

$$P_K\left[y;\alpha,\tau\right]=\int_{-\infty}^{\infty}M_K\left[y(t);\tau\right]e^{-j\alpha t}dt \qquad [12.13]$$

It allows the polynomial coefficients corresponding to the PPS $y(t)$ to be estimated as indicated below:

$$a_K = \frac{1}{K!\tau^{K-1}} \arg\max_{\alpha} \left| P_K \left[y;\alpha,\tau \right] \right| \qquad [12.14]$$

This result means that a constant amplitude signal can be characterized by a Kth order polynomial phase if its Kth order HAF is a pure spectral component for any positive τ. The HAF is thus an appropriate tool for the PPS analysis, since all the coefficients of the phase function can be estimated by its successive application.

The discrete HIM and the discrete HAF are defined by the following relationships:

$$DHIM_K = DM_K = \prod_{q=0}^{K-1} \left[y_{n-q\tau}^{(*q)} \right]^{C_{K-1}^{q}} \qquad [12.15]$$

$$DHAF_K = DP_K \left[y;\alpha,\tau \right] = \sum_{n=-\infty}^{\infty} DM_K \left[y_n;\tau \right] e^{-j\alpha\Delta n} \qquad [12.16]$$

where τ is a positive integer and Δ stands for the sampling period.

In the case of finite length signals, the sum in equation [12.16] is finite and the discrete HAF calculation is performed by FFT.

COMMENTS

a. The choice of τ is arbitrary, but it is decisive for the accuracy of the coefficients estimation. A small value of τ increases the number of admissible frequency values, but decreases the resolution. The asymptotic analysis of the algorithm performance as a function of τ was carried out by Peleg. He stated that the lowest asymptotic variance of $\hat{a}_K$ is obtained for:

– $\tau = N/K$ if K = 2 or 3,

– $\tau = N/(K+2)$ if $4 \leq K \leq 10$.

The last result is based on a numerical evaluation of a certain function and has not been demonstrated for any K so far. In the same way, the optimal choice of τ for $k < K$ has not been analytically determined yet. Nevertheless, many simulations indicated that the choice $\tau = N/k$ is close to the optimal value. Furthermore, some authors state that this choice yields the best estimation accuracy for the signal polynomial coefficients.

b. Since the discrete HAF is periodic with the period $2\pi/\Delta$, coefficient a_k can be unambiguously expressed if and only if:

$$|a_k| < \frac{\pi}{k!\Delta(\Delta\tau)^{k-1}} \qquad [12.17]$$

This constraint results in a Nyquist type condition on the value of Δ.

Multi-lags HAF concept (ml-HAF)

For a Kth order PPS, $x(t)$, the Kth order ml-HIM is defined in the following sequential manner:

$$\begin{aligned} &\mathbf{x}_1(t) = x(t) \\ &\mathbf{x}_2(t;T_1) = \mathbf{x}_1(t+\tau_1)\mathbf{x}_1^*(t-\tau_1) \\ &\vdots \\ &\mathbf{x}_K(t;T_{K-1}) = \mathbf{x}_{K-1}(t+\tau_{K-1};T_{K-2})\mathbf{x}_{K-1}^*(t-\tau_{K-1};T_{K-2}) \end{aligned} \qquad [12.18]$$

where:

$$T_i = (\tau_1, \tau_2, ..., \tau_i) \qquad [12.19]$$

It can be seen that the ml-HIM is equal to the HIM if $\tau_1 = \tau_2 = ... = \tau_{K-1} = \tau$. If the signal $s(t)$ contains M K^{th} order PPS:

$$s(t) = \sum_{k_1=1}^{M} A_{k_1} \exp\left(j2\pi \sum_{k=0}^{K} a_{k_1,k} t^k \right) \qquad [12.20]$$

then the corresponding ml-HIM is also a mc-PPS:

$$\mathbf{s}_K(t;T_{K-1}) = \sum_{k_1,...,k_{2^{K-1}}=1}^{M} A_{k_1} \cdots A_{k_{2^{K-1}}} \exp\left(j2\pi \sum_{k=0}^{K} b_{k_1,...,k_{2^{K-1}};k}^{(K)} t^k \right) \qquad [12.21]$$

Coefficients $b_{k_1,...,k_{2^{K-1}};k}^{(K)} t^k$ are calculated using following iterative rule:

$$b_{k_1;k}^{(1)} = a_{k_1;k}, \quad k = 0..K \qquad [12.22a]$$

$$b_{k_1,k_2;k}^{(2)} = \sum_{l=0}^{K-k} C_{k+l}^{k} \left[b_{k_1;k+l}^{(1)} - (-1)^l \, b_{k_2;k+l}^{(1)} \right] \tau_1^l \qquad [12.22b]$$

$$b^{(K)}_{k_1,\dots,k_{2^{K-1}};k} = \sum_{l=0}^{K-k} C^k_{k+l}\left[b^{(K-1)}_{k_1,\dots,k_{2^{K-2}};k+l} - (-1)^l\, b^{(K-1)}_{k_{2^{K-2}+1},\dots,k_{2^{K-1}};k+l}\right]\tau^l_{K-1} \qquad [12.22c]$$

The coefficients expressed by [12.22a] correspond to ($k_1 = k_2 = \dots = k_{2^{K-1}}$) and are called *auto-terms*, while the others are called *interference or cross-terms*.

For example, for K = 2, the 2nd order ml-HIM is the sum of quadratic phase signals whose 1st and 2nd order coefficients are given by:

$$\begin{aligned} b^{(2)}_{k_1,k_2;2} &= b^{(1)}_{k_1;2} - b^{(1)}_{k_2;2} \\ b^{(2)}_{k_1,k_2;1} &= b^{(1)}_{k_1;1} - b^{(1)}_{k_2;1} + 2\left(b^{(1)}_{k_1;2} + b^{(1)}_{k_2;2}\right)\tau_1 \end{aligned} \qquad [12.23]$$

The 2nd order ml-HIM is then expressed as:

$$\begin{aligned} s_2(t;T_1) = {} & \underbrace{A_1^2 \exp\left(j2\pi \sum_{k=0}^{2} b^{(2)}_{1,1;k} t^k\right) + A_2^2 \exp\left(j2\pi \sum_{k=0}^{2} b^{(2)}_{2,2;k} t^k\right)}_{\text{auto-terms}} \\ & \underbrace{+ A_1 A_2 \exp\left(j2\pi \sum_{k=0}^{2} b^{(2)}_{2,1;k} t^k\right) + A_1 A_2 \exp\left(j2\pi \sum_{k=0}^{2} b^{(2)}_{1,2;k} t^k\right)}_{\text{cross-terms}} \end{aligned} \qquad [12.24]$$

The polynomial coefficients involved in equation [12.24] are indicated in the table below.

	k=0	***k=1***	***k=2***
$b^{(2)}_{1,1;k}$	$2a_{11}\tau_1$	$4a_{12}\tau_1$	0
$b^{(2)}_{2,2;k}$	$2a_{21}\tau_1$	$4a_{22}\tau_1$	0
$b^{(2)}_{2,1;k}$	$a_{10} - a_{20} + (a_{11} + a_{21})\tau_1 + (a_{12} - a_{22})\tau_1^2$	$(a_{11} - a_{21}) + 2(a_{12} + a_{22})\tau_1$	$a_{12} - a_{22}$
$b^{(2)}_{1,2;k}$	$a_{20} - a_{10} + (a_{11} + a_{21})\tau_1 + (a_{22} - a_{12})\tau_1^2$	$(a_{21} - a_{11}) + 2(a_{12} + a_{22})\tau_1$	$a_{22} - a_{12}$

Table 12.1. *The polynomial coefficients of the 2nd order ml-HIM for a PPS having 2 components*

The following notes can be made using the above equations:

a. A 2nd order coefficient of the auto-terms is zero, while the 1st order coefficients are given by $b^{(2)}_{k_1,k_2;1} = 4a_{k_1;2}\tau_1$. Consequently, the auto-terms are represented by spectral peaks.

b. The cross-terms generally have a 2nd order instantaneous phase if $a_{k_1;2} \neq a_{k_2;2}$ for $k_1 \neq k_2$. However, the cross-terms are also sinusoids if the 2nd order coefficients are the same ($a_{k_1;2} = a_{k_2;2}$ for $k_1 \neq k_2$). In this case, the cross-terms are also represented by spectral peaks localized at $b^{(2)}_{k_1,k_2;1} = b^{(1)}_{k_1;1} - b^{(1)}_{k2;1} + 4b^{(1)}_{k_1;2}\tau_1$.

The useful peaks (associated with the auto-terms) are related to the time-delays or lags by the following relationship:

$$f = 2^{K-1} K! \left(\prod_{i=1}^{K-1} \tau_i \right) a_K \qquad [12.25]$$

The cross-terms can thus be canceled using several lag values. For $K = 2$ or 3, the optimal lag is $\tau = N / K$, where N stands for the number of samples and K is the polynomial order.

12.1.3. *Time-frequency representations based on warping operators*

The family of modulation operators is formed by all the linear operators $\mathbf{M}_\beta$ defined on a Hilbertian space as indicated below:

$$\left(\mathbf{M}_\beta s\right)(t) = e^{j2\pi\beta m(t)} s(t) \qquad [12.26]$$

where $m(t)$ represents the irreversible modulation function, β stands for the modulation rate and s is the analyzed signal.

The eigenfunctions and eigenvalues of the operator $\mathbf{M}_\beta$ are given by:

$$\begin{aligned} u_a^{\mathbf{M}_\beta}(t) &= \sqrt{\left|m'(t)\right|}\,\delta\left(m(t) - a\right) \\ \lambda_a^{\mathbf{M}_\beta} &= e^{j2\pi\beta a} \end{aligned} \qquad [12.27]$$

where a denotes the variable of the new domain from the application of this operator.

The family of *warping operators* is formed by all linear operators $\mathbf{W}_\alpha$ defined on a Hilbertian space as indicated below:

$$(\mathbf{W}_\alpha s)(t) = \sqrt{w'_\alpha(t)}\, s\left(w_\alpha(t)\right) \qquad [12.28]$$

where $w_\alpha(t)$ represents the warping function corresponding to the time axis.

This function depends on parameter α and is assumed differentiable, irreversible and satisfactory for the composition property:

$$w_{\alpha_1}\left(w_{\alpha_2}(t)\right) = w_{\alpha_1 \bullet \alpha_2}(t) \qquad [12.29]$$

where • is the operation corresponding to the composition rule: $\mathbf{A}_{\alpha_1}\mathbf{A}_{\alpha_2} = \mathbf{A}_{\alpha_1 \bullet \alpha_2}$.

If $m(t)$ is an irreversible and differentiable modulation function, then the associated warping function is given by:

$$w_\alpha(t) = m^{-1}\left(m(t) \bullet \alpha^{-1}\right) \qquad [12.30]$$

This result allows the unambiguous design of the warping operator associated with a given modulation function. The design methodology steps will be described in the following.

Warping function calculation

Warping function $w_\alpha(t)$ and warping operator $\mathbf{W}_\alpha$ are calculated using function $m(t)$ and equations [12.30] and [12.28].

Prewarping

This step consists of changing the time or frequency axis using the previously designed operator. The application of a TFR from the Cohen class leads to a new joint distribution, which represents the signal in the time-frequency plane $(\tilde{t}, \tilde{f})$:

$$\begin{aligned} &\tilde{t} = w_\alpha(t) \;\; ; \;\; \tilde{f} = f\dot{w}_\alpha^{-1}\left(w_\alpha(t)\right) - \text{time operator} \\ &\tilde{t} = t\dot{\hat{w}}_\alpha^{-1}\left(\hat{w}_\alpha(f)\right) \;\; ; \;\; \tilde{f} = \hat{w}_\alpha(f) - \text{frequency operator} \end{aligned} \qquad [12.31]$$

The application of a TFR from the Cohen class to the prewarped signal leads to the linearization of its instantaneous frequency in the $(\tilde{t}, \tilde{f})$ plane.

Unwarping

The result obtained after the previous step is a two-dimensional distribution, without cross-terms, of a linear structure. The matched non-linear TFR corresponding to the analyzed signal can be obtained using the inverse transformation indicated below:

$$(a,b)=\Gamma^{-1}\left(\tilde{t},\tilde{f}\right)$$

$$a=w_{\alpha}^{-1}\left(\tilde{t}\right);b=\frac{\tilde{f}}{\dot{w}^{-1}\left(\tilde{t}\right)} \quad \text{– time unwarping} \qquad [12.32]$$

$$a=\frac{\tilde{t}}{\dot{\hat{w}}^{-1}\left(\tilde{f}\right)};b=\hat{w}_{\alpha}^{-1}\left(\tilde{f}\right) \quad \text{– frequency unwarping}$$

Application to the hyperbolic TFR class

This class is represented by a non-linear TFR set, derived from the Altes-Marinovich distribution, which was designed to match the hyperbolic chirp class. This signal class is defined by:

$$H_c(f)=\frac{1}{\sqrt{f}}\exp\left(-j2\pi c\ln\frac{f}{f_r}\right)u(f) \qquad [12.33]$$

where $u(f)$ stands for the Heaviside function and f_r is a fixed "reference" frequency chosen so that the Nyquist condition is met. A hyperbolic chirp is characterized by an energy spectral density $\left|H_c(f)\right|^2=u(f)/f$ and a group delay $t=c/f$.

Let us consider f_r = 1 for simplicity. The frequency warping operator, which matches H_c and transforms it into a stationary structure, is then obtained as follows:

$$\left(\hat{\mathbf{W}}_{\mathbf{log}}H_c\right)(f)=\sqrt{\left|\hat{w}'_{\log}(f)\right|}H_c\left(\hat{w}_{\log}(f)\right)=e^{-j2\pi cf}$$

$$\Leftrightarrow e^{-j2\pi c\ln\hat{w}_{\log}(f)}=e^{-j2\pi cf}\Rightarrow\hat{w}_{\log}(f)=e^{f} \qquad [12.34]$$

Thus, if this operator is used for the prewarping step, a Dirac pulse is obtained in the time domain. The application of the DWV to the prewarped signal leads to the linear time-frequency structure plotted in Figure 12.2b.

The hyperbolic TFR class can be obtained from this distribution using the inverse transform (unwarping). Therefore, a non-linear (hyperbolic) TFR matching signal h_c (hTFR) (Figure 12.2c) in the plane $(\tilde{t}, \tilde{f})$ is obtained. The axes of the latest TFR are obtained as a particular case of equation [12.31] for $\hat{w}_{\log}(f)$.

$$\begin{aligned} \tilde{t} &= t\exp(\ln f) = tf \\ \tilde{f} &= \ln f \end{aligned} \qquad [12.35]$$

Thus, the Atlas-Marinovich distribution (AMD) can be also defined by the following equation:

$$AMD_x(t,f) = DWV_{\mathbf{W}_{\log}x}(tf, \ln f) \qquad [12.36]$$

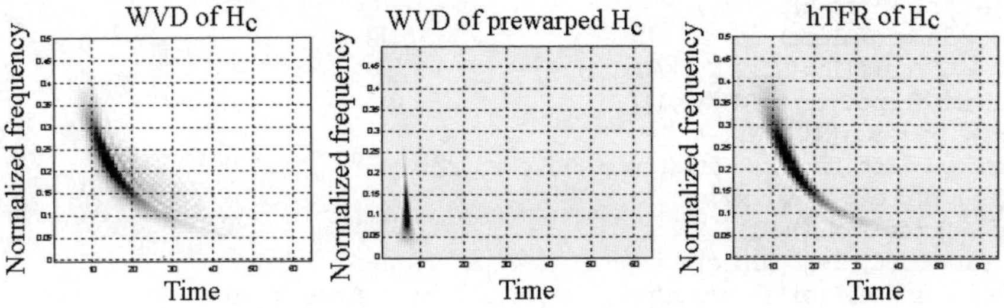

Figure 12.2. *Illustration of the hTFR construction*

The general form of the hyperbolic TFRs class can be obtained if a generic Cohen class distribution replaces the DWV in equation [12.36]. The hyperbolic TFRs have the same properties as the Cohen class TFRs.

The most important property of the hTFR is that it is perfectly focused on the group delay *c/f* of signals h_c (given by the inverse Fourier transform of the H_c function):

$$hTFR_{h_c}(t,f) = \frac{1}{f}\delta\left(t - \frac{c}{f}\right) \qquad [12.37]$$

12.2. Solved exercises

EXERCISE 12.1.

Write a MATLAB function to calculate the FRFT of a signal for a given order.

```
function FRFT = FracFR(x0,a);
% Inputs: x0   - Original signal
```

```
%           a     - FRFT order
% Output: FRFT - Transformed signal
deltax = sqrt(length(x0)/2);
phi= a*pi/2; N = fix(length(x0)/2);
deltax1 = 2*deltax;
alpha = 1/tan(phi);
beta = 1/sin(phi);
% Kernel implementation
T = [-N:N-1]/deltax1;
T = T(:);x0=x0(1:2*N);
f1 = exp(-i*pi*tan(phi/2)*T.*T);
f1 = f1(:);x = x0.*f1;clear T;
t = [-2*N+1:2*N-1]/deltax1;
hlptc = exp(i*pi*beta*t.*t);
clear t; hlptc=hlptc(:);
N2 = length(hlptc);
N3 = 2^(ceil(log(N2+2*N-1)/log(2)));
hlptcz=[hlptc;zeros(N3*N2,1)];
xz = [x;zeros(N3-2*N,1)];
Hcfft = ifft(fft(xz).*fft(hlptcz));
clear hlptcz; clear xz;
Hc = Hcfft(2*N:4*N-1);
clear Hcfft;clear hlptc;
Aphi = exp(-i*(pi*sign(sin(phi))/4-phi/2))/sqrt(abs(sin(phi)));
xx = [-N:N-1]/deltax1;
f1 = f1(:); FRFT = (Aphi*f1.*Hc)/deltax1;
```

EXERCISE 12.2.

This exercise aims to estimate the linear modulation rate of a chirp signal using the FRFT.

The MATLAB code below generates the analyzed signal.

```
t=0:255;
x0=[ zeros(1,128),exp(j*2*pi*(.1*t+.0007*t.^2)),zeros(1,128)].';
```

The instantaneous frequency law and the modulation rate corresponding to this signal are calculated as follows:

```
LFI=.1+.0014*t;
c=4*(LFI(256)-LFI(1)) ; % the modulation rate
plot(t,LFI);
xlabel('Temps');
ylabel('Normalized frequency');
title('Instantaneous frequency law');
```

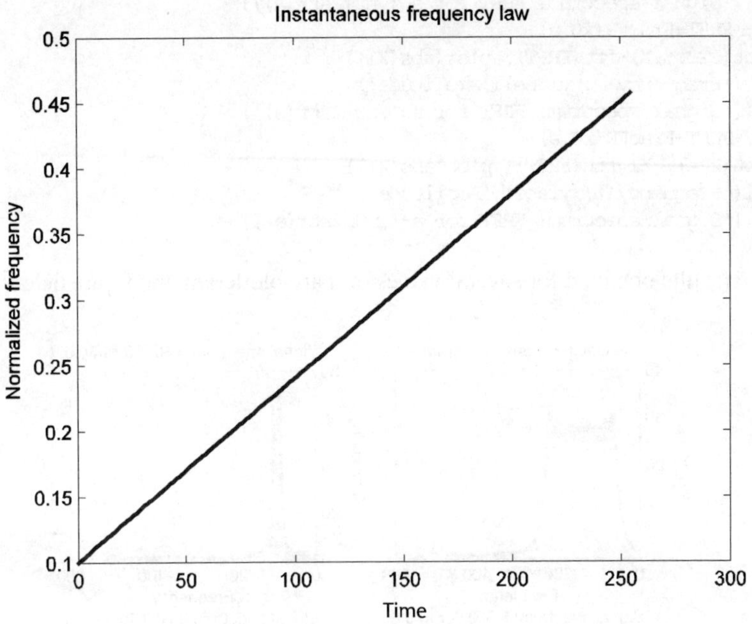

Figure 12.3. *Instantaneous frequency law for a chirp signal*

The modulation rate estimate can be found by applying the FRFT to the signal, for different values of its order a. Thus, if $\hat{a}$ is the order leading to the best concentration of the spectral energy for the signal obtained by FRFT, then the modulation rate estimate is given by $c = \tan(\hat{a}\pi / 2)$.

```
figure
for a=0:.01:1
     disp('Order value a = '); a
     c=tan(a*pi/2) % modulation rate value
     FRFT=FracFR(x0,a);
     X1=fft(FRFT); plot(abs(X1))
     xlabel('Frequency'); ylabel('Amplitude')
     title(['Transformed signal spectrum: FRFT for a=',num2str(a)])
     pause
end
FRFT=FracFR(x0,a);
subplot(221); X=fft(x0); plot(abs(X))
xlabel('Frequency');ylabel('Amplitude')
title('Original signal spectrum')
a=0.5; FRFT=FracFR(x0,a);
subplot(222); X1=fft(FRFT); plot(abs(X1))
xlabel('Frequency'); ylabel('Amplitude')
```

```
title(['Signal spectrum: FRFT for a=',num2str(a)])
a=0.6; FRFT=FracFR(x0,a);
subplot(223); X1=fft(FRFT); plot(abs(X1))
xlabel('Frequency'); ylabel('Amplitude')
title(['Signal spectrum: FRFT for a=',num2str(a)])
a=0.8; FRFT=FracFR(x0,a);
subplot(224); X1=fft(FRFT); plot(abs(X1))
xlabel('Frequency'); ylabel('Amplitude')
title(['Signal spectrum: FRFT for a=',num2str(a)])
```

The results obtained for several values of *a* are plotted on the figure below.

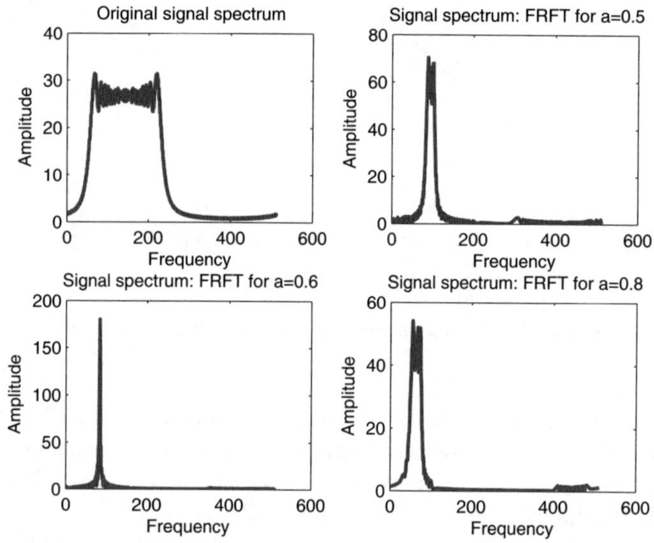

Figure 12.4. *Application of the fractional Fourier transform*

Note that the best concentration of the spectral energy is obtained for $a = 0.6$. This value corresponds to a modulation rate of 1.3764, which is close to the theoretical value, 1.42. An even better estimation accuracy could be obtained using more values for *a*.

EXERCISE 12.3.

Write a MATLAB function to calculate the Mth order moment of an input signal, for a given lags vector.

```
function h=him(x,l,M);
% Inputs:
% x - the signal
```

```
% l - lags vector
% M - HIM order
% Output:
% h - M-th order HIM
h = x;
for i=2:M
    hpp= [h,zeros(1,l(i-1))];
    hmm= [zeros(1,l(i-1)),h];
    h=hpp.*conj(hmm);
end
```

EXERCISE 12.4.

Estimate the polynomial coefficients of the signal below.

$$s(t) = \exp\left[j2\pi\left(0.25t - 9.76\cdot 10^{-4} t^2 + 3.204\cdot 10^{-6} t^3\right)\right], t = 0..255$$

```
% Signal generation
t=0:255;
a=[0.25,-9.76*10^-4,3.204*10^-6 ];
s=exp(j*2*pi*(a(1)*t+a(2)*t.^2+a(3)*t.^3));
LFI=a(1)+2*a(2)*t+3*a(3)*t.^2;
plot(LFI);
title('Instantaneous frequency law');
axis([1 256 0 0.5]);
xlabel('Time');
ylabel('Normalized frequency')
```

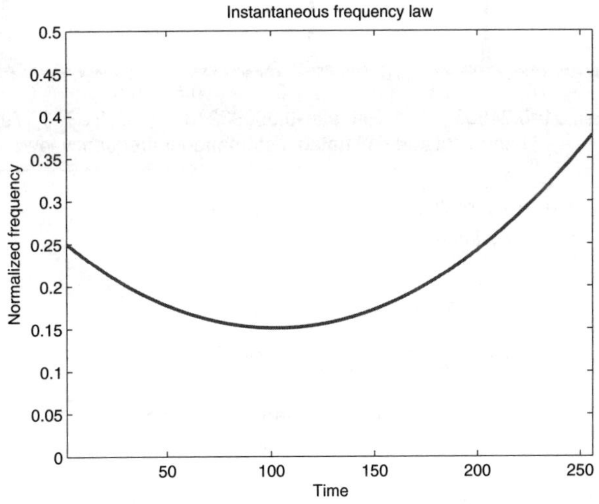

Figure 12.5. *Instantaneous frequency law of the signal from exercise 12.4*

```
% Estimation of the polynomial coefficients
N = length(s); K = 3;
Nfft=2^12; F=-0.5:1/Nfft:0.5-1/Nfft;
a_est=zeros(1,K); tau = N/K ;
for i=K:-1:1
        h=him(s,tau*ones(1,3),i);
        hf=abs(fft(h,Nfft));
        hf=[hf(length(hf)-Nfft/2+1:length(hf)),hf(1:Nfft/2)];
        subplot(2,K,i);plot(F,hf);
        title(['HAF: order ',num2str(i)])
        [ym,zm]=max(hf); Fmax=(zm-Nfft/2-1)/Nfft;
        a_est(i)=Fmax/(factorial(i)*tau^(i-1));
        xlabel(['Est. a',num2str(i),'=',num2str(a_est(i))]);
          s=s.* exp(-j*2*pi*a_est(i)*t.^(i));
end
LFI_est=a_est(1)+2*a_est(2)*t+3*a_est(3)*t.^2 ;
subplot(212); plot(LFI,'--');
hold on; plot(LFI_est,'r');
legend('Theoretical','Estimated');
xlabel('Time'); ylabel('Normalized frequency')
title('Theoretical and estimated instantaneous frequency laws')
```

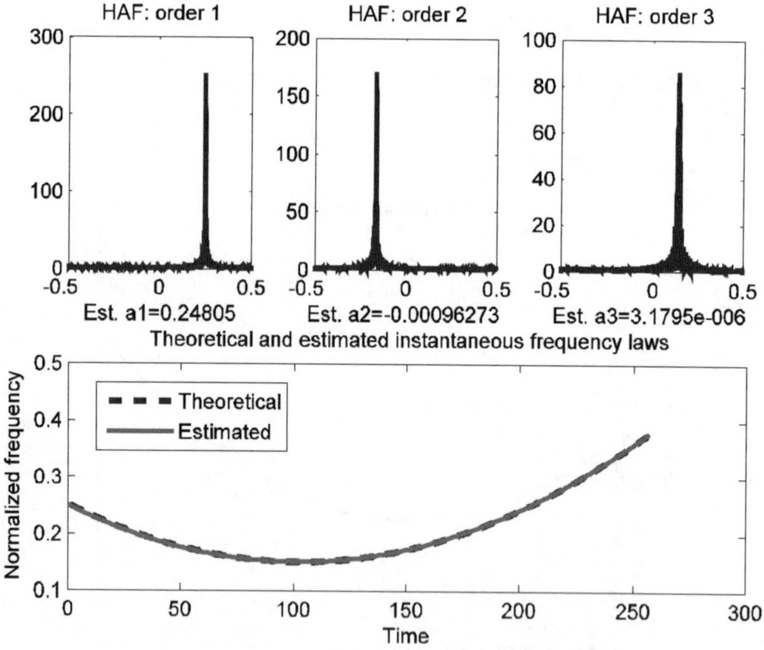

Figure 12.6. *Polynomial coefficients estimated using the HAF and the recovered instantaneous frequency law*

Note that the polynomial coefficients are well recovered and lead to an accurate estimation of the instantaneous frequency law.

EXERCISE 12.5.

The noise-free and single component signal case was considered in exercise 12.4. Let us now suppose the case of a signal having two components ($s_1(t)$), and then the case of a noisy signal ($s_2(t)$), where $b(t)$ is a white noise.

$$s_1(t) = \exp\left[j2\pi\left(0.25t - 9.76\cdot 10^{-4}t^2 + 3.204\cdot 10^{-6}t^3\right)\right]$$

$$+\exp\left[j2\pi\left(0.12t\text{-}4.687\cdot 10^{-4}t^2 + 1.525\cdot 10^{-6}t^3\right)\right];$$

$$s_2(t) = \exp\left[j2\pi\left(0.25t - 9.76\cdot 10^{-4}t^2 + 3.204\cdot 10^{-6}t^3\right)\right] + b(t),\ t = 0..255$$

The polynomial modeling of the phase of these signals can be performed using the following MATLAB code:

```
% For the signal s1
t=0:255;
a=[0.25,-9.76*10^-4,3.204*10^-6 ;.12,-4.687*10^-4,1.525*10^-6];
s1=exp(j*2*pi*(a(1,1)*t+a(1,2)*t.^2+a(1,3)*t.^3)) +
exp(j*2*pi*(a(2,1)*t+a(2,2)*t.^2+a(2,3)*t.^3));
N = length(s1);
K = 3;
Nfft=2^12;
F=-0.5:1/Nfft:0.5-1/Nfft;
a_est=zeros(1,K);
tau = N/K ;
for i=K:-1:1
     h=him(s1,tau*ones(1,3),i);
     hf=abs(fft(h,Nfft));
     hf=[hf(length(hf)-Nfft/2+1:length(hf)),hf(1:Nfft/2)];
     subplot(2,K,i);
     plot(F,hf);
     title(['HAF: order ',num2str(i)])
     [ym,zm]=max(hf);
     Fmax=(zm-Nfft/2-1)/Nfft;
     a_est(i)=Fmax/(factorial(i)*tau^(i-1));
     xlabel(['Est. a',num2str(i),'=',num2str(a_est(i))]);
     s1=s1.* exp(-j*2*pi*a_est(i)*t.^(i));
end
LFI1= a(1,1)+2*a(1,2)*t+3*a (1,3)*t.^2;
LFI2= a(2,1)+2*a(2,2)*t+3*a (2,3)*t.^2 ;
LFIe= a_est(1,1)+2*a_est(1,2)*t+3*a_est (1,3)*t.^2 ;
subplot(212)
plot(LFI1,'--'); hold on;
```

```
plot(LFI2,':g');
plot(LFIe,'r');
legend('Theoretical 1st cmp.','Theoretical 2nd cmp.','Estimated');
xlabel('Time');
ylabel('Normalized frequency')
title('Theoretical and estimated instantaneous frequency laws')
```

Note that the HAF contains cross-terms because of the multicomponent nature of the analyzed signal and therefore the polynomial modeling of the signal phase becomes less accurate in this case.

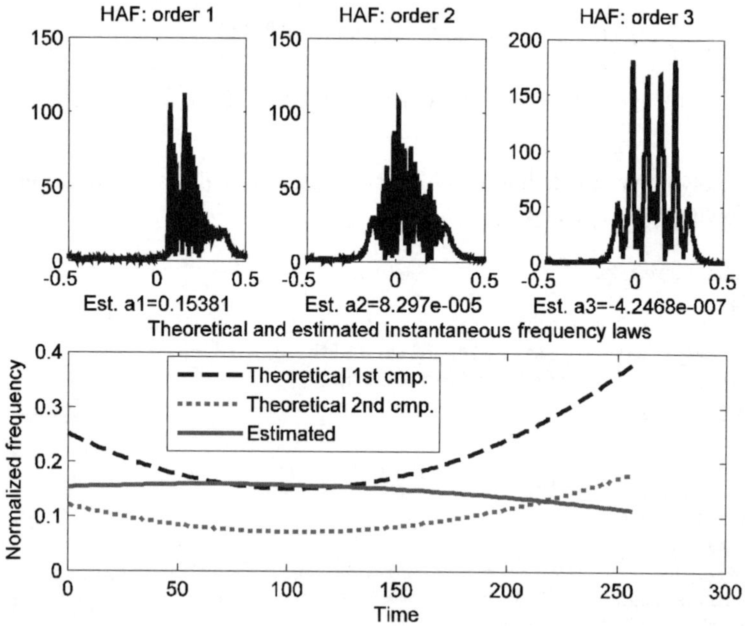

Figure 12.7. *Polynomial coefficients estimated using the HAF and the recovered instantaneous frequency law for the signal $s_1(t)$*

```
% For the signal s2
t=0:255;
a=[0.25,-9.76*10^-4,3.204*10^-6];
s2=exp(j*2*pi*(a(1,1)*t+a(1,2)*t.^2+a(1,3)*t.^3));
s2= s2+ 0.7*hilbert(randn(1,256));
N = length(s2); K = 3;
Nfft=2^12;
F=-0.5:1/Nfft:0.5-1/Nfft;
a_est=zeros(1,K);
tau = N/K;
```

```
for i=K:-1:1
    h=him(s2,tau*ones(1,3),i);
    hf=abs(fft(h,Nfft));
    hf=[hf(length(hf)-Nfft/2+1:length(hf)),hf(1:Nfft/2)];
    subplot(2,K,i); plot(F,hf);
    title(['HAF: order ',num2str(i)])
     [ym,zm]=max(hf);
    Fmax=(zm-Nfft/2-1)/Nfft;
    a_est(i)=Fmax/(factorial(i)*tau^(i-1));
    xlabel(['Est. a',num2str(i),'=',num2str(a_est(i))]);
    s2=s2.* exp(-j*2*pi*a_est(i)*t.^(i));
end
LFI = a(1,1)+2*a(1,2)*t+3*a (1,3)*t.^2;
LFIe = a_est(1,1)+2*a_est(1,2)*t+3*a_est (1,3)*t.^2;
subplot(212); plot(LFI,'--');
hold on; plot(LFIe,'r');
legend('Theoretical','Estimated');
xlabel('Time');
ylabel('Normalized frequency')
title('Theoretical and estimated instantaneous frequency laws')
```

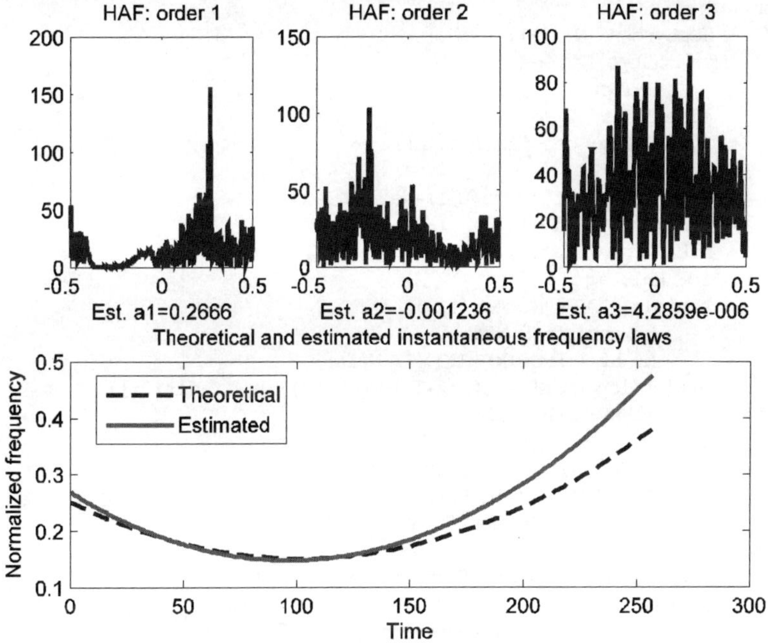

Figure 12.8. *Polynomial coefficients estimated using the HAF and the recovered instantaneous frequency law for the signal $s_2(t)$*

The result, depicted in Figure 12.8, illustrates the high noise sensitivity of the HAF.

The two examples considered in this exercise highlight the HAF limitations in the case of a multicomponent of noisy signals.

EXERCISE 12.6.

Repeat exercise 12.5 but using the ml-HAF instead of the HAF for the analysis of signal $s_1(t)$.

```
t=1:256;
a=[0.25,-9.76*10^-4,3.204*10^-6;.12,-4.687*10^-4,1.525*10^-6];
s1=exp(j*2*pi*(a(1,1)*t+a(1,2)*t.^2+a(1,3)*t.^3))+
exp(j*2*pi*(a(2,1)*t+a(2,2)*t.^2+a(2,3)*t.^3));
Ncomp = 2; Lag = 30;
N = length(s1); K = 3;
Nfft = 2^14; F = -0.5:1/Nfft:0.5-1/Nfft;
s = s1; a_est = zeros(Ncomp,K); warning off;
for icomp=1:Ncomp
   for i=K:-1:2
          Topt=round(N/(i));
        if fix(Topt/Lag)<=0
               L=Topt-1; dec=1;
          else
               L=Lag; dec=fix(Topt/L);
          end
          LAG=round(Topt)*ones(L,1)-dec*(0:L-1)';
          PROD=(LAG).^(i-1);
          mlHAF=ones(1,Nfft);
          for k=1:Lag
          h=him(s1,LAG(k)*ones(1,i-1),i);
               hf=abs(fft(h,Nfft));
               hf=hf(1:PROD(k)/PROD(1):Nfft);
          hf=[hf(length(hf)-Nfft/2+1:length(hf)),hf(1:Nfft/2)];
               mlHAF=mlHAF.*hf;
          end
          subplot(Ncomp+1,K,K*(icomp-1)+i);
          plot(F,mlHAF/max(mlHAF));
          title(['ml-HAF: order ',num2str(i)])
        [ym,zm]=max(mlHAF); Fmax=(zm-Nfft/2-1)/Nfft;
          a_est(icomp,i)=Fmax/(factorial(i)* PROD(1));
          xlabel(['Est. a',num2str(i),'=',num2str(a_est(icomp,i))]);
          s1=s1.* exp(-j*2*pi*a_est(icomp,i)*t.^(i));
   end
   hf=fftshift(abs(fft(s1,Nfft)));
   subplot(Ncomp+1,K,K*(icomp-1)+1);
   plot(F,hf);
   title(['ml-HAF: order ',num2str(1)]);grid
```

```
   [ym,xm]=max(hf);
   fmax=(xm-Nfft/2-1)/Nfft;
   a_est(icomp,1)=fmax;
   xlabel(['Est. a',num2str(1),'=',num2str(a_est(icomp,1))]);
   s1=s-exp(j*2*pi*(a_est(1,1)*t+a_est(1,2)*t.^2+a_est(1,3)*t.^3));
end
LFI1= a(1,1)+2*a(1,2)*t+3*a (1,3)*t.^2;
LFI2= a(2,1)+2*a(2,2)*t+3*a (2,3)*t.^2;
LFIe1= a_est(1,1)+2*a_est(1,2)*t+3*a_est (1,3)*t.^2;
LFIe2= a_est(2,1)+2*a_est(2,2)*t+3*a_est (2,3)*t.^2;
subplot(313);
plot(LFI1,'--b'); hold on;
plot(LFIe1,'r');
plot(LFI2,'--b');
plot(LFIe2,'r');
legend('Theoretical','Estimated');
xlabel('Time');
ylabel('Normalized frequency')
title('Theoretical and estimated instantaneous frequency laws')
```

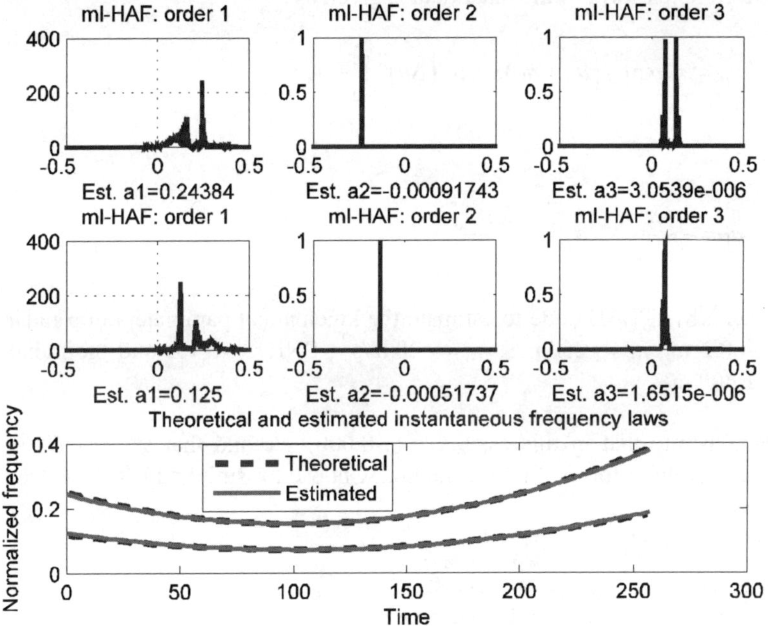

Figure 12.9. *Polynomial coefficients estimated using the ml-HAF and the recovered instantaneous frequency law for the signal $s_1(t)$*

It can be seen on the figure obtained that the ml-HAF approach yields an accurate estimation of the polynomial coefficients corresponding to this signal mixture. Even better results could be obtained by increasing the number of lags and the number of calculation points for the FFT.

EXERCISE 12.7.

Suppose a radar target having a constant radial acceleration during the observation time. Demonstrate that the distance between the radar and the target has in this case the following expression:

$$r(t) = r_0 + \dot{r}_0 t + 0.5\ddot{r}t^2$$

where r_0, $\dot{r}_0$, $\ddot{r}_0$ are the distance, the radial speed and the radial acceleration at the beginning of the observation period.

Denote by λ the wavelength and by Δ the sampling period. Demonstrate that the discrete complex form of the radar data is given by:

$$y_n = A\exp\left[j\left(a_0 + a_1\Delta n + a_2\left(\Delta n\right)^2\right)\right] + u_n$$

where:

$$a_1 = -\frac{4\pi \cdot \dot{r}_0}{\lambda},\ a_2 = -\frac{2\pi \cdot \ddot{r}_0}{\lambda}$$

Write a MATLAB code to estimate the kinematical parameters of a radar target for $\lambda = 0.2$ m, $\dot{r}_0 = -250$ m/s, $\ddot{r}_0 = -20\ \text{m/s}^2$, SNR = –3 dB and the radial speed range ± 400 m/s.

Demonstrate that in this case Δ = (1/8,000) sec and that at least N_{fft} = 4,000 points are required for the FFT calculation. Choose for simplicity N_{fft} = 4,096.

```
% Radar parameters
L=0.2;                    % wavelength
Range_v=400;          % speed range
PRI=L/(4*Range_v); % sampling period
Nfft=4096;
% Target parameters
vt=-250;                  % radial speed
at=-20;                   % radial acceleration
% Received signal simulation
t=0:Nfft-1; A=1;          % amplitude
SNR=-3; a0=0;
```

```
a1=-4*pi*vt/L;a2=-2*pi*at/L;
x=A*exp(j*(a0+a1*PRI*t+a2*(PRI*t).^2));
B=sqrt(A^2*10^(-SNR/10));
b=B*randn(1,Nfft); y=x+b;
% 3-rd order HAF analysis
Lag=20;                          % lags number
N=length(y);  K=3;               % polynomial modeling order
Nfft=4096;                       % FFT calculation points
F=-0.5:1/Nfft:0.5-1/Nfft;        % frequencies vector
a_est=zeros(1,K);           % coefficients vector
warning off; figure
s2 = y; tau = N/K ;
for i=K:-1:1
        h=him(s2,tau*ones(1,3),i);
        hf=abs(fft(h,Nfft));
        hf=[hf(length(hf)-Nfft/2+1:length(hf)),hf(1:Nfft/2)];
        subplot(2,K,i); plot(F,hf);
        title(['HAF: order ',num2str(i)])
        [ym,zm]=max(hf); Fmax=(zm-Nfft/2-1)/Nfft;
        a_est(i)=Fmax/(factorial(i)*tau^(i-1));
        xlabel(['Est. a',num2str(i),'=',num2str(a_est(i))]);
         s2=s2.* exp(-j*2*pi*a_est(i)*t.^(i));
end
% 3-rd order ml-HAF analysis
a_est=zeros(1,K);
for i=K:-1:2
     Topt=round(N/(i));
    if fix(Topt/Lag)<=0
         L=Topt-1; dec=1;
     else
         L=Lag; dec=fix(Topt/L);
     end
     LAG=round(Topt)*ones(L,1)-dec*(0:L-1)';
     PROD=(LAG).^(i-1); mlHAF=ones(1,Nfft);
     for k=1:Lag
         h=him(y,LAG(k)*ones(1,i-1),i);
         hf=abs(fft(h,Nfft));
         hf=hf(1:PROD(k)/PROD(1):Nfft);
         hf=[hf(length(hf)-Nfft/2+1:length(hf)),hf(1:Nfft/2)];
         mlHAF=mlHAF.*hf;
     end
     subplot(2,K,i+K);plot(F,mlHAF/max(mlHAF));
     title(['ml-HAF: order ',num2str(i)])
     [ym,zm]=max(mlHAF); Fmax=(zm-Nfft/2-1)/Nfft;
     a_est(i)=Fmax/(factorial(i)* PROD(1));
     xlabel(['Est. a',num2str(i),'=',num2str(a_est(i))]);
     y=y.* exp(-j*2*pi*a_est(i)*t.^(i));
end
hf=fftshift(abs(fft(y,Nfft)));
subplot(2,K,1+K); plot(F,hf);
```

```
title(['ml-HAF: order ',num2str(1)]); grid; [ym,xm]=max(hf);
fmax=(xm-Nfft/2-1)/Nfft; a_est(1)=fmax;
xlabel(['Est. a',num2str(1),'=',num2str(a_est(1))]);
```

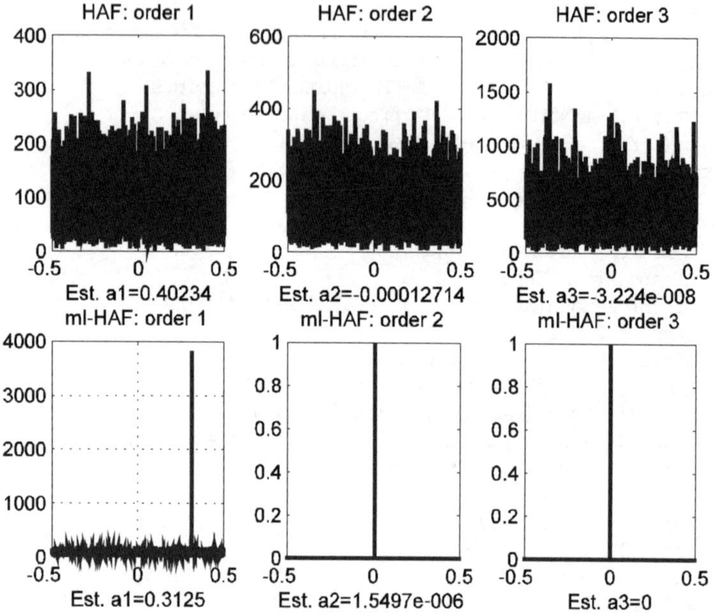

Figure 12.10. *Estimation of the polynomial coefficients using the HAF (top) and the ml-HAF (bottom)*

The polynomial coefficients estimated using the HAF and the ml-HAF are shown in Figure 12.10. Note that despite the noise, the ml-HAF approach yields a zero value for the third coefficient, since the analyzed signal is a 2nd order PPS. This result is due to the calculation of the HAF for several lags, which makes the polynomial modeling less sensitive to the order overestimation.

The ml-HAF based estimation obviously outperforms the HAF approach. Indeed, unlike the last case, the spectral peaks are clearly identified in the first case, so that the target motion parameters can be accurately estimated (see the MATLAB code below).

```
% Estimation of the target motion parameters
c2=2*pi*a_est(2)/(PRI^2)
a2
c1=2*pi*a_est(1)/(PRI)
a1
```

EXERCISE 12.8.

Consider the signal $s(t) = \exp\left[j2\pi \cdot 0.02 \cdot t^{1.5} \right], t = 0..255$. Estimate its instantaneous frequency law using the warping operators.

The ideal instantaneous frequency law and the signal Wigner-Ville distribution are generated by the MATLAB code given below.

```
t=0:255; k=1.5;
f=0:0.5/256:0.5-0.5/256;
s0=exp(j*2*pi*0.02*t.^k);
LFI=.02*k*t.^(k-1);
figure; plot(LFI);
title('Instantaneous frequency law')
xlabel('Time');
ylabel('Normalized frequency');
axis([0 256 0 .5])
tfr=tfrwv(s0.');
figure; imagesc(t,f,tfr);
axis xy; colormap(flipud(jet))
title('Wigner-Ville distribution')
xlabel('Time');ylabel('Normalized frequency');
```

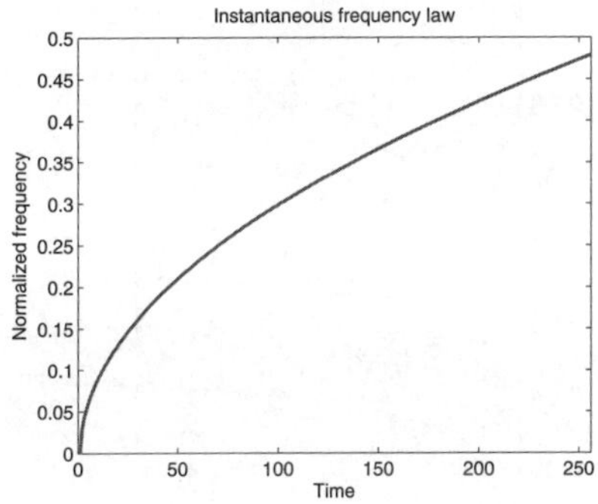

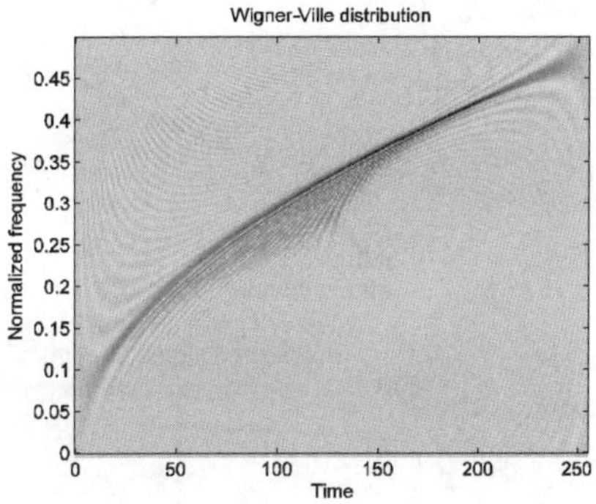

Figure 12.11. *Instantaneous frequency law (left) and Wigner-Ville distribution (right) for the signal from exercise 12.8*

The first two steps to evaluate the non-linear matched TFR are implemented by the MATLAB code given below:

```
% 1st step: warping operator calculation
u=3; Tu=0:1/u:255-1/u;
fu=0:0.5/256:u*0.5-0.5/256;
fu=fu(1:length(Tu)); w=Tu.^(1/(k));
% 2nd step: Prewarping
```

```
tw=w; fw=fu.*(Tu.^(1-k))/k; w=Tu.^(1/(k));
s1=interp1(t,s0,tw,'cubic');
subplot(221);plot(Tu,w);
title('Warping law');
xlabel('Time'); ylabel('Warped time')
subplot(122); tfr=tfrwv(s1.');
imagesc(t,f,tfr); axis xy
colormap(flipud(hot));
axis([0 256 0 0.1])
title('Warped signal WVD'); xlabel('Time');
ylabel('Normalized frequency');
subplot(223);S1=fft(s1);
ff=-0.5:1/length(S1):.5-1/length(S1);
plot(ff,abs(fftshift(S1)));
title('Prewarped signal spectrum')
xlabel('Normalized frequency');
ylabel('Amplitude')
```

Figure 12.12 illustrates the linearization of the time-frequency content performed by the warping operator.

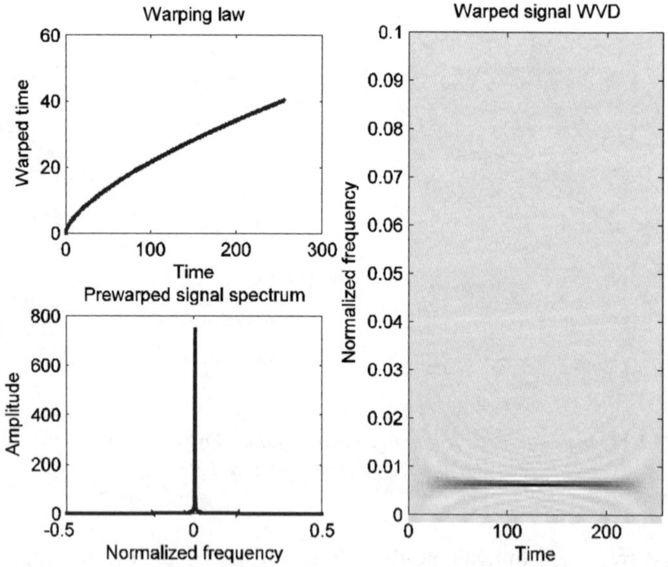

Figure 12.12. *Influence of the warping operator on the signal spectral features*

The prewarped signal, obtained after the first two steps, is equivalent to a sinusoid, whose frequency is related to the non-linear modulation rate. In the case of this exercise we obtain $6.5 \cdot 10^{-3}$. This value is multiplied by 3 (upsampling rate) and finally yields 0.0195, which is close to the theoretical value, 0.02. An even better accuracy could be obtained by increasing the upsampling rate.

The last processing step consists of calculating the Wigner-Ville distribution of the prewarped signal. The associated MATLAB code is given below.

```
% 3rd step: Inverse transformation
[N,N]=size(tfr);
% Find the frequency of the prewarped signal = max(tfr)
[M,C]=max(tfr(:,ceil(N/2)));
TFR=zeros(N,N);
% Calculate the inverse coordinates
Fs1=fw*N/fw(end);Fs1(1)=1;
for i=1:N-1,
        TFR(ceil(Fs1(i)),i)=tfr(C,i);
end
% Retrieve the original time-frequency coordinates
t1=0:1/u:256-1/u;f1=0:0.5/(256):0.5-.5/256;
figure; imagesc(t1,f1,log2(TFR));
axis xy ; title('Non-linear TFR');
colormap(flipud(hot)); xlabel('Time');
ylabel('Normalized frequency');
figure; tfr=tfrwv(s0.');
imagesc(t,f,tfr); axis xy ;
title('Wigner-Ville distribution')
xlabel('Time'); colormap(flipud(jet))
ylabel('Normalized frequency');
```

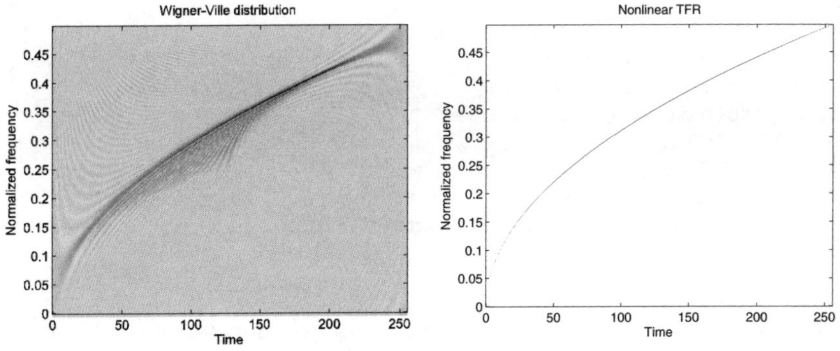

Figure 12.13. *Wigner-Ville distribution (left) and non-linear matched TFR (right) of the signal from exercise 12.8*

Note that the non-linear TFR of the signal has no cross-terms. Furthermore, its resolution is better than that obtained in the case of the WVD.

The "dotted-line" aspect of the non-linear TFR is due to the non-linear distribution of the time-frequency coordinates. This problem could be overcome by using a more effective interpolation technique.

Consequently, the non-linear TFRs yield a perfect image of the signal time-frequency content provided that its modulation law is known. If not, it can be estimated using a bank of warping operators, as it is shown in exercise 12.10.

The methodology illustrated by this exercise also allows the calculation of the matched TFRs for some standard signal classes. An example is provided by exercise 12.9, in the case of the hyperbolic modulations.

EXERCISE 12.9.

Determine the operator which insures the best time-frequency representation of the hyperbolic modulations. Illustrate the application of this operator in the case of a hyperbolic chirp.

The following MATLAB code calculates the matched TFR corresponding to a hyperbolic chirp, defined by equation [12.33].

```
N=128;N=N/2;
c=4 ; % hyperbolic modulation rate
df=1/N;f=df:df:1; inverse=1./sqrt(f);
HM = inverse.*exp(-j*2*pi*c*log(f));
HM=[HM.*hamming(N)' zeros(1,N)]; x=ifft(HM);

% Calculate the warping operator + prewarping
u = 8; L = length(x); N = u*L;
inputzero = [x zeros(size(1:N-L))];
fftinsig = fft(inputzero);
npos = L/2; M=2*L-6;
dv=log((L/2)/(L/2 -1));
df=((N/2)/(N/2 -u))^M * u/(N/2) ;
fftinsig(1:npos*u)=fftinsig(1:npos*u).*sqrt(df.*[0:npos*u-1]);
for count=0:M-1
    est = exp(count*dv)/df*u;%
    estint = floor(est);
fftwarp0(count+1)=(fftinsig(estint+1)-fftinsig(estint))*(est-
estint)+fftinsig(estint);
end
fftwarp0(M+1) = fftinsig(npos*u);
fftwarp0(M+2:4*npos)=zeros(size(M+2:4*npos));
fftwarp(1:npos*2)=fftwarp0(M+2-npos*2:M+1);
fftwarp(npos*2+1:npos*8)=zeros(size(1:npos*8-npos*2));
fftwarp(npos*8-M+npos*2:npos*8)=fftwarp0(1:M-npos*2+1);
warpsig = ifft(fftwarp); L2=npos*4*2; tfr=tfrwv(warpsig.');
```

```
% Unwarping
L=L2/4;
for count=0:L2-1
   countd=count+1;
   tfr(L,countd)=0;
   for count2=1:L-1
      est=log(count2*df/2)/dv;
      est=2*(est+L2/4-M); est=est+L2/2;
      estint=floor(est);
       est=L2-est+1; estint=L2-estint+1;
      count1=L-count2+1; tfr(count1-1, count+1)=0;
          if estint>0
               if estint<=L2
diff=(tfr(estint-1, count+1)-tfr(estint, count+1))*(est-estint);
                tfr(count1-1,count+1) = diff+tfr(estint-1,count+1);
               end
          end
   end
end
tfr1=zeros(L,L);
for count1=1:L
   for count2=1:L
      freqcnt=L-count2+1;
      expand=freqcnt/L*L2/L;
      arg = (count1-1)*expand+1;
      intarg = floor(arg);
      if intarg>=1
         if intarg<=L2
diff=(tfr(count2,intarg+1)-tfr(count2,intarg))*(arg-intarg);
            tfr1(count2, count1) = diff+tfr(count2,intarg);
         end
      end
      if intarg<1
         tfr1(count2,count1)=tfr(count2,1);
      end
   end
end
tx=0:1/u:128-1/u;
fy=0:0.5/(128):0.5-.5/128;
imagesc(tx,fy,tfr1);
axis xy ; title('Hyperbolic TFR');
colormap(flipud(hot)); xlabel('Time');
ylabel('Normalized frequency');
```

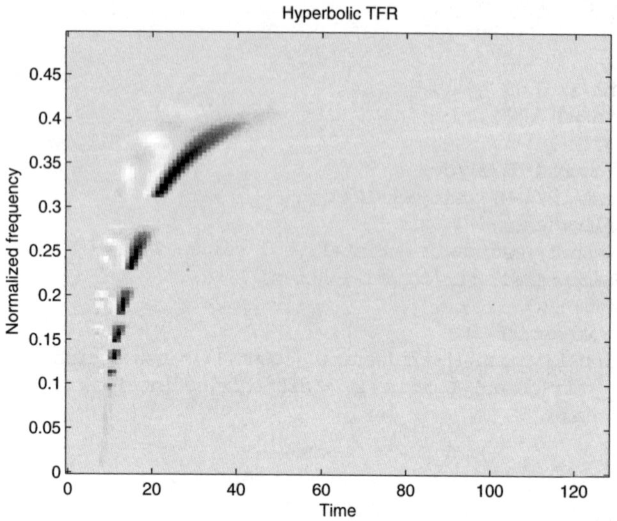

Figure 12.14. *Matched TFR corresponding to a hyperbolic chirp*

EXERCISE 12.10.

The warping concept is useful to precisely estimate the type of non-linearity of the time-frequency content. The basic idea is to use a bank of warping operators and to select the one which minimizes the width of the spectral peak corresponding to the prewarped signal. Its order k thus characterizes the signal modulation.

Illustrate this concept for the following mixture of two signals:

$$s(t) = \exp\left[j2\pi 0.02 \cdot t^{1.5} \right] + \exp\left[j2\pi 10^{-6} \cdot t^{3} \right], t = 0..255$$

```
t=0:255; f=0:0.5/256:0.5-0.5/256;
s0=exp(j*2*pi*0.02*t.^1.5)+exp(j*2*pi*0.1*10.^-5*t.^3);
LFI1=.02*1.5*t.^(1.5-1); LFI2=3*10^-6*t.^2;
subplot(221); plot(LFI1); hold on; plot(LFI2);
title('Instantaneous frequency laws');
xlabel('Time'); ylabel('Normalized frequency');
axis([0 256 0 .5])
subplot(222); tfr=tfrwv(s0.'); imagesc(t,f,tfr); axis xy;
title('Wigner-Ville distribution');
xlabel('Time');ylabel('Normalized frequency');
u=10; Tu=0:1/u:255-1/u;
fu=0:0.5/256:u*0.5-0.5/256; fu=fu(1:length(Tu));
S=zeros(length([1:.25:3]),length(Tu));
i=1;
```

```
for k=1:.25:3,
    w=Tu.^(1/(k)); tw=w;fw=fu.*(Tu.^(1-k))/k;
    s1=interp1(t,s0,tw,'cubic');
    subplot(212); S1=fft(s1);
    ff=-0.5:1/length(S1):.5-1/length(S1);
    S(i,:)=abs(fftshift(S1))/max(abs(S1));
    plot(ff,S(i,:)); i=i+1; pause
end
k=1:.25:3; imagesc(ff,k,S); axis xy;
title('Prewarped signal spectrum for several values of k')
xlabel('Normalized frequency'); ylabel('k');
axis([-5*10^-3 5*10^-3 1 3.2]); colormap(flipud(jet));
```

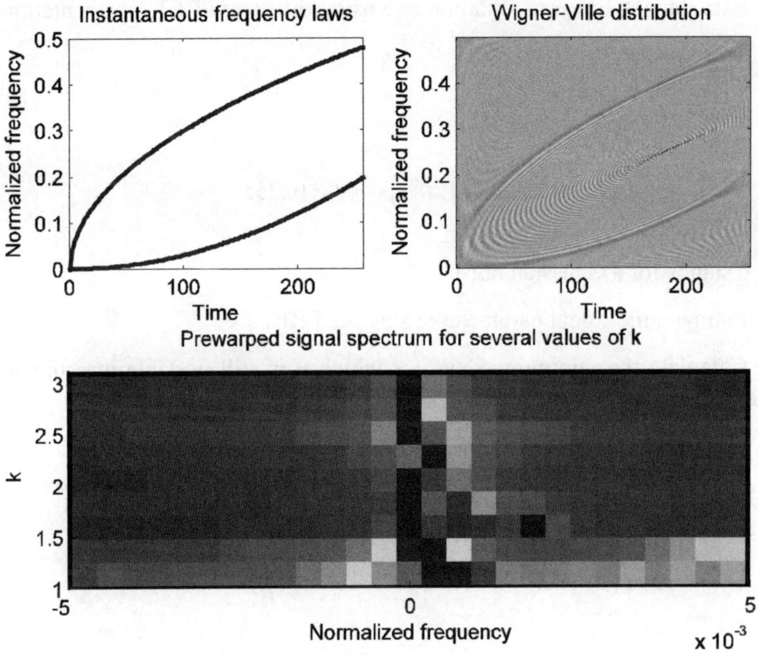

Figure 12.15. *Illustration of the warping operator bank concept*

The two-dimensional structure *S* containing the prewarped signal spectrum for several warping operators $t^{1/k}$, with *k* between 1 and 3, is plotted in Figure 12.15 (bottom). Note that the best spectral concentrations are obtained for k = 1.5 and 3, which correspond to the modulation laws of the analyzed signal components.

12.3. Exercises

EXERCISE 12.11.

Calculate the FRFT to the sum of two *chirps* given below:

$$s(t)=\exp\left(j2\pi\left(0.1t+0.0012t^2\right)\right)+\exp\left(j2\pi\left(0.4t-0.0012t^2\right)\right), t=0..255$$

for an order sweeping the range between 0 and 2 with an increment of 0.1.

a. Plot the instantaneous frequency law for each component of the signal.

b. Estimate the linear modulation rate using different FRFT representations.

EXERCISE 12.12.

Consider the signal:

$$s(t)=\exp\left(j2\pi\left(0.1t+0.0012t^2\right)\right)+b, t=0..255$$

where b stands for a Gaussian noise.

a. Estimate the signal parameters using the FRFT.

b. Calculate the minimum SNR, for which it is still possible to estimate these parameters.

EXERCISE 12.13.

Consider the following signal:

$$s(t)=\exp\left(j2\pi\left(0.1t+.0012t^2+.13\cdot 10^{-6}t^3\right)\right), t=1..256.$$

Express the time-frequency content of this signal as the sum of two chirps using the FRFT.

EXERCISE 12.14.

Consider the signal below:

$$s(t)=\exp\left(j2\pi\left(0.1t+0.0012t^2\right)\right)+\exp\left(j2\pi\left(0.12t+0.0012t^2\right)\right), t=0..255$$

a. Plot the instantaneous frequency law for each component of the signal.

b. Estimate the parameters of each component using the FRFT.

c. Propose a solution to increase the obtained resolution.

EXERCISE 12.15.

Let us consider the following sinusoidal frequency modulation:

$$s(t)=\exp\left[j2\pi\left(0.3t+0.1\sin\left(2\pi\cdot 0.02t\right)\right)\right],\ t=0..255.$$

a. Model the phase of this signal by means of the HAF.

b. Comment on the influence of the modeling order on the estimation quality. Find the most appropriate order value.

EXERCISE 12.16.

Repeat exercise 12.15 but using ml-HAF and discuss the influence of the number of lags on the modeling quality.

EXERCISE 12.17.

Consider the noisy signal given below:

$$s(t)=\exp\left[j2\pi\left(0.25t-9.76\cdot 10^{-4}t^2+3.204\cdot 10^{-6}t^3\right)\right]+b(t),\ t=0..255.$$

Find the minimum SNR for which the ml-HAF still provides appropriate results. Propose an objective criterion for the assessment of the results.

EXERCISE 12.18.

Consider the following 6th order polynomial phase signal:

$$s(t)=\exp\left[j2\pi\begin{pmatrix}0.17\cdot t-9.7\cdot 10^{-4}\cdot t^2-2.35\cdot 10^{-7}\cdot t^3\\ +3.8\cdot 10^{-8}\cdot t^4+2.8\cdot 10^{-10}\cdot t^5-3.29\cdot 10^{-13}\cdot t^6\end{pmatrix}\right]$$

Modelize the phase of this signal by means of the ml-HAF for several orders (4, 5, 6, 7, 8). Comment on the results obtained.

EXERCISE 12.19.

Consider the received signal from a mobile underwater source, which moves away with speed v = 6 m/s and acceleration a = 0.07 m/s^2. The transmitted signal is a chirp given by $\exp[j2\pi(410t+2.48t^2)]$. The received signal is sampled with the sampling frequency $F_s = 1{,}000$ Hz.

Give the theoretical expression of the received signal phase (the speed of the sound in the water is considered equal to 1,500 m/s). Estimate the signal phase using the ml-HAF.

EXERCISE 12.20.

The chirp signal below:

$$s(t) = \exp\left[j2\pi\left(0.25t - 9.76\cdot10^{-4}t^2 + 3.204\cdot10^{-6}t^3\right)\right], t = 0..255$$

propagates in a multipath channel characterized by the following impulse response:

$$h(t) = \delta(t) + 0.8\delta(t-24) + 0.4\delta(t-64)$$

a. Demonstrate that the received signal is expressed as the sum of three 3rd order PPSs. Calculate the polynomial coefficients corresponding to this mixture.

b. Modelize this mixture by means of the ml-HAF and find out the channel parameters.

c. Comment on the performance of this method and indicate its possible applications.

EXERCISE 12.21.

Consider the following signal:

$$s(t) = \exp[j2\pi 3\ln t] + \exp[j2\pi 4\ln t]$$

a. Compare the Wigner-Ville distributions corresponding to this signal and to its prewarped version obtained using the hyperbolic warping operator.

b. Analyze the errors and identify the cross-terms.

c. Indicate the type of errors which are removed by the warping operator.

d. Propose a solution for the time-frequency representation of the prewarped signal in order to obtain a TFR without cross-terms.

EXERCISE 12.22.

Consider a signal with two similar components:

$$s(t)=\exp\left[j2\pi\cdot 0.02\cdot t^{1.5}\right]+\exp\left[j2\pi\cdot 0.01\cdot t^{1.5}\right], t=0..255.$$

Estimate the modulation rate of each component using the warping principle. Calculate the non-linear matched TFR corresponding to this signal. Compare the results obtained to the instantaneous frequency laws of the two components.

EXERCISE 12.23.

Consider a signal with two different components:

$$s(t)=\exp\left[j2\pi\cdot 0.02\cdot t^{1.5}\right]+\exp\left[j2\pi\cdot 3\ln(t)\right], t=0..255.$$

a. Extract each component of this mixture by projecting the signal on the function sets $\exp(at^k)$ and $\exp(b\ln(t))$.

b. Calculate the matched TFR corresponding to each component.

c. Find the non-linear matched TFR corresponding to signal $s(t)$ by superposing the two previously calculated TFRs.

Chapter 13

Supervised Statistical Classification

13.1. Theoretical background

13.1.1. *Introduction*

Pattern recognition, one of the most important aspects of artificial intelligence, is an appropriate field for the development, validation and comparison of different learning techniques: statistical or structural, supervised or unsupervised, inductive or deductive, etc.

Patterns are general concepts describing an object (mechanical part, obstacle, human face, etc.) or a phenomenon (disease, system operating state, emotion, etc.). They are provided by a sensor or transducer to a recognition system in the form of a data set. The classification is possible only if characteristic information about the observed pattern is confined in this data set, which is then called a *pattern signature* or *fingerprint*.

Since the signatures are generally measured, they have a random nature. Thus, the statistical approach, which is described in this chapter, is the most used for pattern recognition. More precisely, we consider the *supervised learning* framework, which requires a database. This database contains labeled patterns belonging to the *M* predefined classes $\omega_1, \omega_2, \ldots, \omega_M$. These patterns are repeatedly presented to the classifier in order to derive a decision rule optimizing a given criterion.

The general structure of a supervised statistical classification chain is presented in Figure 13.1. This process aims at classifying any unknown pattern using its signature and consists of the following steps:

– *feature vector extraction* from the recorded signatures; the feature vector confines the discriminant information, which allows the different classes to be separated;

– *data analysis*, which is useful for reducing the initial space dimension while keeping the discriminant properties of the extracted features;

– *classification*, which is the decision step and associates a label with an unknown input pattern.

The first step may involve many different techniques: statistical, spectral, time-frequency, etc.). However, there is no rule for finding the best salient features while their nature essentially depends on the considered application.

The two last steps are, from this point of view, easier to implement, since they make use of some standard techniques, which will be briefly described in this chapter.

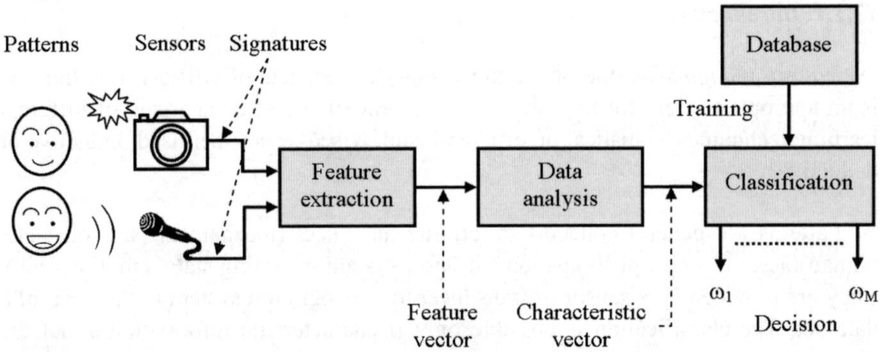

Figure 13.1. *Supervised statistical classification flowchart*

13.1.2. *Data analysis methods*

The data analysis is a key step of any pattern recognition process, which is closely related to the classifier performance and complexity. It is also the interface between the signal processing and the information processing domains. In fact, if the features extracted from recorded signatures still have a physical mean (amplitude, frequency, contour, etc.), the characteristics obtained after the data analysis step loose it in the new representation space.

Two data analysis approaches are generally used for reducing the initial space dimension. The first consists of selecting from among the extracted features the best subset using sequential methods, such as SFS (*sequential forward selection*), SBS (*sequential backward selection*), etc. However, these methods are suboptimal and the convergence rate toward the final solution depends on the extracted feature set. Furthermore, this approach does not completely remove the feature redundancy.

The second approach, which will be described in this section, deals with projection methods, such as the principal component analysis, the discriminant factor analysis, the Sammon method, etc. The projection is performed by a linear or non-linear transformation, which aim to optimize a given criterion: variance maximization, distance conservation, class separability maximization, etc.

The linear transformations are the most used thanks to their simplicity and calculation speed. They are defined by the following relationship:

$$\mathbf{y} = \mathbf{K} \cdot \mathbf{x} \qquad [13.1]$$

where $\mathbf{K}$ is the transformation matrix, $\mathbf{x} = [x_1\ x_2\ \ldots\ x_n]^T$ denotes the feature vector and $\mathbf{y} = [y_1\ y_2\ \ldots\ y_m]^T$ is the characteristic vector ($m < n$).

Principal component analysis

The most well known linear method for data analysis is the Karhunen-Loève transform or *the principal component analysis method (PCA)*, which maximizes the variance of the projected vectors.

It is defined by a matrix having as rows the eigenvectors of the feature space covariance matrix $\mathbf{\Sigma_x}$. The main steps for the implementation of this transform are indicated in Figure 13.2.

The PCA removes any redundancy between the components of the projected vectors, since the covariance matrix in the transformed space becomes diagonal:

$$\mathbf{\Sigma_y} = \mathbf{K}\mathbf{\Sigma_x}\mathbf{K}^T = \mathrm{diag}[\lambda_1\ \ \lambda_2\ \ \ldots\ \ \lambda_n] \qquad [13.2]$$

where $\{\lambda_i\}_{i=1..n}$ stand for the eigenvalues of matrix $\mathbf{\Sigma_x}$.

The mean square error associated with this transform is minimal among all the linear transforms, and is given by:

$$EQM = \sum_{i=m+1}^{n} \lambda_i \qquad [13.3]$$

This result shows that the contribution of each projection space dimension to the mean square error is proportional to the associated eigenvalue. Consequently, the selected components correspond to the most important eigenvalues.

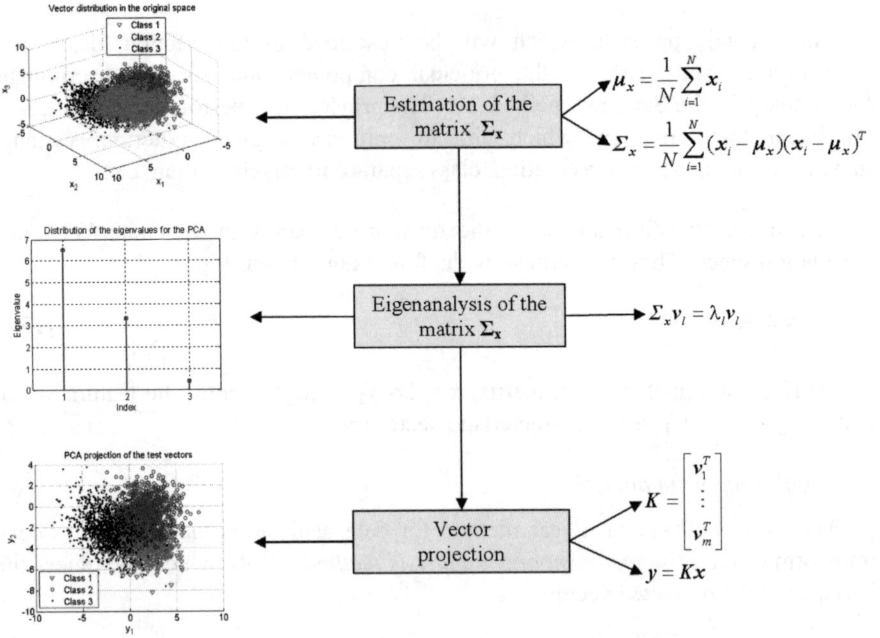

Figure 13.2. *Main steps of the PCA implementation*

The PCA performs the vector projection without any knowledge of their labels. It is therefore an unsupervised data analysis method.

Linear discriminant analysis

Unlike the PCA, the *linear discriminant analysis* (LDA) is a supervised projection method, which aims to maximize the separability of the projected classes. It is defined by the corresponding projection matrix according to equation [13.1].

The LDA objective is to minimize the distances among the vectors belonging to the same class and to maximize the distances among the class centers. These distances are represented by matrices $\mathbf{S_W}$ and $\mathbf{S_B}$ respectively.

The rows of matrix $\mathbf{K}$ are the eigenvectors of matrix $\mathbf{\Sigma_{LDA}}$ in this case. This last matrix is obtained as shown in Figure 13.3 using the two matrices $\mathbf{S_W}$ and $\mathbf{S_B}$. Note that the number of non-zero eigenvalues of matrix $\mathbf{\Sigma_{LDA}}$ is the number of classes minus 1. All the eigenvalues are comprised between 0 and 1 and indicate the discriminant capability of the axes represented by the corresponding eigenvectors.

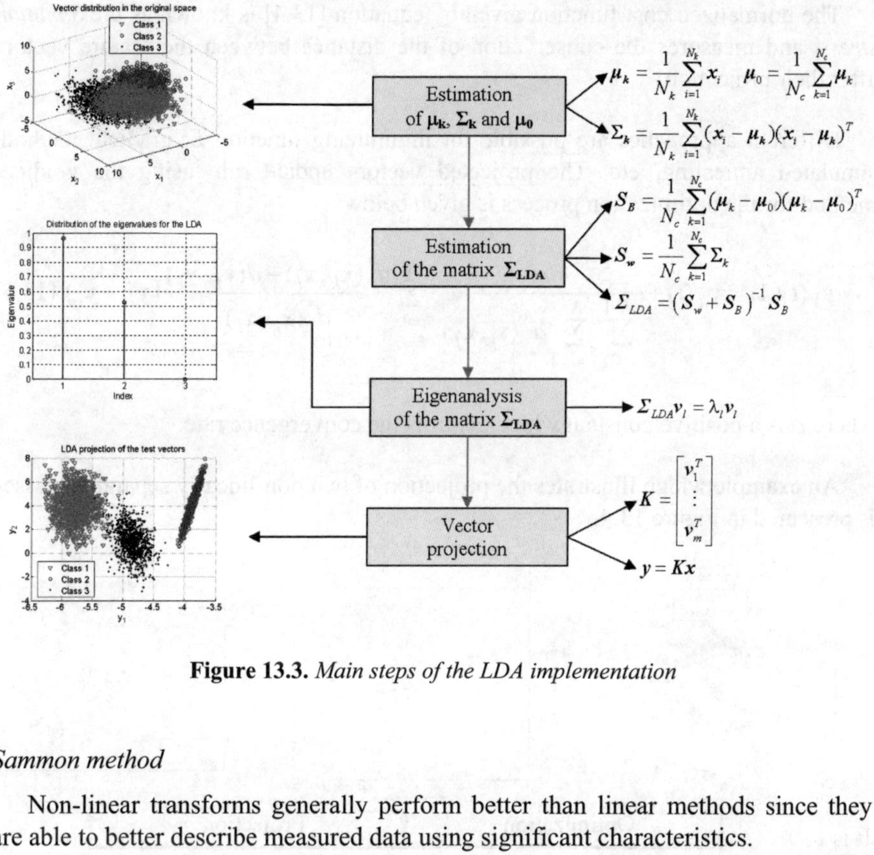

Figure 13.3. *Main steps of the LDA implementation*

Sammon method

Non-linear transforms generally perform better than linear methods since they are able to better describe measured data using significant characteristics.

A particularly interesting non-linear transform is the Sammon method, which tries to ensure the best conservation of the distances among the vectors. This is equivalent to maintaining the same neighborhood relationship as in the initial space. Therefore, the transformed space will be the closest image of the initial space in a lower dimension.

The Sammon method minimizes the following cost function:

$$E = \frac{1}{\sum_{i=1}^{N_v-1}\sum_{j=i+1}^{N_v} d^*(\mathbf{x}_i,\mathbf{x}_j)} \cdot \sum_{i=1}^{N_v-1}\sum_{j=i+1}^{N_v} \frac{\left[d^*(\mathbf{x}_i,\mathbf{x}_j) - d(\mathbf{y}_i,\mathbf{y}_j)\right]^2}{d^*(\mathbf{x}_i,\mathbf{x}_j)} \qquad [13.4]$$

where N_v is the number of vectors, while $d^*(\mathbf{x}_i,\mathbf{x}_j)$ and $d(\mathbf{y}_i,\mathbf{y}_j)$ stand for the distances between a couple of vectors in the initial and transformed space respectively.

The normalized cost function given by equation [13.4] is known as the *Sammon stress*, and measures the conservation of the distance between the feature vectors after their projection.

Different approaches are possible for minimizing function E: gradient method, simulated annealing, etc. The projected vectors update rule using the gradient method for the optimization process is given below:

$$\mathbf{y}_k(t+1)=\mathbf{y}_k(t)+\frac{\alpha}{\sum_{i=1}^{N-1}\sum_{j=i+1}^{N} d^*(\mathbf{x}_i,\mathbf{x}_j)}\cdot\sum_{\substack{i=1\\ i\neq k}}^{N}\frac{d^*(\mathbf{x}_k,\mathbf{x}_i)-d(\mathbf{y}_k,\mathbf{y}_i)}{d^*(\mathbf{x}_k,\mathbf{x}_i)}(\mathbf{y}_k-\mathbf{y}_i) \quad [13.5]$$

where α is a positive constant which controls the convergence rate.

An example which illustrates the projection of two non-linearly separable classes is presented in Figure 13.4.

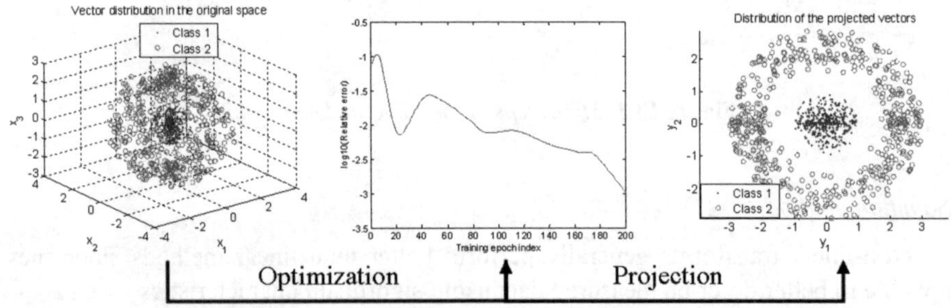

Figure 13.4. *Application of the Sammon method to the case of two classes*

The most recent approach makes use of neural networks to generalize the optimization process result. The projection of a new vector can thus be performed by a trained neural network without repeating the optimization process.

13.1.3. *Supervised classifiers*

The classifier is the decision element of a pattern recognition system. The previous two phases (feature extraction and data analysis) aimed to extract the essential information about the observed patterns and present it in the most appropriate and condensed form to the classifier. From this point, the quality of decision made depends only on the classifier choice and its implementation.

The general structure of a supervised classifier is illustrated in Figure 13.5. Implementing such a classifier is equivalent to finding out, during the training phase, and to using, during the test phase, the discriminant functions $g_1, g_2, \ldots, g_M$, so that:

$$\mathbf{x} \in \omega_i \Rightarrow g_i(\mathbf{x}) = \max_{j=1..M} \left[g_j(\mathbf{x}) \right] \qquad [13.6]$$

The classifier is *optimal* according to the mean error probability P_{err_mean} minimization criterion if:

$$g_i(\mathbf{x}) = f(\mathbf{x} \mid \omega_i) \cdot P(\omega_i) \qquad [13.7]$$

In the above equation, $P(\omega_i)$ is the *a priori* probability of class ω_i, while $f(x|\omega_i)$ denotes the conditional probability density function of the vectors belonging to this class.

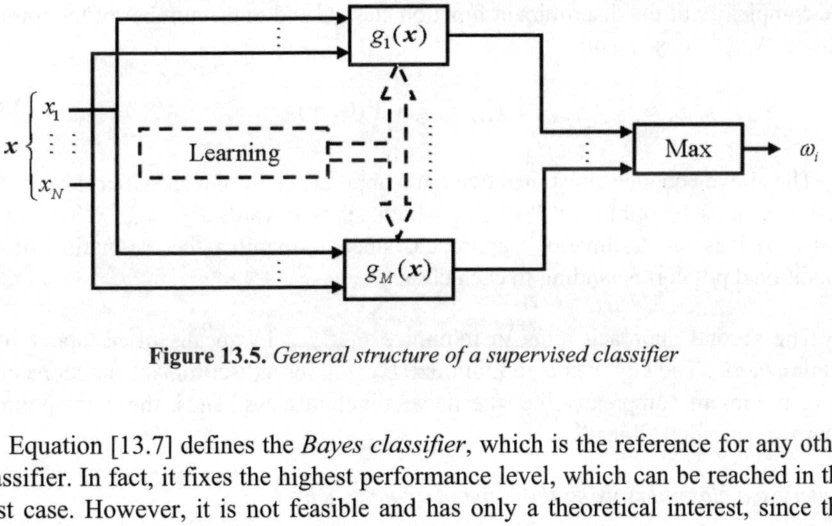

Figure 13.5. *General structure of a supervised classifier*

Equation [13.7] defines the *Bayes classifier*, which is the reference for any other classifier. In fact, it fixes the highest performance level, which can be reached in the best case. However, it is not feasible and has only a theoretical interest, since the exact expressions of $P(\omega_i)$ and $f(x|\omega_i)$ are unknown in real situations.

If P_{err_emp} is an empirical mean error probability evaluated over a training set containing N_{train} vectors, thus:

$$\lim_{N_{train} \to \infty} P_{err\_emp} = P_{err\_mean}$$

However, the minimization of P_{err_emp} does not automatically lead to a high generalization capability, since the training vector set is rather limited in practical situations. The discriminant functions calculated in this way are also linked to some particular aspects of the training vector distribution.

To avoid the data overfitting two solutions are possible. The first is empirical and consists of introducing a third vector set, which is used during the training phase only to evaluate the system performance. The classification rate corresponding to this set measures the generalization capability. Thus, the training phase is finished when it reaches a maximum value.

The start point for the second solution is the relationship between the discriminant function complexity and the data overfitting risk. The key idea is thus to use the simplest functions which are able to fit the data, according to the *Occam razor principle*.

It is therefore possible to define an upper limit for the mean error probability, which will be minimized by the training process. This limit depends on the empirical mean error probability P_{err_emp} and on a *penalty term* Ψ. Ψ is defined with respect to the complexity of the discriminant function class G and to the number of the training vectors N_{appr}. Consequently:

$$P_{err\_Bayes} \leq P_{err\_mean} \leq P_{err\_emp} + \Psi(G, N) \qquad [13.8]$$

The above equation illustrates two main approaches for the classifier design. The first one aims to minimize P_{err_mean}, which tends towards P_{err_Bayes}. This can be considered as an "estimation" approach, since it requires the estimation of the conditional pdf corresponding to each class.

The second approach aims to minimize P_{err_mean} by means of its upper limit minimization. The key idea is to minimize P_{err_emp} for a discriminant functions class G of minimum complexity, i.e. the linear function class. Thus, the corresponding approach is called "linear".

Supervised classifiers using the estimation approach

The methods derived from this approach rely on either the explicit (parametric or non-parametric) estimation of the classes' distributions or their implicit estimation (neural networks).

The *parametric estimation* assumes that the analytic form of these distributions is known up to a given number of parameters, which will then be estimated using the training vector set. Although this approach is close to the original Bayes theory, it has some shortcomings:

– most theoretical distributions are unimodal, unlike the real data distributions, which are generally multimodal,

– even if the analytical forms of the data distributions can be identified, the estimation of their parameters require a large number of training vectors; this approach is therefore inappropriate whenever the number of available training vectors is reduced.

The *non-parametric estimation* does not consider any assumptions on the analytical form of the different classes' distributions. They are locally estimated using the training vector set.

KNN (K nearest neighbor) classifiers

KNN classifiers are the most used non-parametric classification techniques. The decision rule is very simple and can be easily generalized to the multiclass case. Thus, if $V_K(\boldsymbol{x})$ is the Kth order neighborhood of the vector $\boldsymbol{x}$ and:

$$K_j(\mathbf{x}) = \text{card}\left\{\mathbf{x}_n \middle| \mathbf{x}_n \in \omega_j \text{ , } \mathbf{x}_n \in V_K(\mathbf{x})\right\} \quad [13.9]$$

thus:

$$K_i(\mathbf{x}) = \max_{j=1..M}\left[K_j(\mathbf{x})\right] \Rightarrow \mathbf{x} \in \omega_i \quad [13.10]$$

This decision rule, illustrated in Figure 13.6, consists of assigning an unknown vector $\boldsymbol{x}$ to the class which has the largest number of vectors among its K nearest neighbors. If $K = 1$ the unknown vector is classified in the same class as its nearest neighbor (NN or nearest neighbor rule).

The performance of the KNN classifier can be increased if some additional information is considered. The new classifier, called fuzzy KNN, takes into account not only the number of neighbors but also their distances to the unknown vector and their membership coefficients calculated for each class.

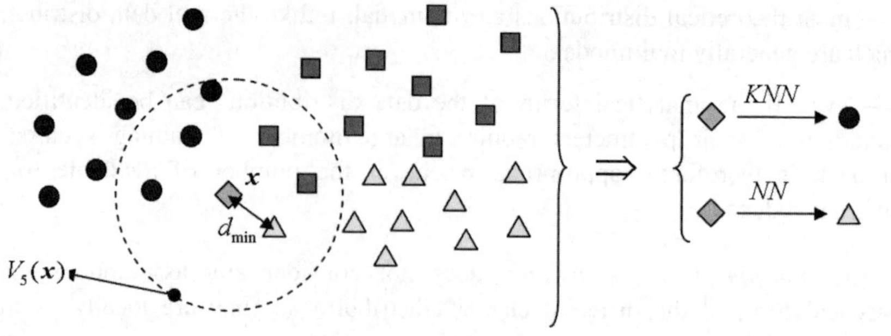

Figure 13.6. *KNN classifier principle*

The membership coefficient of an element (vector) to a set (class) is the central concept of the fuzzy logic theory. It can take any value between 0 and 1, provided that the sum of all the membership coefficients associated with a given vector is equal to 1. This membership definition is much more flexible than the standard definition, which is based on a binary rule, so that a vector belongs to a class or not. Furthermore, the fuzzy membership concept is also more appropriate for real world problems, where the transition from one pattern to another is often continuous.

The fuzzy KNN algorithm involves the three main steps shown in Figure 13.7: fuzzification, calculation of unknown vector membership coefficients, and unfuzzification.

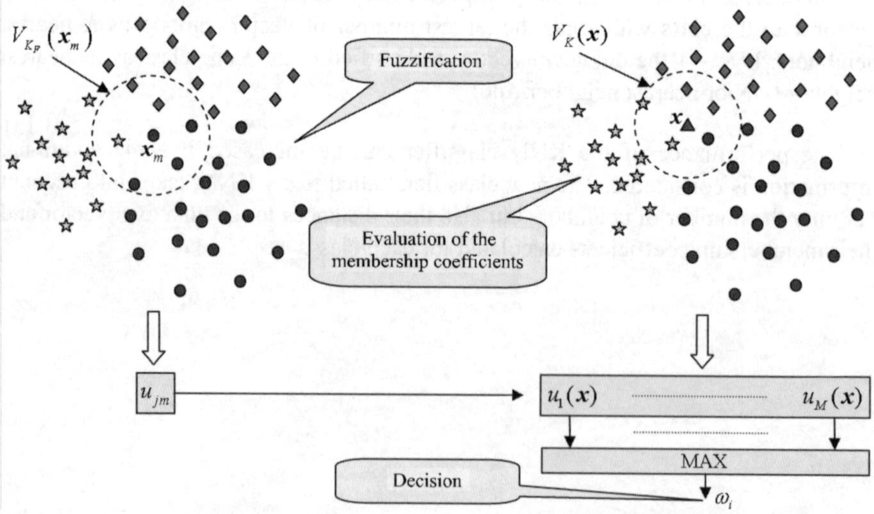

Figure 13.7. *Fuzzy KNN classifier principle*

The first step consists of calculating the membership coefficients of each learning vector corresponding to all classes. A commonly used fuzzification rule is indicated below:

$$u_{jm} = \frac{K_j^{(m)}}{K_F}, \quad K_j^{(m)} = \text{card}\left\{ \mathbf{x}_n^{(j)} \mid \mathbf{x}_n^{(j)} \in V_{K_F}\left(\mathbf{x}_m\right) \right\} \qquad [13.11]$$

In the above equation, u_{jm} is the membership coefficient of the training vector $\mathbf{x}_m$ corresponding to the class ω_j, $K_j^{(m)}$ stands for the number of its nearest neighbors belonging to this class and K_F denotes the total number of nearest neighbors considered in the fuzzification step. Coefficient u_{jm} can be seen as a measure of the matching between the components of vector $\mathbf{x}_m$ and the characteristics of class ω_j.

In the next step, the unknown vector membership coefficients corresponding to each class are given by the following relationship:

$$u_j(\mathbf{x}) = \frac{\sum\limits_{\mathbf{x}_m \in V_K(\mathbf{x})} u_{jm} \left\| \mathbf{x} - \mathbf{x}_m \right\|^{-2/\delta}}{\sum\limits_{\mathbf{x}_m \in V_K(\mathbf{x})} \left\| \mathbf{x} - \mathbf{x}_m \right\|^{-2/\delta}} \qquad [13.12]$$

where K denotes the total number of nearest neighbors considered in this phase and δ is an appropriately chosen constant (usually $\delta = 1$).

Finally, a decision is made in the third step about the classification of the unknown vector according to the following rule:

$$u_i(\mathbf{x}) = \max_j \left\{ u_j(\mathbf{x}) \right\} \Rightarrow \mathbf{x} \in \omega_i \qquad [13.13]$$

Furthermore, these membership coefficients can also be helpful to provide a confidence measure for the classification result.

The main drawback of the KNN classifier comes from the fact that all the training vectors are used in the classification phase for calculating their distances to the unknown vector and selecting its nearest neighbors. Thus, the use of this classifier for real-time applications is inappropriate in most cases.

Neural networks

The connectionist approach, represented by the *neural networks,* relies on the parallel-distributed processing performed by simple elementary structures called neurons. Their weights are randomly initialized and then repeatedly updated during the training phase using different recursive learning algorithms. The neural network final configuration is highly adapted to the structure of the training vectors' space.

There is a large variety of neural networks for the supervised classification: multilayer perceptron, radial basis function (RBF) networks, learning vector quantization (LVQ) networks, etc. The basic component of each of them is the neuron, which generally calculates a scalar product or a distance.

The neural networks are trained using a learning rule of *biological inspiration* (such as the Hebb rule) or of *mathematical inspiration* (such as the generalized delta rule or the backpropagation algorithm). These rules calculate automatically the network weights, which confine the information about the distribution of the vector classes. In other words, the network weights generate borders which separate the classes and approximate the discriminant functions of the Bayes classifier.

No *a priori* knowledge is needed about the classes' statistics (as in the case of the non-parametric classifiers) and a significant compression of the database volume is obtained (as in the case of the parametric classifiers), since the whole information is concentrated in the network weights.

The neural networks are very fast classifiers, since the classification time of an unknown vector is the propagation time through the network layers. Furthermore, they are much less sensitive than other classifiers to local failures (one or even several defective neurons) due to the information distribution all over the network.

Finally, a very important property of neural networks is that if the training set is *statistically characteristic* for the input vectors' distribution and if the network is *properly trained,* its outputs approximate the *a posteriori* probabilities associated with the considered classes. As the unknown vector is assigned to the class whose corresponding output is maximal, a neural network makes the same optimal decision as the Bayes classifier.

However, due to the finite number of training vectors, a real neural network minimizes the mean square error (MSE) instead of the mean error probability. The MSE is expressed in this case as follows:

$$MSE = \frac{1}{N_v} \sum_{i=1}^{M} \sum_{j=1}^{N_v} \left[o_i(\mathbf{x}_j) - d_i(\mathbf{x}_j) \right]^2 \qquad [13.14]$$

where N_v is the number of training vectors, while $o_i(\boldsymbol{x}_j)$ and $d_i(\boldsymbol{x}_j)$ denote the actual and target outputs of the neural network corresponding to class ω_i and training vector $\boldsymbol{x}_j$.

The most well known neural network for the supervised classification is the *multilayer perceptron* (MLP). Its general structure is illustrated in Figure 13.8. The neurons are organized in an input layer, an output layer and one or two hidden layers. The output of each neuron is connected to the input of all neurons from the next layer (fully connected forward network).

The MLP neuron performs the scalar product between the input vector and its weight vector. Its output is then calculated by means of a differentiable non-linear function *f*(*net*), called the activation function.

Its most commonly used forms are indicated below together with the corresponding derivatives:

– *sigmoid* function:

$$f(net) = \left[1 + \exp(-\alpha net)\right]^{-1} \Rightarrow f'(net) = \alpha f(net)\left[1 - f(net)\right]$$

– *tanh* function:

$$f(net) = \tanh(-\alpha net) \Rightarrow f'(net) = \alpha\left[1 - f^2(net)\right]$$

The constant α controls the variation rate of function *f*(*net*).

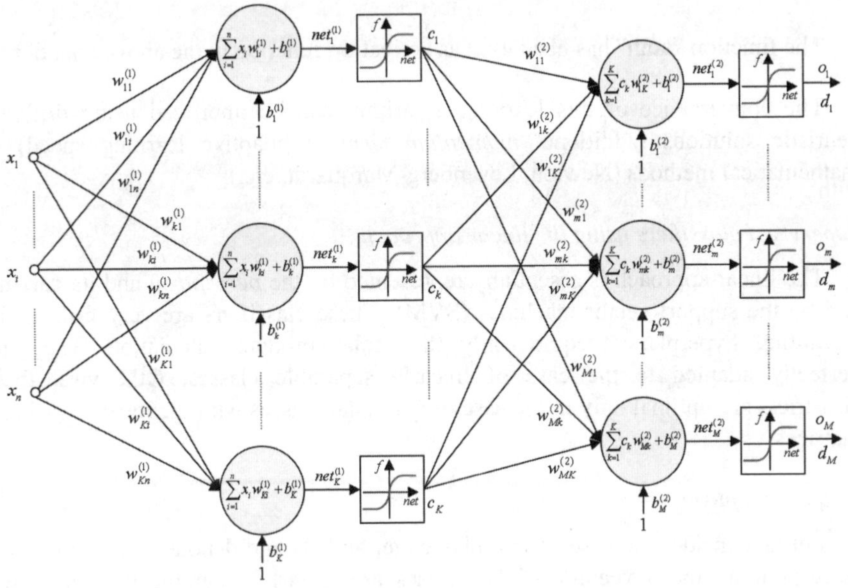

Figure 13.8. *Structure of a multilayer perceptron*

The neurons' weights are the network free parameters. They are randomly initialized and reach stable values at the end of the training process.

During the training process all the couples training vector – target outputs are repeatedly presented to the neural network. At each iteration, called an epoch, the neurons' weights change according to the backpropagation learning algorithm in order to minimize the error function defined by equation [13.14]. The output layer neurons' weights are updated using the following relationships:

$$w_{mk}^{(2)}(t+1) = w_{mk}^{(2)}(t) + \eta\delta_m^{(2)}(t)c_k(t) \qquad [13.15]$$

$$\delta_m^{(2)}(t) = \alpha\left[1 - o_m(t)^2\right]\left[d_m - o_m(t)\right] \qquad [13.16]$$

where η is a constant which controls the learning rate.

For the hidden layers, the updating rule is given below:

$$w_{ki}^{(1)}(t+1) = w_{ki}^{(1)}(t) + \eta\delta_k^{(1)}(t)x_i \qquad [13.17]$$

$$\delta_k^{(1)}(t) = \alpha\left[1 - c_k(t)^2\right]\sum_{m=1}^{M} w_{mk}^{(2)}(t)\delta_m^{(2)}(t) \qquad [13.18]$$

The function "tanh" has been used as activation function in the above equations.

The convergence of this learning algorithm can be improved using different heuristic solutions (additional *momentum term* or adaptive learning speed) or mathematical methods (Newton, Levenberg-Marquardt, etc.).

Supervised classifiers using the linear approach

The linear approach is essentially represented by the *perceptron* and its variants and by the support vector machines (SVM). These classifiers are very fast, as the separating hyperplanes require only the scalar product calculation. They are perfectly adapted to the case of linearly separable classes. Otherwise, these classifiers are optimal only in the case of Gaussian classes with identical covariance matrices.

Fuzzy perceptron

Let us consider the case of two classes ω_1 and ω_2 and denote by μ_1 and μ_2 the corresponding mean vectors. If the classes are linearly separable the perceptron having the structure illustrated in Figure 13.9 is able to find the equation of the separating hyperplane in a finite number of epochs.

Its learning rule is given below:

$$\tilde{\mathbf{w}}(t+1) = \tilde{\mathbf{w}}(t) + (d_k - o_k)\tilde{\mathbf{x}}_k \qquad [13.19]$$

where the target output d_k is considered equal to 1 for the training vectors belonging to ω_1 and 0 for those belonging to ω_2.

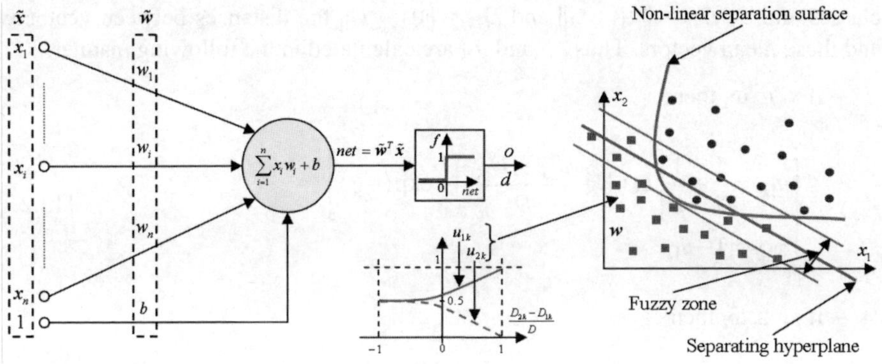

Figure 13.9. *Fuzzy perceptron structure and principle*

The actual output value o_k can be either 0 or 1 due to the Heaviside activation function. At the end of the training process, all the training vectors are well classified. However, if the classes are not linearly separable, as illustrated in Figure 13.9, the learning algorithm does not converge.

The fuzzy perceptron has the same structure as the perceptron, but its learning rule is different:

$$\tilde{\mathbf{w}}(t+1) = \tilde{\mathbf{w}}(t) + \eta \left| u_{1k} - u_{2k} \right|^{\delta} (d_k - o_k)\tilde{\mathbf{x}}_k \qquad [13.20]$$

where u_{1k} and u_{2k} are the membership coefficients of vector $\boldsymbol{x}_k$ relative to the two classes, η stands for the learning rate and δ is a positive appropriately chosen constant.

The membership coefficients of each training vector have to be inside the interval [0.5, 1] for its own class, and inside the interval [0, 0.5] for the other class. They are defined so that the following conditions are met:

– the membership coefficient of a training vector to its own class is equal to 1 if it is the mean vector of its class and is equal to 0.5 if it is the mean vector of the other class;

– the membership coefficients of a vector equally spaced with respect to the mean vectors of the two classes is close to 0.5;

– the membership coefficients vary exponentially with the distance to each class mean vector.

Let us denote by $D = \|\mu_1 - \mu_2\|$ the distance between the mean vectors of the two classes and by $D_{1k} = \|\mu_1 - \mathbf{x}_k\|$ and $D_{2k} = \|\mu_2 - \mathbf{x}_k\|$ the distances between vector $\boldsymbol{x}_k$ and these mean vectors. Thus u_{1k} and u_{2k} are calculated in the following manner:

– if $\mathbf{x}_k \in \omega_1$ then:

$$\begin{cases} u_{1k} = \dfrac{1}{2} + \dfrac{1}{A}\left[\exp\left(g\dfrac{D_{2k} - D_{1k}}{D}\right) - \exp(-g)\right] \\ u_{2k} = 1 - u_{1k} \end{cases} \qquad [13.21]$$

– if $x_k \in \omega_2$ then:

$$\begin{cases} u_{1k} = 1 - u_{2k} \\ u_{2k} = \dfrac{1}{A}\left[\exp\left(g\dfrac{D_{1k} - D_{2k}}{D}\right) - \exp(-g)\right] \end{cases} \qquad [13.22]$$

where $A = 2[\exp(g)\text{-}\exp(\text{-}g)] = 4\text{sh}(g)$ and g is a positive constant which controls the variation rate of the membership coefficients.

The learning rate defined by equation [13.20] reduces the influence of the training vectors whose class membership is ambiguous. These vectors define the fuzzy zone illustrated in Figure 13.9, which corresponds to a membership coefficient value around 0.5:

$$0.5 - \beta < u_{ik} < 0.5 + \beta \qquad [13.23]$$

The width of the fuzzy region is determined by the largest membership coefficient of the training vectors equally spaced with respect to the mean vectors of the two classes, denoted by β. Thus, the largest membership coefficients of the training vectors outside this zone have to be higher than β. Thus, the β value is calculated as follows:

$$\beta = \frac{1-\exp(-g)}{2[\exp(g)-\exp(-g)]} + \varepsilon, \quad \varepsilon \geq 0 \qquad [13.24]$$

where constant ε is related to the convergence rate (typically $\varepsilon = 0.02$).

The learning process is stopped when, during a complete epoch, the fuzzy perceptron weight vector is updated only by the training vectors from the fuzzy zone. Therefore, the fuzzy perceptron is able to find a separating hyperplane in a finite number of epochs, even if the classes are not linearly separable.

SVM classifiers

The key idea of the SVM classifiers is to find a separating hyperplane between two classes (h: $\mathbf{w}^T\mathbf{x} + b = 0$), so that its minimal distance with respect to the training vectors, called the margin, is maximum.

The optimal solution is obtained when this hyperplane is located in the middle of the distance between the convex envelopes of the two classes. This distance is denoted by d_m in Figure 13.10, and is expressed as follows:

$$d_m = \frac{2}{\|\mathbf{w}\|} \qquad [13.25]$$

The *support vectors* are situated on the margins of the two classes. If the training vectors' membership is defined by:

$$u_k = \begin{cases} 1 & \text{if} \quad \mathbf{x}_k \in \omega_1 \\ -1 & \text{if} \quad \mathbf{x}_k \in \omega_2 \end{cases} \qquad [13.26]$$

then the support vectors' set can be written in the form $\Omega_s = \{\mathbf{x}_k \mid u_k(\mathbf{w}^T\mathbf{x}_k + b) = 1\}$.

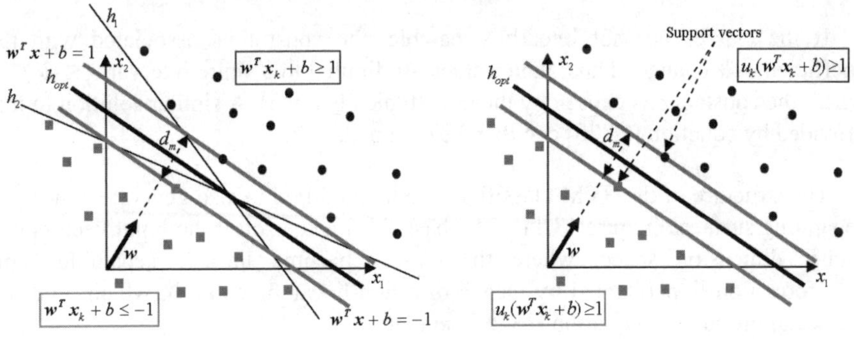

Figure 13.10. *SVM classifier basic principle*

Consequently, the optimal hyperplane separates the vectors of the two classes so that the margin is maximized. This is equivalent to finding the solution of the following constrained optimization problem:

$$\begin{cases} \max\limits_{\mathbf{w}} \left[\dfrac{2}{\| \mathbf{w} \|} \right] \\ u_k \left(\mathbf{w}^{\mathbf{T}} \mathbf{x_k} + b \right) \geq 1 \end{cases} \Leftrightarrow \begin{cases} \min\limits_{\mathbf{w}} \left[\dfrac{1}{2} \| \mathbf{w} \|^2 \right] \\ u_k \left(\mathbf{w}^{\mathbf{T}} \mathbf{x_k} + b \right) - 1 \geq 0 \end{cases} \quad [13.27]$$

Since both the function to be optimized and the associated constraints are convex, solution (**w**, b) can be determined using the Lagrange multipliers:

$$\begin{cases} \mathbf{w} = \sum\limits_{k} u_k \lambda_k \mathbf{x_k} = \sum\limits_{\mathbf{x_k} \in \Omega_s} u_k \lambda_k \mathbf{x_k} \\ u_k \left(\mathbf{w}^{\mathbf{T}} \mathbf{x_k} + b \right) \geq 1 \end{cases} \quad [13.28]$$

The coefficients λ_k are obtained by solving the following optimization problem:

$$\begin{cases} \max\limits_{\lambda} \left[\sum\limits_{k} \lambda_k - \dfrac{1}{2} \sum\limits_{j} \sum\limits_{k} u_j u_k \lambda_j \lambda_k \mathbf{x}_j^T \mathbf{x}_k \right] \\ \sum\limits_{k} u_k \lambda_k = 0 \\ \lambda_k \geq 0 \quad \left(\lambda_k = 0 \text{ if } u_k \left(\mathbf{w}^{\mathbf{T}} \mathbf{x_k} + b \right) - 1 > 0 \right) \end{cases} \quad [13.29]$$

Note that the separating hyperplane requires only the calculation of the scalar product between the input space vectors. The solution thus depends only on the number of support vectors and does not depend on the input space dimension.

If the classes are not linearly separable, the constraints associated with the coefficients λ_k change. Thus, their values are limited to a finite interval $0 \leq \lambda_k \leq c$, where the constant c is chosen by the user (typically $c = 5$). A similar solution to that provided by equation [13.28] can then be obtained.

The structure of the SVM classifiers can be modified to also generate non-linear separating surfaces (Figure 13.11). The basic idea is to project the input vectors in a higher dimension space, where the classes become linearly separable. This transformation is performed by means of a non-linear function Φ, which modifies the scalar products in equation [13.29], so that:

$$\left.\begin{array}{l}\mathbf{x_k} \rightarrow \Phi(\mathbf{x_k}) \\ \mathbf{x_j} \rightarrow \Phi(\mathbf{x_j})\end{array}\right\} \Rightarrow \mathbf{x}_j^T \mathbf{x_k} \rightarrow \Phi(\mathbf{x_k})^T \Phi(\mathbf{x_j}) \quad [13.30]$$

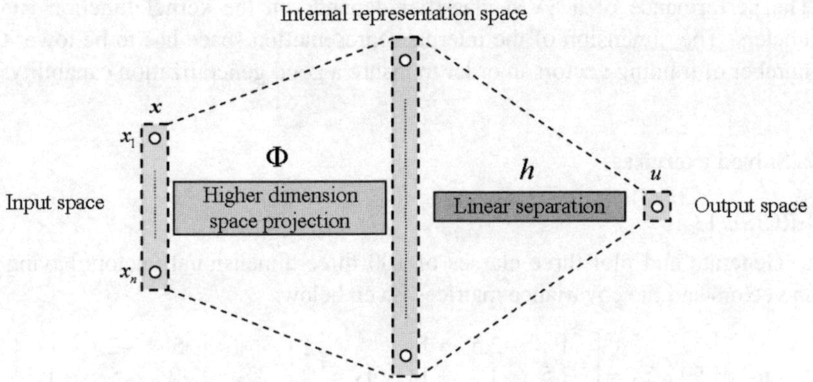

Figure 13.11. *SVM classification using non-linear separating surfaces*

Function Φ does not appear explicitly in the expression of the final solution. In fact, it is replaced by a symmetric and separable function, called *kernel*[1] in order to overcome the problem related to the scalar product calculation in a higher dimension space. The kernel function is defined as follows:

$$\Delta : \mathbf{R}^n \times \mathbf{R}^n \rightarrow \mathbf{R}, \Delta(\mathbf{x_k}, \mathbf{x_j}) = \Phi(\mathbf{x_k})^T \Phi(\mathbf{x_j}) \quad [13.31]$$

The solution is then obtained in the following form:

$$\mathbf{w}^T \mathbf{x} + b = \sum_{\mathbf{x_k} \in \Omega_s} u_k \lambda_k \Delta(\mathbf{x_k}, \mathbf{x}) + b \quad [13.32]$$

Some normal kernel functions are provided below, α and a being positive constants:

– linear kernel: $\Delta(\mathbf{x_k}, \mathbf{x_j}) = \mathbf{x}_j^T \mathbf{x_k}$,

– polynomial kernel: $\Delta(\mathbf{x_k}, \mathbf{x_j}) = \left(\alpha \mathbf{x}_j^T \mathbf{x_k} + 1\right)^n$,

– Gaussian RBF kernel: $\Delta(\mathbf{x_k}, \mathbf{x_j}) = \exp\left(-\alpha \| \mathbf{x_k} - \mathbf{x_j} \|^2\right)$,

1 The kernel function has to meet Mercer's condition: $\int_{-\infty}^{\infty} \Delta(x, y) g(x) g(x) dx dy \geq 0$ for any finite energy function $g(x)$.

– exponential RBF kernel: $\Delta(\mathbf{x_k},\mathbf{x_j}) = \exp\left(-\alpha \| \mathbf{x_k} - \mathbf{x_j} \|\right)$,

– sigmoid kernel: $\Delta(\mathbf{x_k},\mathbf{x_j}) = \tanh\left(\alpha \mathbf{x}_j^T \mathbf{x_k} - a\right)$.

The performance of a SVM classifier depends on the kernel function and its parameters. The dimension of the internal representation space has to be lower than the number of training vectors in order to insure a good generalization capability.

13.2. Solved exercises

EXERCISE 13.1.

a. Generate and plot three classes of 100 three-dimensional vectors having the mean vectors and the covariance matrices given below:

$$\boldsymbol{\mu}_1 = \begin{bmatrix} 3 \\ 5 \\ 2 \end{bmatrix}, \ \boldsymbol{\Sigma}_1 = \begin{bmatrix} 1 & -2.5 & -1 \\ -2.5 & 1 & -1 \\ -1 & -1 & 1 \end{bmatrix}, \ \boldsymbol{\mu}_2 = \begin{bmatrix} 1 \\ 3 \\ 3 \end{bmatrix}, \ \boldsymbol{\Sigma}_2 = \begin{bmatrix} .5 & -2 & -2 \\ -2 & .5 & -1.5 \\ -2 & -1.5 & .5 \end{bmatrix},$$

$$\boldsymbol{\mu}_3 = \begin{bmatrix} 5 \\ 2 \\ 1 \end{bmatrix}, \ \boldsymbol{\Sigma}_3 = \begin{bmatrix} 1 & -2 & -1 \\ -2 & 3 & -1.5 \\ -1 & -1.5 & 1 \end{bmatrix}$$

b. Project all these vectors in a two-dimensional space using the principal component analysis. What is the resulting relative error?

c. Perform the linear discriminant analysis on the same vectors' set. Conclude about the interest of this method for better organizing the data before the classification phase.

a.

The following function can be used for generating Gaussian classes:

```
function matr_gen =
generation_class_gauss(mean_vect,cov_matr,vect_nbr)
% Generation of a Gaussian class
% matr_gen=generation_class_gauss(mean_vect,cov_matr,vect_nbr);
% mean_vect  - mean vector
% cov_matr   - covariance matrix
% vect_nbr - number of vectors
% matr_gen   - matrix of generated vectors
nbre_var=length(mean_vect); matr_init=randn(nbre_var,vect_nbr);
cov_matr_init=cov(matr_init');
matr_transf=real((cov_matr^.5)*(cov_matr_init^-.5));
matr_gen_centered=matr_transf*matr_init ;
matr_gen=repmat(mean_vect,1,vect_nbr)+matr_gen_centered;
```

The MATLAB code below allows then generating and plotting the three classes:

```
mean_vect1=[3;5;2]; cov_matr1=[1 -2.5 -1;-2.5 1 -1;-1 -1 1];
mean_vect2=[1;3;3]; cov_matr2=[.5 -2 -2;-2 .5 -1.5;-2 -1.5 .5];
mean_vect3=[5;2;1]; cov_matr3=[1 -2 -1;-2 3 -1.5;-1 -1.5 1];
vect_nbr=100;
class1=generation_class_gauss(mean_vect1,cov_matr1,vect_nbr);
class2=generation_class_gauss(mean_vect2,cov_matr2,vect_nbr);
class3=generation_class_gauss(mean_vect3,cov_matr3,vect_nbr);
figure; plot3(class1(1,:),class1(2,:),class1(3,:),'vr')
hold on; plot3(class2(1,:),class2(2,:),class2(3,:),'ob')
plot3(class3(1,:),class3(2,:),class3(3,:),'.k')
xlabel(,x_1'); ylabel(,x_2'); zlabel(,x_3');
title('Distribution of the vectors belonging to the three classes')
legend('Class 1','Class 2','Class 3',0); grid
```

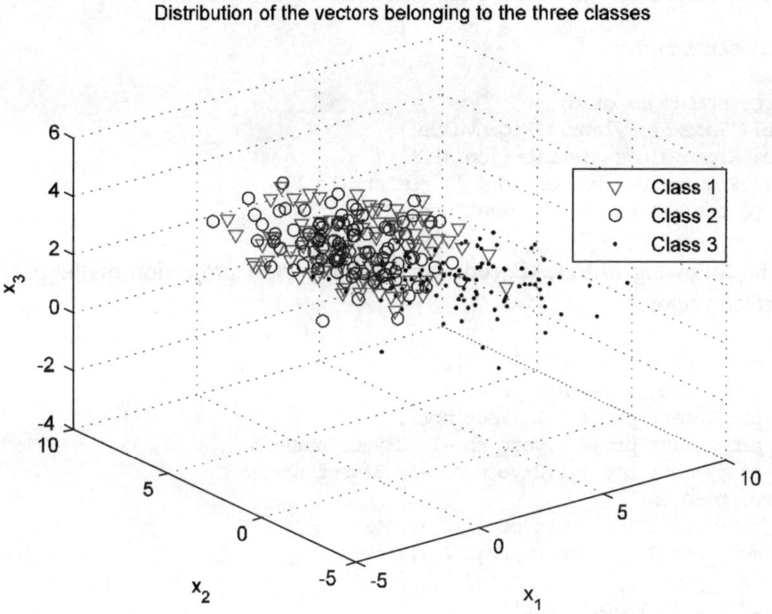

Figure 13.12. *Distribution of the vectors belonging to the three classes*

b.

An example of a function implementing the principal component analysis is provided below.

```
function [kmat,pes]=pca(pen,threshold)
```

```
% Principal component analysis
% [kmat,pes]=pca(pen,threshold);
% pen - matrix of input vectors
% threshold - if higher than 1, stands for the projection space dimension; if
lower than 1, stands for the percentage of the preserved variance
% kmat - projection matrix
% pes - matrix of projected vectors
moy=mean(pen,2); sigx=cov(pen');
[vectp,valp]=eig(sigx);
vlpr=abs(diag(valp));
[ld,idx]=sort(vlpr);
idx=flipud(idx);
ld=flipud(ld);
vectps=vectp(:,idx);
if threshold<1
    ldn=(norm(ld))^2; ldns=cumsum(ld.^2);
    ldns=ldns/ldn; idxc=find(ldns>=threshold); ns=idxc(1);
else
    ns=threshold;
end
figure; stem(ld); grid;
xlabel('Index'); ylabel('Eigenvalue');
title('Eigenvalues distribution (PCA)')
disp(['Eigenvalues vector ld = ', num2str(ld')])
kmat=vectps(:,1:ns).'; pes=kmat*pen;
```

The following MATLAB code performs the PCA projection of the previously generated vectors:

```
matr_vect=[class1 class2 class3];
[kmat,matr_proj]=pca(matr_vect,2);
matr_proj1=matr_proj( :,1 :vect_nbr);
matr_proj2=matr_proj( :,vect_nbr+1 :2*vect_nbr);
matr_proj3=matr_proj( :,2*vect_nbr+1 :3*vect_nbr);
figure; hold on
plot(matr_proj1(1,:),matr_proj1(2,:),'vr')
plot(matr_proj2(1,:),matr_proj2(2,:),'ob')
plot(matr_proj3(1,:),matr_proj3(2,:),'.k')
xlabel('y_1'); ylabel('y_2');
title('PCA projection')
legend('Class 1','Class 2','Class 3',0);
```

The PCA yields the following eigenvalues: 6.77, 3.16 and 0.42. Thus, equation [13.3] leads to the relative error calculated below:

$$\varepsilon_r = \frac{\lambda_3}{\sum_{i=1}^{3} \lambda_i} = \frac{0.42}{6.77+3.16+0.42} = 0.04$$

Consequently, there is a minor information loss associated with this projection. The space dimension reduction from 3 to 2 almost without any information loss was possible due to the initial space redundancy.

The projection result is illustrated in Figure 13.13. The axes of the projection plane correspond to the maximum variance directions in the initial space.

c.

The following function performs the linear discriminant analysis.

```
function [kmat,pes]=lda(pen,threshold)
% Linear disriminant analysis
% [kmat,pes]=lda(pen,threshold);
% pen - matrix of input vectors (the last row contains the labels)
% threshold - if higher than 1, stands for the projection space dimension; if
lower than 1, stands for the percentage of the preserved variance
% kmat - projection matrix
% pes - matrix of projected vectors (the last row contains the labels)
x=pen(1:size(pen,1)-1,:);
labelx=pen(size(pen,1),:);
nc=max(labelx); [np,nvt]=size(x);
for k=1:nc
    ic=num2str(k);  idxc=find(labelx==k);
    eval(['x',ic,'=x(:,idxc);']);
    eval(['xr',ic,'=x',ic,''';']);
end
moy0=zeros(np,1); sw=zeros(np,np); sb=zeros(np,np);
for l=1:nc
    eval(['moy',num2str(l),'=mean(xr',num2str(l),')'';']);
    eval(['moy0=moy0+(1/nc)*moy',num2str(l),';']);
    eval(['vary',num2str(l),'=cov(xr',num2str(l),');']);
    eval(['sw=sw+(1/nc)*vary',num2str(l),';']);
end
for l=1:nc
    eval(['dif=moy',num2str(l),'-moy0;']);
    eval(['sb=sb+(1/nc)*dif*dif'';']);
end
s=(inv(sw+sb)*sb);
[vectpr,valpr]=eig(s);
ld=abs(diag(valpr));
[ld,idxvp]=sort(ld);
idxvp=flipud(idxvp);
ld=flipud(ld);
vectprs=vectpr(:,idxvp);
if threshold<1
    ldn=(norm(ld))^2; ldns=cumsum(ld.^2);
    ldns=ldns/ldn; idxc=find(ldns>=threshold); ns=idxc(1);
else
```

```
    ns=threshold;
end
figure; stem(ld);
xlabel('Index'); ylabel('Eigenvalue');
title('Eigenvalues distribution (LDA)')
disp(['Eigenvalues vector ld = ', num2str(ld')])
kmat=vectprs(:,1:ns).';
y=kmat*x; pes=[y;labelx];
```

The following MATLAB code performs the LDA projection of the previously generated vectors:

```
etiq1=ones(1,vect_nbr);
vect_labeluettes=[ etiq1 2*etiq1 3*etiq1];
matr_vect=[matr_vect ;vect_labeluettes]; [kmat,matr_proj]=lda(matr_vect,2);
matr_proj1=matr_proj( :,1 :vect_nbr);
matr_proj2=matr_proj( :,vect_nbr+1 :2*vect_nbr);
matr_proj3=matr_proj( :,2*vect_nbr+1 :3*vect_nbr);
figure;
hold on;
plot(matr_proj1(1,:),matr_proj1(2,:),'vr')
plot(matr_proj2(1,:),matr_proj2(2,:),'ob');
plot(matr_proj3(1,:),matr_proj3(2,:),'.k')
xlabel('y_1');
ylabel('y_2');
title('LDA projection')
legend('Class 1','Class 2','Class 3',0)
```

The LDA yields the following normalized eigenvalues: 0.96, 0.55 and 0. The first eigenvalue is close to 1 and indicates that the corresponding eigenvector has a high discriminant capability.

The LDA leads to a maximum separability of the projected classes, so that it makes the classifier task much easier.

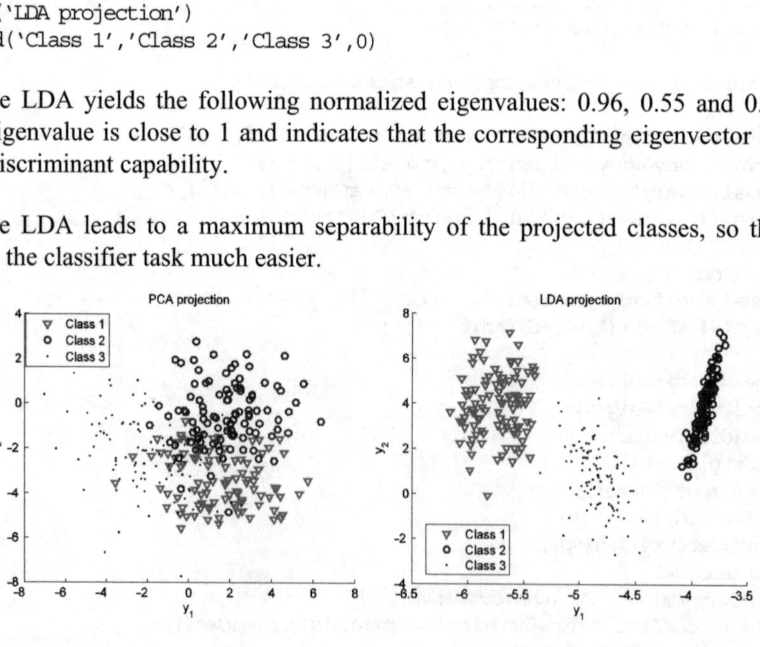

Figure 13.13. *PCA (left) and LDA (right) projection of the three Gaussian classes*

EXERCISE 13.2.

a. Generate two Gaussian classes of 200 two-dimensional vectors with the mean vectors and the covariance matrices given below:

$$\boldsymbol{\mu}_1 = \begin{bmatrix} 3 \\ 5 \end{bmatrix},\ \boldsymbol{\Sigma}_1 = \begin{bmatrix} 4 & 6 \\ 6 & 12 \end{bmatrix},\ \boldsymbol{\mu}_2 = \begin{bmatrix} 15 \\ 4 \end{bmatrix},\ \boldsymbol{\Sigma}_2 = \begin{bmatrix} 12 & -4 \\ -4 & 8 \end{bmatrix}$$

Then divide each class randomly into a training subset and a test subset containing the same number of vectors.

b. The *a priori* probabilities of the two classes are considered equal. Perform the classification of the test subset using a direct approximation of the Bayes classifier obtained from the training subset.

c. Repeat the classification using the fuzzy KNN method. Plot on the same figure the test vectors and the separating surfaces generated by the two classifiers.

d. Evaluate the classification rate for the two classifiers by averaging their performances on 1,000 outcomes.

a.

The vectors of the two classes can be generated using the following MATLAB code (the function `randperm.m` performs the random permutation of the integer from 1 to n):

```
class_nbr=2; vect_nbr=200;
mean_vect1=[5;10]; cov_matr1=[4 6;6 12];
class1=generation_class_gauss(mean_vect1,cov_matr1,vect_nbr);
mean_vect2=[10;20]; cov_matr2=[12 -4;-4 8];
class2=generation_class_gauss(mean_vect2,cov_matr2,vect_nbr);
idx=randperm(vect_nbr);
vect_nbr_test=fix(vect_nbr/2);
idx_test=idx(1:vect_nbr_test);
idx_train=idx(vect_nbr_test+1:end);
vect_label_train1=ones(1,vect_nbr-vect_nbr_test);
vect_label_test1=ones(1,vect_nbr_test);
vect_label_train2=2* vect_label_train1;
vect_label_test2=2* vect_label_test1;
vect_train1=class1( :,idx_train);
vect_test1=class1(:,idx_test);
vect_train2=class2( :,idx_train);
vect_test2=class2(:,idx_test);
matr_train=[vect_train1 vect_train2];
vect_label_train=[vect_label_train1 vect_label_train2];
matr_test=[vect_test1 vect_test2];
```

```
vect_label_test=[vect_label_test1 vect_label_test2];
```

b.

Equation [13.7] can be used for implementing the Bayes classifier with the class distributions given by the following 2D Gaussian functions:

$$f(\mathbf{x} \mid \omega_i) = \frac{1}{2\pi \left|\det\left(\boldsymbol{\Sigma}_i\right)\right|^{1/2}} \exp\left[-\frac{1}{2}(\mathbf{x}-\boldsymbol{\mu}_i)^T \boldsymbol{\Sigma}_i^{-1} (\mathbf{x}-\boldsymbol{\mu}_i)\right]$$

The logarithmic form of the Bayes classifier discriminant functions has been preferred in the MATLAB code provided below.

```
mean_vect1_est=mean(vect_train1,2);
cov_matr1_est=cov(vect_train1');
matr_moy1=repmat(mean_vect1_est,1,2*vect_nbr_test);
mean_vect2_est=mean(vect_train2,2);
cov_matr2_est=cov(vect_train2');
matr_moy2=repmat(mean_vect2_est,1,2*vect_nbr_test);
restest=zeros(2,2*vect_nbr_test);
restest(1,:)=vect_label_test;
vect_fun_discrim1=log(abs(det(cov_matr1_est)))+sum(((matr_test-
matr_moy1)'*inv(cov_matr1_est))'.*(matr_test-matr_moy1));
vect_fun_discrim2=log(abs(det(cov_matr2_est)))+sum(((matr_test-
matr_moy2)'*inv(cov_matr2_est))'.*(matr_test-matr_moy2));
vect_fun_discrim=vect_fun_discrim1-vect_fun_discrim2;
restest(2,vect_fun_discrim<=0)=1;
restest(2,vect_fun_discrim>0)=2;
conf_matr=zeros(class_nbr);
for k1=1:class_nbr
   for k2=1:class_nbr
    conf_matr(k1,k2)=length(find((restest(1,:)==k1)&(restest(2,:)==k2)));
   end
end
conf_matr
vect_vect_nbr=sum(conf_matr,2);
vect_prob_classif=diag(conf_matr)./vect_vect_nbr
maxx=max(matr_test(1,:)); maxx=maxx+.1*abs(maxx);
maxy=max(matr_test(2,:)); maxy=maxy+.1*abs(maxy);
minx=min(matr_test(1,:)); minx=minx-.1*abs(minx);
miny=min(matr_test(2,:)); miny=miny-.1*abs(miny);
xv=linspace(minx,maxx,100);
yv=linspace(miny,maxy,100);
matr_x,matr_y]=meshgrid(xv,yv);
matr_test0=[matr_x(:)';matr_y(:)'];
vect_nbr_test0=size(matr_test0,2);
matr_moy1=repmat(mean_vect1_est,1,vect_nbr_test0);
matr_moy2=repmat(mean_vect2_est,1,vect_nbr_test0);
```

```
vect_fun_discrim1=log(abs(det(cov_matr1_est)))+sum(((matr_test0-
matr_moy1)'*inv(cov_matr1_est))'.*(matr_test0-matr_moy1));
vect_fun_discrim2=log(abs(det(cov_matr2_est)))+sum(((matr_test0-
matr_moy2)'*inv(cov_matr2_est))'.*(matr_test0-matr_moy2));
vect_fun_discrim=vect_fun_discrim1-vect_fun_discrim2;
matr_fun_discrim=reshape(vect_fun_discrim,size(matr_x,1),size(matr_x,2));
figure; imagesc(xv,yv,matr_fun_discrim);
colormap(flipud(spring));
colorbar; hold on;
xlabel(,x_1'); ylabel(,x_2');
contour(xv,yv,matr_fun_discrim,[0 0],'-k','LineWidth',[3]);
plot(vect_test1(1,:),vect_test1(2,:),'db','Markersize',[8])
plot(vect_test2(1,:),vect_test2(2,:),'og','Markersize',[8])
title('Classification of the test vectors using the Bayes rule')
legend('S_s_e_p_a_r_a_t_i_o_n','Class 1','Class 2')
```

c.

The Fuzzy KNN classifier has been implemented using equation [13.11], [13.12] and [13.13]. The result is shown in Figure 13.14.

```
kn=3; kf=10;
% Calculation of the training vectors' membership coefficients
vect_nbr_train=size(matr_train,2);
vect_nbr_test=size(matr_test,2);
matr_dist=dist(matr_train',matr_train);
matr_max=max(max(matr_dist))*eye(vect_nbr_train);
matr_dist=matr_dist+matr_max;
[matr_dists,matr_idxs]=sort(matr_dist);
matr_label=vect_label_train(matr_idxs); matr_label_red=matr_label(1 :kf,:);
matr_hist_label=histc(matr_label_red,[.5 :1 :class_nbr+.5]);
matr_hist_label=matr_hist_label(1 :class_nbr,:);
matr_coeff_appart=matr_hist_label/kf;
% Fuzzy KNN classification
matr_sort=zeros(class_nbr,vect_nbr_test); mc=2;
matr_dist=dist(matr_train',matr_test);
matr_dist_inv=matr_dist.^(-2/mc);
[matr_dists,matr_idxs]=sort(matr_dist);
matr_label=vect_label_train(matr_idxs);
matr_label_red=matr_label(1 :kn,:);
matr_idxs_red=matr_idxs(1 :kn,:); matr_dist_inv_red=zeros(kn,vect_nbr_test);
for k=1 :vect_nbr_test
    matr_dist_inv_red(1 :kn,k)=matr_dist_inv(matr_idxs_red( :,k),k);
end
vect_denom=sum(matr_dist_inv_red); matr_sort=zeros(class_nbr,vect_nbr_test);
for k1=1:vect_nbr_test
    denom=vect_denom(k1);
    for k2=1:class_nbr
     numer=sum(matr_dist_inv_red(:,k1).*matr_coeff_appart(k2,matr_idxs_red(:,
k1))');
```

```
    matr_sort(k2,k1)=numer/denom;
    end
end
[valmax,idxmax]=max(matr_sort);
restest=[vect_label_test;idxmax];
conf_matr=zeros(class_nbr);
for k1=1:class_nbr
    for k2=1:class_nbr
     conf_matr(k1,k2)=length(find((restest(1,:)==k1)&(restest(2,:)==k2)));
    end
end
conf_matr
vect_vect_nbr=sum(conf_matr,2);
vect_prob_classif=diag(conf_matr)./vect_vect_nbr
matr_sort0=zeros(class_nbr,vect_nbr_test0);
matr_dist=dist(matr_train',matr_test0);
matr_dist_inv=matr_dist.^(-2/mc);
[matr_dists,matr_idxs]=sort(matr_dist);
matr_label=vect_label_train(matr_idxs);
matr_label_red=matr_label(1:kn,:);
matr_idxs_red=matr_idxs(1:kn,:);
matr_dist_inv_red=zeros(kn,vect_nbr_test0);
for k=1:vect_nbr_test0
    matr_dist_inv_red(1:kn,k)=matr_dist_inv(matr_idxs_red(:,k),k);
end
vect_denom=sum(matr_dist_inv_red);
for k1=1:vect_nbr_test0
    denom=vect_denom(k1);
    for k2=1:class_nbr
     numer=sum(matr_dist_inv_red(:,k1).*matr_coeff_appart(k2,matr_idxs_red(:,
k1))');
    matr_sort(k2,k1)=numer/denom;
    end
end
vect_fun_discrim=matr_sort(2,:)-matr_sort(1,:);
matr_fun_discrim=reshape(vect_fun_discrim,size(matr_x,1),size(matr_x,2));
figure;imagesc(xv,yv,matr_fun_discrim);
colormap(flipud(spring)); colorbar; hold on
contour(xv,yv,matr_fun_discrim,[0 0],'-k','LineWidth',[3]);
plot(vect_test1(1,:),vect_test1(2,:),'db','Markersize',[8])
plot(vect_test2(1,:),vect_test2(2,:),'og','Markersize',[8])
xlabel(,x_1'); ylabel(,x_2');
title('Classification of the test vectors using the Fuzzy KNN rule')
legend('S_s_e_p_a_r_a_t_i_o_n','Class 1','Class 2')
```

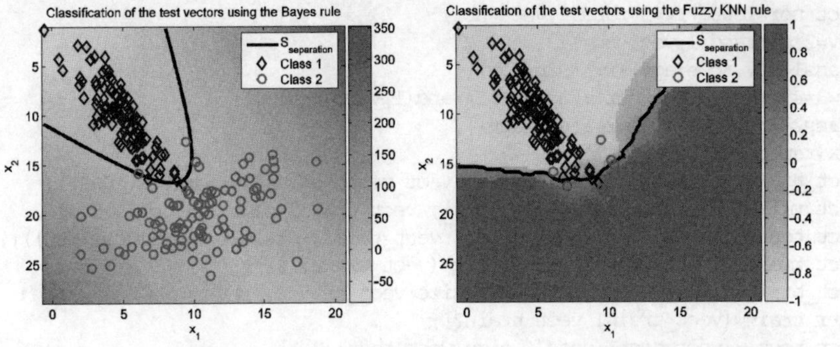

Figure 13.14. *Classification of the test vectors using the Bayes and Fuzzy KNN rules*

d. The averaged classification rates are provided in the table below.

Classifier	Class 1	Class 2
Bayes	0.969	0.979
Fuzzy KNN	0.967	0.973

Table 13.1. *Mean classification rates obtained using the Bayes and Fuzzy KNN classifiers*

EXERCISE 13.3.

a. Generate two classes of 1,500 vectors uniformly distributed inside a disk with the radius 3 (first class) and inside a circular zone whose radius varies from 3 to 6 (second class). The two classes are centered at the origin.

Then divide each class randomly into three subsets (training, test and validation) containing the same number of vectors. Plot the training vectors of the two classes.

b. Use the first subset to train a multilayer perceptron and the third subset to decide the end of this process when its generalization capability becomes maximal. Consider the backpropagation learning algorithm with *momentum term* and variable learning rate. Classify the test vectors using trained multilayer perceptron.

c. Plot the outputs of the multilayer perceptron corresponding to the zone occupied by the two classes and highlight their link with the vectors' distributions.

a.

The MATLAB code below generates the vectors of the two classes and divides them to form the training, test and validation subsets.

```
vect_nbr=1500; rv=3*rand(1,vect_nbr);
agv=2*pi*rand(1,vect_nbr);
class1=[rv.*cos(agv);rv.*sin(agv)];
rv=3+3*rand(1,vect_nbr); agv=2*pi*rand(1,vect_nbr);
class2=[rv.*cos(agv);rv.*sin(agv)];
idx=randperm(1500);
vect_train1=class1(:,idx(1:500));  vect_test1=class1(:,idx(501:1000));
vect_valid1=class1(:,idx(1001:1500)); vect_train2=class2(:,idx(1:500));
vect_test2=class2(:,idx(501:1000)); vect_valid2=class2(:,idx(1001:1500));
vect_train1=[vect_train1;ones(1,size(vect_train1,2))];
vect_train2=[vect_train2;2*ones(1,size(vect_train2,2))];
matr_train=[vect_train1 vect_train2];
vect_test1=[vect_test1;ones(1,size(vect_test1,2))];
vect_test2=[vect_test2;2*ones(1,size(vect_test2,2))];
matr_test=[vect_test1 vect_test2];
vect_valid1=[vect_valid1;ones(1,size(vect_valid1,2))];
vect_valid2=[vect_valid2;2*ones(1,size(vect_valid2,2))];
matr_valid=[vect_valid1 vect_valid2];
figure; hold on;
plot(vect_train1(1,:),vect_train1(2,:),'xr','Markersize',[8])
plot(vect_train2(1,:),vect_train2(2,:),'ob','Markersize',[8])
xlabel('x_1'); ylabel('x_2');
title('Distribution of the training vectors')
legend('Class 1','Class 2'); axis equal
```

b.

The multilayer perceptron is trained using the MATLAB code given below. The test vectors and the separation surface between the two classes are plotted in Figure 13.15.

```
vect_ettiq_train=matr_train(end,:);
matr_train=matr_train(1 :2,:);
vect_ettiq_test=matr_test(end,:);
matr_test=matr_test(1 :2,:);
vect_ettiq_valid=matr_valid(end,:);
matr_valid=matr_valid(1 :2,:);
class_nbr=max(vect_ettiq_train);
vect_nbr_train=size(matr_train,2);
vect_nbr_test=size(matr_test,2);
vect_nbr_valid=size(matr_valid,2);
PR_matr=minmax(matr_train); nbre_neuro=[10 10 class_nbr];
net=newff(PR_matr, nbre_neuro,{'logsig' 'logsig' 'logsig'});
target_train=zeros(class_nbr,vect_nbr_train);
target_test=zeros(class_nbr,vect_nbr_test);
target_valid=zeros(class_nbr,vect_nbr_valid);
for k=1:class_nbr
    idx=find(vect_ettiq_train==k);
    target_train(k,idx)=ones(1,length(idx));
    idx=find(vect_ettiq_test==k);
```

```
    target_test(k,idx)=ones(1,length(idx));
    idx=find(vect_ettiq_valid==k);
    target_valid(k,idx)=ones(1,length(idx));
end
net.trainFcn='traingdx'; net.trainParam.epochs=1000;
net.trainParam.goal=0; net.trainParam.lr=0.01;
net.trainParam.lr_inc=1.05; net.trainParam.lr_dec=0.7;
net.trainParam.max_fail=10; net.trainParam.max_perf_inc=1;
net.trainParam.mc=0.9; net.trainParam.min_grad=0;
net.trainParam.show=50; net.trainParam.time=inf;
VV.P=matr_valid;VV.T=target_valid;
TV.P=matr_test;TV.T=target_test;
net = train(net,matr_train,target_train,[],[],VV,TV);
Yr = sim(net,matr_test);
[valmax,idxmax]=max(Yr);
restest=zeros(2,vect_nbr_test);
restest(1,:)=vect_ettiq_test;
restest(2,:)=idxmax;
for k1=1:class_nbr
    for k2=1:class_nbr
     conf_matr_nnet(k1,k2)=length(find((restest(1,:)==k1)&(restest(2,:)==k2))
);
    end
end
conf_matr_nnet
vect_vect_nbr=sum(conf_matr_nnet,2);
vect_prob_classif_nnet=(diag(conf_matr_nnet)./vect_vect_nbr)'
maxx=max(matr_train(1,:)); maxx=maxx+.1*abs(maxx);
maxy=max(matr_train(2,:)); maxy=maxy+.1*abs(maxy);
minx=min(matr_train(1,:)); minx=minx-.1*abs(minx);
miny=min(matr_train(2,:)); miny=miny-.1*abs(miny);
xv=linspace(minx,maxx,100); yv=linspace(miny,maxy,100);
[matr_x,matr_y]=meshgrid(xv,yv);
matr_xy=[matr_x(:)'; matr_y(:)'];
nv=size(matr_xy,2);
sortie_pmc = sim(net,matr_xy);
g1_pmc=reshape(sortie_pmc(1,:),size(matr_x,1),size(matr_x,2));
g2_pmc=reshape(sortie_pmc(2,:),size(matr_x,1),size(matr_x,2));
figure; contour(xv,yv,g1_pmc-g2_pmc,[0 0],'-k','LineWidth',[3]);
hold on;plot(vect_test1(1,:),vect_test1(2,:),'xr','Markersize',[8])
plot(vect_test2(1,:),vect_test2(2,:),'ob','Markersize',[8])
xlabel(,x_1'); ylabel(,x_2');
title('Classification using a MLP')
legend('S_s_e_p_a_r_a_t_i_o_n','Class 1','Class 2'); axis equal
```

The confusion matrix and the classification rates corresponding to the two classes are obtained as follows.

conf_matr_nnet =

497 3

2 498

vect_prob_classif_nnet =

0.9940 0.9960

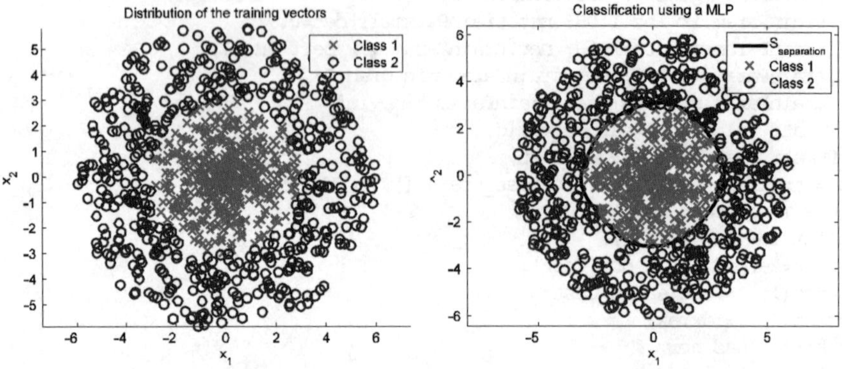

Figure 13.15. *Distribution of the training vectors and classification of the test vectors using a multilayer perceptron*

c.

The following MATLAB code plots the outputs of the multilayer perceptron corresponding to the zone occupied by the vectors of each class.

```
figure; imagesc(xv,yv,g1_pmc); hold on;
contour(xv,yv,g1_pmc);
axis xy; colormap(spring)
xlabel('x_1'); ylabel('x_2');
colorbar; axis equal
title('MLP output corresponding to the class 1')
figure; imagesc(xv,yv,g2_pmc); hold on;
contour(xv,yv,g2_pmc);
axis xy; colormap(spring)
xlabel('x_1'); ylabel('x_2');
colorbar; axis equal
title('MLP output corresponding to the class 2')
```

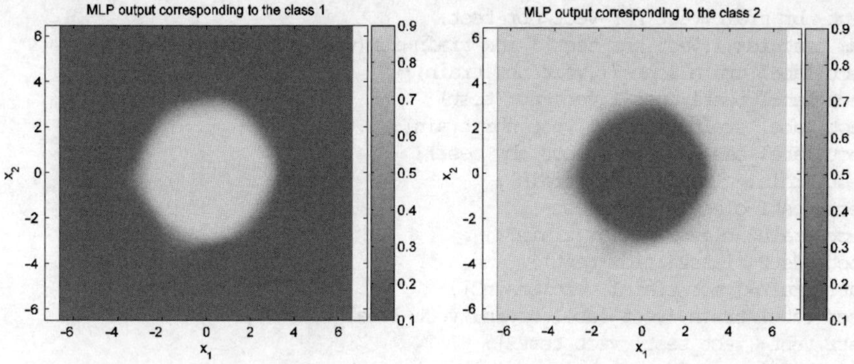

Figure 13.16. *MLP outputs corresponding to the vectors of the two classes*

It can be seen that at the end of the training phase, the two outputs of the multilayer perceptron approximate the statistical distributions of the two classes. The decision is therefore optimal according to the MAP (maximum *a posteriori*) criterion and the trained MLP approximates the Bayes classifier.

EXERCISE 13.4.

a. Generate two classes of 200 two-dimensional Gaussian vectors having the following mean vectors and covariance matrices:

$$\boldsymbol{\mu}_1 = \begin{bmatrix} 5 \\ 10 \end{bmatrix}, \ \boldsymbol{\mu}_2 = \begin{bmatrix} 12 \\ 10 \end{bmatrix}, \ \boldsymbol{\Sigma}_1 = \boldsymbol{\Sigma}_2 = \begin{bmatrix} 4 & 0 \\ 0 & 4 \end{bmatrix}$$

Then divide each class randomly into two subsets (training and test) containing the same number of vectors. Normalize the two subsets so that for each variable the mean is 0 and the standard deviation is 1.

b. Indicate the optimal separation surface in this case. Classify the test vectors using a fuzzy perceptron and a SVM classifier and compare the obtained results.

a.

The generation of the two classes is the same as for the previous exercises.

```
class_nbr=2; vect_nbr=200;
mean_vect1=[5;10]; cov_matr1=[4 0;0 4];
class1=generation_class_gauss(mean_vect1,cov_matr1,vect_nbr);
mean_vect2=[10;5]; cov_matr2=cov_matr1;
class2=generation_class_gauss(mean_vect2,cov_matr2,vect_nbr);
idx=randperm(vect_nbr); vect_nbr_test=fix(vect_nbr/2);
```

```
vect_nbr_train=vect_nbr-vect_nbr_test;
idx_test=idx(1:vect_nbr_test); idx_train=idx(vect_nbr_test+1:end);
vect_label_train1=ones(1,vect_nbr_train);
vect_label_test1=ones(1,vect_nbr_test);
vect_label_train2=-ones(1,vect_nbr_train);
vect_label_test2=-ones(1,vect_nbr_test);
vect_train1=class1( :,idx_train);
vect_test1=class1(:,idx_test);
vect_train2=class2( :,idx_train);
vect_test2=class2(:,idx_test);
matr_train=[vect_train1 vect_train2];
vect_label_train=[vect_label_train1 vect_label_train2];
matr_test=[vect_test1 vect_test2];
vect_label_test=[vect_label_test1 vect_label_test2];
matr_train = prestd(matr_train);  matr_test = prestd(matr_test);
vect_train1=matr_train( :,1 :vect_nbr_train);
vect_train2=matr_train( :,vect_nbr_train+1 :end);
```

b.

Since the two classes are Gaussian, with the same covariance matrices, the optimal separating surface is a line. The MATLAB code below is able to find an approximation of this line by training a fuzzy perceptron and a SVM classifier (see Figure 13.17). In the second case, the results were obtained using the functions provided by "SVM-KM, *SVM and Kernel Methods* MATLAB *Toolbox*", created by Stéphane Canu and Alain Rakotomamonjy (Perception Systems and Information Laboratory, Rouen, France).

```
% Calculation of the membership coefficients for the training vectors
moy1=mean(vect_train1,2);
moy2=mean(vect_train2,2);
D=dist(moy1',moy2);
epsy=0.02; g=2; eta=.1;
delta=.1; A=4*sinh(g);
bety=epsy+(1-exp(-g))/A
matr_coeff_train1=zeros(2,vect_nbr_train);
matr_coeff_train2=zeros(2,vect_nbr_train);
for k=1 :vect_nbr_train
    xk=vect_train1( :,k);
    D1k=(dist(moy1',xk));
    D2k=(dist(moy2',xk));
    u1k=1/2+(1/A)*(exp(g*(D2k-D1k)/D)-exp(-g));
    matr_coeff_train1(1,k)=u1k;
    matr_coeff_train1(2,k)=1-u1k;
    xk=vect_train2( :,k);
    D1k=(dist(moy1',xk));
    D2k=(dist(moy2',xk));
    u2k=1/2+(1/A)*(exp(g*(D1k-D2k)/D)-exp(-g));
    matr_coeff_train2(2,k)=u2k;
```

```
    matr_coeff_train2(1,k)=1-u2k;
end
% Training
maxx=max(matr_train(1,:));
maxx=maxx+.1*abs(maxx);
maxy=max(matr_train(2,:));
maxy=maxy+.1*abs(maxy);
minx=min(matr_train(1,:));
minx=minx-.1*abs(minx);
miny=min(matr_train(2,:));
miny=miny-.1*abs(miny);
matr_train=[matr_train;ones(1,2*vect_nbr_train)];
matr_test=[matr_test;ones(1,2*vect_nbr_test)];
target_output=[ones(1,vect_nbr_train) zeros(1,vect_nbr_train)];
matr_coeff_train=[matr_coeff_train1 matr_coeff_train2];
idx_ext_fuzzy_zone=find((matr_coeff_train(1,:)<0.5-
bety)|(matr_coeff_train(1,:)>0.5+bety));
idx_fuzzy_zone=find((matr_coeff_train(1,:)>=0.5-bety)
&(matr_coeff_train(1,:)<=0.5+bety));
plot(matr_train(1,idx_fuzzy_zone),matr_train(2,idx_fuzzy_zone),'xr','Markersi
ze',[8])
label_ext_fuzzy_zone=vect_label_train(idx_ext_fuzzy_zone);
vect_diff_uk=(abs(matr_coeff_train(1,:)-matr_coeff_train(2,:))).^delta;
weight_vect=0.1*randn(3,1); stop_criterion=1;
while stop_criterion~=0
    vect_net=weight_vect'*matr_train;
    actual_output=(1+sign(vect_net))/2;
    dec_ext_fuzzy_zone=actual_output(idx_ext_fuzzy_zone);
    dec_ext_fuzzy_zone(dec_ext_fuzzy_zone==0)=-1;
    stop_criterion=sum(abs(dec_ext_fuzzy_zone-label_ext_fuzzy_zone))
    xhp=[minx maxx];
     yhp=-(weight_vect(1)/weight_vect(2))*xhp-
(weight_vect(3)/weight_vect(2));
    hh=plot(xhp,yhp,'r-','LineWidth',[3]);
    axis([minx maxx miny maxy]); drawnow
    weight_vect=weight_vect+eta*matr_train*((target_output-
actual_output).*vect_diff_uk)';
    pause;  set(hh,'Visible','off');
    legend('Class 1','Class 2','Vect. zone floue','S_s_e_p_a_r_a_t_i_o_n')
end
set(hh,'Visible','on');
legend('Class 1','Class 2','Vect. zone floue','S_s_e_p_a_r_a_t_i_o_n')
vect_net_test=weight_vect'*matr_test; restest=zeros(2,2*vect_nbr_test);
restest(1,:)=vect_label_test;
restest(2,:)=sign(vect_net_test);
restest(restest==-1)=2;
for k1=1:class_nbr
    for k2=1:class_nbr
     conf_matr_fuzzy_perceptron(k1,k2)=length(find((restest(1,:)==k1)&(restes
t(2,:)==k2)));
```

```
    end
end
conf_matr_fuzzy_perceptron
vect_vect_nbr=sum(conf_matr_fuzzy_perceptron,2);
vect_prob_classif_fuzzy_perceptron=(diag(conf_matr_fuzzy_perceptron)./vect_ve
ct_nbr)'
xv=linspace(minx,maxx,100);
yv=linspace(miny,maxy,100);
[matr_x,matr_y]=meshgrid(xv,yv); matr_train0=[matr_x(:)';matr_y(:)'];
matr_train0=[matr_train0; ones(1,size(matr_train0,2))];
vect_nbr_train0=size(matr_train0,2);
vect_net=weight_vect'*matr_train0;
matr_net=reshape(vect_net,size(matr_x,1),size(matr_x,2));
figureimagesc(xv,yv,matr_net);
axis xy; colormap(flipud(spring));
colorbar; hold on
plot(xhp,yhp,'k-','LineWidth',[3]);
axis([minx maxx miny maxy]);
plot(vect_train1(1,:),vect_train1(2,:),'sb','Markersize',[8])
plot(vect_train2(1,:),vect_train2(2,:),'og','Markersize',[8])
plot(matr_train(1,idx_fuzzy_zone),matr_train(2,idx_fuzzy_zone),'xk','Markersi
ze',[8])
xlabel(,x_1'); ylabel(,x_2');
title('Fuzzy perceptron based classification')
legend('S_s_e_p_a_r_a_t_i_o_n','Class 1','Class 2','Fuzzy zone vectors')
xapp= matr_train(1 :2,:)';
yapp= vect_label_train';
xtest= matr_test (1 :2,:)';
matr_train=[xapp';vect_label_train];
matr_test=[xtest';vect_label_test];
C=10000000; verbose=0; lambda=1e-7;
kernel='poly'; kerneloption=1;
[xsup,w,b,pos,timeps,alpha,obj]=svmclass(xapp,yapp,C,lambda,kernel,kernelopti
on,verbose);
vect_support= matr_train (1 :2,pos);
ypred = svmval(xtest,xsup,w,b,kernel,kerneloption);
ypred(ypred>=0)=1; ypred(ypred<0)=2;
vect_label_test(vect_label_test==-1)=2;
restest = zeros(2,size(matr_test,2));
restest(1,:) = vect_label_test;
restest(2,:) = ypred;
for k1=1:class_nbr
    for k2=1:class_nbr
     conf_matr_svm(k1,k2)=length(find((restest(1,:)==k1)&(restest(2,:)==k2)))
;
    end
end
conf_matr_svm
vect_prob_classif_svm=(diag(conf_matr_svm)./vect_vect_nbr)'
[matr_x,matr_y]=meshgrid(xv,yv);
```

```
matr_test_g =[matr_x(:);matr_y(:)];
vect_classif_g = svmval(matr_test_g,xsup,w,b,kernel,kerneloption);
matr_classif_g=reshape(vect_classif_g, size(matr_x,1),size(matr_x,2));
figure
imagesc(xv,yv,matr_classif_g);
axis xy; colormap(flipud(spring));
colorbar; hold on
contour(xv,yv,matr_classif_g,[0 0],'-k','LineWidth',[3]);
plot(vect_train1(1,:),vect_train1(2,:),'sb','Markersize',[8])
plot(vect_train2(1,:),vect_train2(2,:),'og','Markersize',[8])
plot(vect_support(1,:),vect_support(2,:),'xk','Markersize',[8])
xlabel(,x_1');
ylabel(,x_2');
title('SVM based classification')
legend('S_s_e_p_a_r_a_t_i_o_n','Class 1','Class 2','Support vectors')
```

Figure 13.17 shows that the obtained solutions are quite similar in the two cases. The vectors from the fuzzy zone play the same role for the first classifier as the support vectors for the SVM.

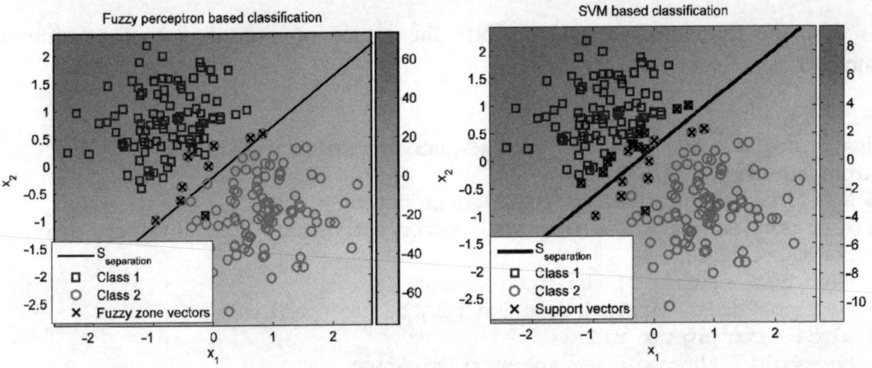

Figure 13.17. *Separation surfaces obtained after training a fuzzy perceptron and a SVM classifier in the case of two classes having the same covariance matrix*

13.3. Exercises

EXERCISE 13.5.

a. Generate two sets of 100 three-dimensional Gaussian vectors having the following mean vector and covariance matrix:

$$\boldsymbol{\mu} = \begin{bmatrix} 3 \\ 5 \\ -2 \end{bmatrix}, \quad \boldsymbol{\Sigma} = \begin{bmatrix} 5 & 7 & -7 \\ 7 & 13 & -12 \\ -7 & -12 & 11 \end{bmatrix}$$

b. Perform the principal component analysis on the first set and then use the obtained result to project the second set. Conclude on the generalization capability of this type of analysis.

c. Repeat the previous analysis after preprocessing the generated vectors, so that the mean of each component is 0 and its standard deviation is 1. Compare to the results previously obtained and conclude about the role of this data preprocessing.

EXERCISE 13.6.

Generate two classes of 500 vectors uniformly distributed inside a sphere of radius 0.3 (first class) and inside a spherical calotte whose radius varies from 0.7 to 1 (second class). The two classes are centered at the origin.

Project the two classes in a two-dimensional space using the PCA, the LDA and the Sammon method randomly initialized. Then repeat the projection using the Sammon algorithm initialized with the PCA solution and conclude on the obtained results.

Use the function below to perform the projection according to the Sammon method.

```
function
[matr_proj,Sammon_stress,esr_vect]=Sammon_method(matr_init,matr_proj_init,
alpha,threshold)
% Non-linear projection using the Sammon method
%[matr_proj,Sammon_stress,esr_vect]=Sammon_method(matr_init,matr_proj_init,
alpha,threshold);
% matr_init - matrix of the input vectors
% matr_proj_init - initial solution for the projected vectors
% alpha - convergence rate
% threshold - threshold for the stop criterion
% matr_proj - matrix of the projected vectors
% Sammon_stress - Sammon stress vector
% esr_vect - relative error vector
matr_dist_init=triu(dist(matr_init',matr_init));
matr_dist_proj_init=triu(dist(matr_proj_init',matr_proj_init));
coeff_sam=1/sum(sum(matr_dist_init));
matr_dist_diff=((matr_dist_init-matr_dist_proj_init).^2)./matr_dist_init;
matr_dist_diff(isnan(matr_dist_diff))=0;
Sammon_stress_old=coeff_sam*sum(sum(matr_dist_diff));
Sammon_stress(1)=Sammon_stress_old;
cont=1; esr=1e3;
vect_nbr=size(matr_proj_init,2);
while esr>threshold
    matr_distc=(matr_dist_init-matr_dist_proj_init)./matr_dist_init;
    matr_distc(isnan(matr_distc))=0;
    for k=1:vect_nbr
```

```
        vectc=matr_proj_init(:,k);
        matref=repmat(vectc,1,vect_nbr-1);
        if k==1
            matrc=matr_proj_init(:,2:vect_nbr);
            vect_distc=matr_distc(1,2:vect_nbr);
        elseif k==vect_nbr
            matrc=matr_proj_init(:,1:vect_nbr-1);
            vect_distc=matr_distc(1:vect_nbr-1,vect_nbr)';
        else
            matrc=[matr_proj_init(:,1:k-1) matr_proj_init(:,k+1:vect_nbr)];
            vect_distc=[matr_distc(1:k-1,k)' matr_distc(k,k+1:vect_nbr)];
        end
        matrc=matref-matrc;
        matr_proj(:,k)=vectc+(4*alpha*coeff_sam)*matrc*vect_distc';
    end
    matr_dist_proj=triu(dist(matr_proj',matr_proj));
    matr_dist_diff=((matr_dist_init-matr_dist_proj).^2)./matr_dist_init;
    matr_dist_diff(isnan(matr_dist_diff))=0;
    Sammon_stress_new=coeff_sam*sum(sum(matr_dist_diff))
    esr=abs((Sammon_stress_old-Sammon_stress_new)/Sammon_stress_old)
    Sammon_stress_old=Sammon_stress_new;
    Sammon_stress(cont+1)=Sammon_stress_old;
     esr_vect(cont)=esr;
    matr_dist_proj_init=matr_dist_proj;
     matr_proj_init=matr_proj;
    cont=cont+1;
end
```

EXERCISE 13.7.

a. Generate three classes of 1,500 three-dimensional Gaussian vectors having the following mean vectors and the covariance matrices:

$$\mu_1 = \begin{bmatrix} 3 \\ 5 \end{bmatrix}, \ \Sigma_1 = \begin{bmatrix} 4 & 6 \\ 6 & 12 \end{bmatrix}, \ \mu_2 = \begin{bmatrix} 18 \\ 10 \end{bmatrix}, \ \Sigma_2 = \begin{bmatrix} 12 & -4 \\ -4 & 8 \end{bmatrix}, \ \mu_3 = \begin{bmatrix} 10 \\ 3 \end{bmatrix}, \ \Sigma_3 = \begin{bmatrix} 4 & 0 \\ 0 & 4 \end{bmatrix}$$

Then divide each class randomly into three subsets (training, test and validation) containing the same number of vectors.

b. Classify the test vectors using the Bayes classifier, the KNN rule and the MLP and compare the results obtained. For the last classifier consider the backpropagation learning algorithm with *momentum term* and variable learning rate. Then use the Levenberg-Marquardt learning algorithm and plot the training results every ten epochs. Conclude on the performance of the three classifiers and on the advantages and the limitations of each of them.

EXERCISE 13.8.

a. Generate three classes of 100 two-dimensional Gaussian vectors having the following mean vectors and the covariance matrices

$$\mu_1 = \begin{bmatrix} -1.5 \\ -1.5 \end{bmatrix},\ \mu_2 = \begin{bmatrix} 0 \\ 2 \end{bmatrix},\ \mu_3 = \begin{bmatrix} 2 \\ -1.5 \end{bmatrix},\ \Sigma_1 = \Sigma_2 = \Sigma_3 = \begin{bmatrix} 0.25 & 0 \\ 0 & 0.25 \end{bmatrix}$$

b. Classify the generated vectors using a SVM classifier. Consider successively a Gaussian and then a polynomial kernel.

c. Repeat the classification, after centering the third class and increasing the number of vectors. Conclude on the SVM classifier limitations.

Chapter 14

Data Compression

14.1. Theoretical background

Data compression aims to reduce the volume of data to be transmitted, processed or recorded, without significant information loss. Although many other quality indices exist, the mean square error defined below will be considered in this chapter to compare the different data compression algorithms:

$$EQM(s, \hat{s}) = \frac{1}{N_d} \|s - \hat{s}\|^2 \qquad [14.1]$$

In the above equation, s is the original signal (1D or 2D) initial, $\hat{s}$ stands for the compressed signal and N_d denotes the signal length (number of samples or number of pixels). Note that in the case of an image compression $\hat{s}$ is not always the best solution from the human eye point of view. For example, the properties of the two-dimensional discrete Fourier transform are not adapted to the human vision system. Thus, it is replaced by the discrete cosine transform in compression schemes.

Most compression algorithms presented in this chapter are defined for the 2-D case (image compression). Compression algorithms for the 1-D case can be easily obtained as particular cases. An image is often cut up into blocks before performing its compression. Their sizes and forms depend on the processing speed, compression rate and memory organization. Furthermore, the spatial correlation of the gray levels, which is usually isotropic, also has to be taken into account. At present, the best trade-off is obtained for square blocks of 4×4 pixels.

The intensities of the pixels belonging to each block form a vector, denoted by S_i. The set of all vectors $\{S_i\}_{i=1..M}$ associated with an image define the original vector space. All the compression methods aim to reduce the dimension of this space.

Some of the most important strategies for the data compression will be presented hereafter.

14.1.1. *Transform-based compression methods*

The key idea in this case is to transform the original signal, so that the main part of its energy is confined to a minimum number of coefficients. Generally, the signal transform is linear in order to insure both the generalization capability of the solution obtained and the signal reconstruction.

The discrete cosine transform (DCT) and Karhunen-Loève transform (KLT) are the best known transforms for data compression. In the case of the DCT-2D the following coefficients are calculated:

$$a_{uv} = \frac{2}{N}\sum_{n=0}^{N-1}\sum_{m=0}^{N-1} s[n,m]\cdot\cos\left[\frac{\pi u}{N}\left(n+\frac{1}{2}\right)\right]\cdot\cos\left[\frac{\pi v}{N}\left(m+\frac{1}{2}\right)\right] \quad [14.2]$$

with $u = 0..N-1$, $v = 0..N-1$.

According to the above equation, the image is projected on the basis of cosine functions. An important property of this transform is that high order coefficients contain only high spatial frequencies. Since the human eye is much less sensitive to these frequencies, the corresponding coefficients can be discarded without any significant information loss.

Actually, this type of compression is equivalent to a lowpass image filtering. The larger the compression rate is, the lower the quality of the reconstructed image is, since fewer details are preserved.

The KLT is an optimal linear transform, because there is no redundancy in the projection space and the obtained coefficients are completely uncorrelated. It consists of projecting the input image on an orthonormal basis formed by its covariance matrix eigenvectors:

$$\Sigma_S = \frac{1}{M}\sum_{i=0}^{M-1}\left(S_i - \overline{S}\right)\cdot\left(S_i - \overline{S}\right)^T \quad [14.3]$$

where S_i stand for vectors derived from the images to be compressed, and $\overline{S}$ is the mean vector.

It is well known that matrix K, having as rows the eigenvectors of matrix Σ_S, is a unitary matrix. Consequently:

$$Y_i = K \cdot S_i, \; i = 0..M-1 \qquad [14.4]$$

where M is the number of blocks issued from the original image and Y_i are the projected vectors.

If the Σ_s matrix eigenvectors are sorted in decreasing order according to the corresponding eigenvalues, then the components of the projected vectors will also be sorted in decreasing order according to their contribution to the original image reconstruction. Image compression is thus possible by preserving only the components which yield an appropriate reconstruction according to some quality criterion.

Despite its optimality, the KLT requires a large number of calculations compared to other data compression methods. Thus, some suboptimal linear transforms, such as the Walsh or Hadamard transforms, are sometimes preferred due to their simplicity and low calculation effort.

14.1.2. *Parametric (predictive) model-based compression methods*

These compression methods also rely on removing the data redundancy, but in a different way. In the case of images, for example, it is often possible to predict the intensity value of a pixel using the intensity values of its neighbors. An AR-2D model is usually preferred in the form:

$$\hat{s}(n,m) = \sum_{(l,k)\in V(n,m)} \sum a_{lk} \cdot s(n-l, m-k) \qquad [14.5]$$

where $V(n,m)$ denote the neighborhood of the pixel (n,m).

The coefficients of this model are obtained, just as in the 1-D case, by minimizing the mean square error:

$$EQM = E\left\{ \left| s(n,m) - \sum_{(l,k)\in V(n,m)} \sum a_{lk} \cdot s(n-l, m-k) \right|^2 \right\} \qquad [14.6]$$

This optimization procedure leads to the Yule-Walker equations, which are expressed as indicated below:

$$\sum_{(i,j)\in V(l,k)} \sum a_{lk} \cdot R(i-l, j-k) = R(i,j) \qquad [14.7]$$

where $R(i,j)$ is the 2-D autocorrelation function of the image $s(n,m)$.

The coefficients a_{lk} issued from the above equations allow removing the data redundancy, since the error, calculated as the difference between the original and the reconstructed images, is equivalent to a white noise. The original image is thus compressed, because it is replaced by the AR model coefficients and the error variance.

14.1.3. *Wavelet packet-based compression methods*

Wavelet packets, introduced by Coifman and Wickerhauser, represent a generalization of the multiresolution analysis described in Chapter 11. They are recursively defined in the form:

$$\begin{cases} W_{2n}(t) = \sqrt{2} \cdot \sum_k h(k) W_n(2t - k) \\ W_{2n+1}(t) = \sqrt{2} \cdot \sum_k g(k) W_n(2t - k) \end{cases} \quad [14.8]$$

$W_0(t)$ corresponds to the scale function $\varphi(t)$, while $W_1(t)$ corresponds to the wavelet function $\psi(t)$. Its scaled and delayed versions form a zero-mean function library, well localized in time and frequency and called wavelet packets.

Each wavelet packet is characterized by three parameters: frequency, scale and position in time. These functions form a set of possible decomposition bases for the analyzed signal.

The wavelet packet decomposition can be represented by means of a table of coefficients. The row index is related to the scale factor, while the column index is related to the frequency and/or the time position, according to the coefficient grouping on each row. This representation is provided in Figure 14.1 in the particular case of a discrete signal, defined on 8 points $\{x_1, x_2, \ldots x_8\}$.

A table cell which is divided into two other cells on the next decomposition level is called the “father cell”, while the two resulting cells are called “children cells”.

The selection of the best decomposition basis is usually performed using a minimal entropy criterion. It is defined in the Shannon-Weaver sense and makes use of the following cost function:

$$\lambda(s) = -\sum_i |s_i|^2 \log\left(|s_i|^2\right) \quad [14.9]$$

x_1	x_2	x_3	x_4	x_5	x_6	x_7	x_8
s_1	s_2	s_3	s_4	d_1	d_2	d_3	d_4
ss_1	ss_2	ds_1	ds_2	sd_1	sd_2	dd_1	dd_2
sss_1	dss_1	sds_1	dds_1	ssd_1	dsd_1	sdd_1	ddd_1

Figure 14.1. *Wavelet packet decomposition in the form of a table of coefficients*

The minimization of this function is equivalent to the entropy minimization, but it is also an additive information measure. After the wavelet packet decomposition, the algorithm for the selection of the best basis compares the entropies of the children cells to the entropy of their father cell, starting from the last decomposition level. The father cell entropy is replaced by the sum of its children entropies if it is higher, otherwise it is preserved.

Only the coefficients above an appropriately chosen threshold are finally selected (for example, the threshold may be calculated so that 90% of the total energy is preserved by the selected coefficients).

The minimum entropy of the best basis means that the signal energy is confined to a minimum number of coefficients. This makes the wavelet packet decomposition particularly suitable to the data compression.

14.1.4. *Vector quantization-based compression methods*

Vector quantization (VQ) is one of the most effective data compression methods. Let us consider an image as a set of q-length vectors, $\{S_j\}_{j=1..M}$. Its VQ is thus defined as an application:

$$Q: R^q \rightarrow W, \quad Q\left(S_j\right) = W_0^{(j)} \qquad [14.10]$$

where $W = \{W_i,\ i = 1..N_c\}$ is the code vector set and $W_0^{(j)}$ denotes the nearest code vector to input vector S_j.

As the code vector set is also available to the decoder, only the indices of the code vectors corresponding to the input vectors have to be conserved. Thus, a high compression rate can be obtained.

The VQ effectiveness depends on the code vector set, which is obtained by means of different clustering methods, such as the *K-means* algorithm. Its main steps are given below:

1. randomly choose $W = \{W_i, i = 1..N_c\}$;

2. update the vector clusters D_i so that:

$$D_i = \left\{ S \,\middle|\, \|S - W_i\| < \|S - W_j\|, \forall\, j \neq i \right\} \qquad [14.11]$$

3. calculate the mean square error:

$$EQM = \frac{1}{N} \sum_{i=1}^{N_c} \sum_{S \in D_i} \|S - W_i\|^2 \qquad [14.12]$$

and stop if:

$$\left| \frac{EQM(t) - EQM(t-1)}{EQM(t-1)} \right| < \varepsilon \qquad [14.13]$$

where ε is a positive constant;

4. replace each code vector W_i by the mean vector of D_i;

5. return to step 2.

14.1.5. *Neural network-based compression methods*

The learning and generalization capabilities of neural networks are also very useful for data compression. Two neural network based compression techniques are illustrated in this chapter: neural KLT and VQ using the Kohonen map.

Neural KLT

The key idea of this method is to train a neural network, having the structure shown in Figure 14.2, for performing the KLT. It can be easily seen that the last layer i^{th} neuron output corresponding to the vector $S_k = [s_1^{(k)} \quad s_2^{(k)} \quad \cdots \quad s_q^{(k)}]^T$ is given by:

$$o_{ki} = W_i \cdot S_k + \sum_{l<i} u_{il} \cdot o_{kl}, \quad i = 1..r \qquad [14.14]$$

where $W_i = [w_{i1}\ w_{i2}\ \ldots\ w_{iq}]^T$ is the weight vector associated with the ith neuron.

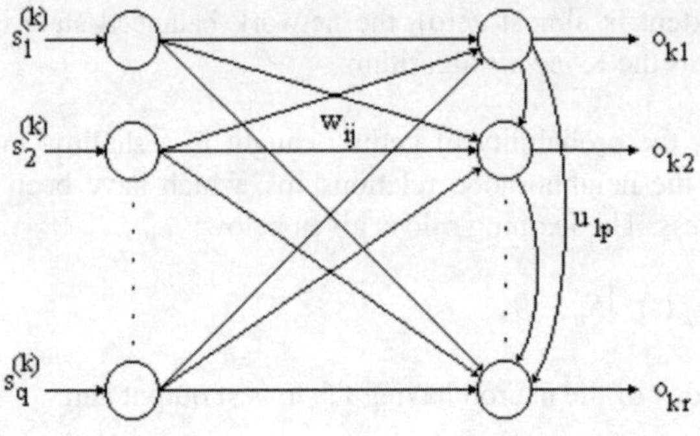

Figure 14.2. *Structure of a neural network for the KLT calculation*

Weights w_{ij} and u_{li} are updated using the Hebb and anti-Hebb rules respectively, with momentum term and variable learning rate:

$$\begin{cases} \Delta w_{ij}(t+1) = \eta(t) \cdot s_j^{(k)} \cdot o_{ki} + \beta(t) \cdot \Delta w_{ij}(t) \\ \Delta u_{li}(t+1) = -\mu(t) \cdot o_{kl} \cdot o_{ki} + \beta(t) \cdot \Delta u_{li}(t) \end{cases} \quad [14.15]$$

Rubner and Tavan have proven that at the end of the training process, the r weight vectors associated with the last layer neurons represent the r eigenvectors of the covariance matrix corresponding to its largest eigenvalues. Consequently, the trained network is able to project any new vector and therefore perform data compression in the KLT sense. However, the projection matrix is coded in the neural network weights, unlike in the case of the KLT, which provides it in an explicit form.

VQ using the Kohonen map

Although easy to implement, standard VQ algorithms sometimes converge to suboptimal solutions, corresponding to a local minimum of the error function. The self-organizing Kohonen map, which involves the neighborhood concept, makes it possible to overcome this problem thanks to its topological properties.

Thus, the self-organizing Kohonen map (whose structure is similar to that of the previously presented neural network) is able to learn both the input space vector distribution and the topology of this space. In other words, two closely spaced neurons will be activated by neighbor zones from the input space.

The main idea is then to progressively reduce the neighborhood extent in the Kohonen model. In this way, at the end of the training process (when the neighborhood extent is almost zero), the network balanced state is defined by the same conditions as the K-*means* algorithm.

Nevertheless, the probability of getting caught in a shallow minimum is much lower thanks to the neighborhood relationships, which have been activated during the training process. The learning rule is given below:

$$\Delta W_i = \alpha_{ik}(t) \cdot [S_n - W_i] \qquad [14.16]$$

where k is the index of the neuron having the lowest output and:

$$\alpha_{ik} = \alpha_0 \cdot e^{-\frac{(i-k)^2}{2\sigma_t^2}}, \ \sigma_t = \sigma_0 \left(\frac{\sigma_T - 1}{\sigma_0}\right)^{\frac{t}{T-1}} \qquad [14.17]$$

In the above equation T is the number of epochs. At the end of the training process, the weights of the neurons on the output layer form the code vector set.

14.2. Solved exercises

EXERCISE 14.1.

Perform the DCT compression of the image "dots", which can be found in the file "imdemos" of the image processing toolbox. Use the MATLAB function `dct2.m`

a. Plot the mean square error variation with the number of conserved coefficients.

b. Perform the image compression by preserving only the 1,024 largest coefficients.

a.

```
load imdemos dots;
[N1,N2]=size(dots);
sl=double(dots); s=zeros(N1,N2);
s=sl; y=dct2(s); [ly,cy]=size(y);
yv=y(:); [yvs,idxs]=sort(abs(yv));
yvs=flipud(yvs); idxs=flipud(idxs);
cont=0;
for r=2.^[0 :14]
      yvr=yv ; yr=y;
      yvr(idxs(r+1 :N1*N2))=zeros(N1*N2-r,1);
      for k=1:N2
            yr(:,k)=yvr((k-1)*N1+1:k*N1,1);
```

```
    end
    sr=idct2(yr); cont=cont+1;
    eqm(cont)=(sum(sum(s-sr).^2))/(N1*N2);
end
plot(eqm);
set(gca,'XTickLabel',2.^(str2num(get(gca,'XTickLabel'))-1))
xlabel('Number of preserved coefficients');
ylabel('Error'); title('Variation of the mean square error')
```

Note that the mean square error decreases very fast with the number of conserved coefficients due to the high redundancy of the original image.

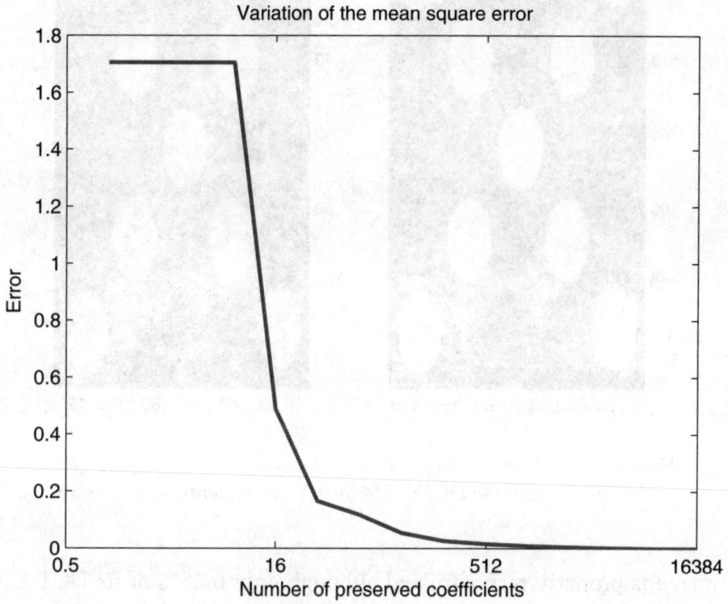

Figure 14.3. *Variation of the mean square error (the horizontal axis is represented in logarithmic scale)*

b.

```
load imdemos dots
r=1024; [N1,N2]=size(dots); s1=double(dots);
s=zeros(N1,N2); s=s1;
y=dct2(s); [ly,cy]=size(y); yv=y(:);
[yvs,idxs]=sort(abs(yv));
yvs=flipud(yvs); idxs=flipud(idxs); cont=0;
yv(idxs(r+1:N1*N2))=zeros(N1*N2-r,1);
for k=1:N2
```

```
    y(:,k)=yv((k-1)*N1+1:k*N1,1);
end
sr=idct2(y); figure
subplot(121); imagesc(s); colormap(gray);
title('Original image')
subplot(122); imagesc(sr); colormap(gray);
title('Reconstructed image')
```

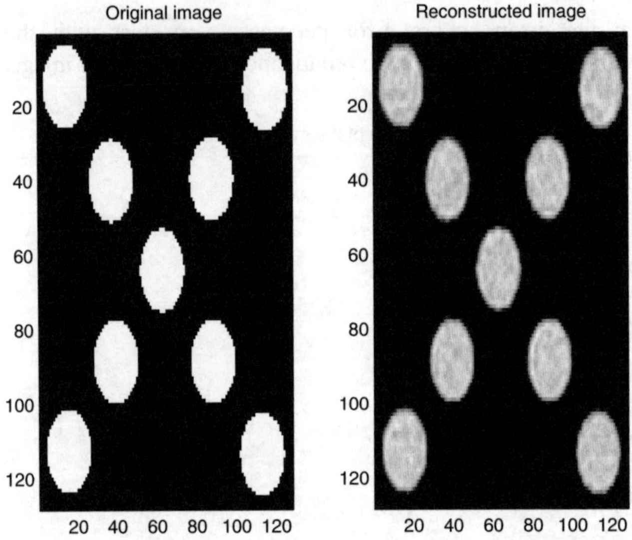

Figure 14.4. *Original image (left) and reconstructed image (right) with the DCT 1,024 largest coefficients*

The image is properly reconstructed although only 6.25% of its DCT have been conserved. The main shortcoming of this method is that its coefficients are not integers, so that they cannot be obtained by simple calculations. This is why other transforms (Walsh, Hadamard, etc.) have been long preferred to the DCT.

Consequently, the coefficients issued from the DCT have to be quantified, according to their dynamic range and weight for image reconstruction. The number of discarded coefficients depends on the required compression rate, the image content and the expected quality of the reconstructed image.

EXERCISE 14.2.

Consider again the image used in the previous exercise. Perform its compression using the KLT of the vectors obtained by cutting up the image into 4×4 pixels blocks.

a. Plot the mean square variation according to the number of conserved coefficients.

b. Perform the compression by preserving only the largest component of each vector in the projection space.

a.

```
load imdemos dots;
[N1,N2]=size(dots);
s1=double(dots);
s=zeros(N1,N2); s=s1;
x=im2col(s,[4 4],'distinct');
xr=x'; sigx=cov(xr);
[vectp,valp]=eig(sigx);
vlpr=diag(valp);
[ld,idx]=sort(vlpr);
idx=flipud(idx);
ld=flipud(ld);
for r=1:16
     vectps='=[';
     for k=1:r
          no=num2str(k);
          eval(['vectps=[vectps ''vectp(:,idx(',no,')) ''];'])
     end
     vectps=[vectps ']'';'];
     eval(['kmat',vectps]);
     kmatc=zeros(16,16);
     kmatc(1:r,:)=kmat;
     sc=kmat*x;
     srcol=kmat'*sc;
     sr=col2im(srcol,[4 4],[N1 N2],'distinct');
     eqm(R)=(sum(sum(s-sr).^2))/(N1*N2);
end
plot(eqm);
xlabel('Number of preserved coefficients')
ylabel('Error');
title('Variation of the mean square error')
```

The high redundancy level of the original image is again clearly visible, since all information is confined by the first 9 coefficients of each data block.

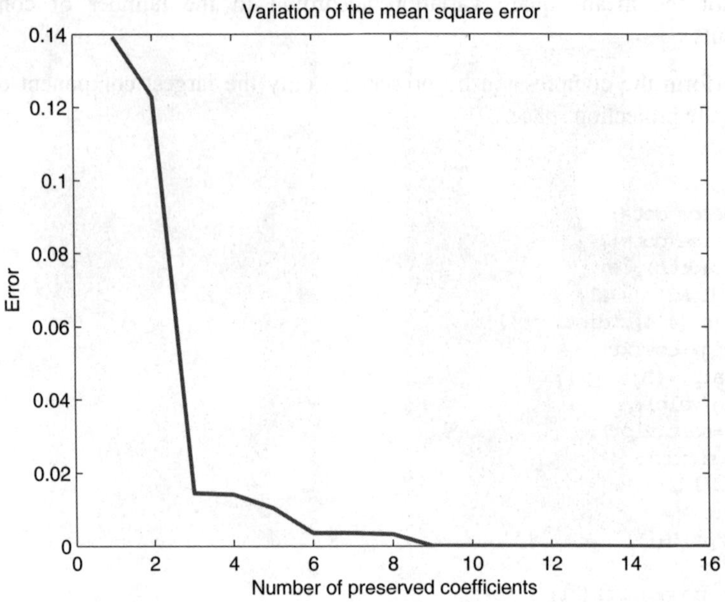

Figure 14.5. *Mean square error variation*

b.

```
load imdemos dots; r=1;[N1,N2]=size(dots);
s1=double(dots); s=zeros(N1,N2); s=s1;
x=im2col(s,[4 4],'distinct');
xr=x'; sigx=cov(xr);
[vectp,valp]=eig(sigx);
vlpr=diag(valp); [ld,idx]=sort(vlpr);
idx=flipud(idx); ld=flipud(ld);
vectps='=[';
for k=1:r
    no=num2str(k);
    eval(['vectps=[vectps ''vectp(:,idx(',no,')) '';']'])
end
vectps=[vectps ']'';'];
eval(['kmat',vectps]);
kmatc=zeros(16,16);kmatc(1:r,:)=kmat;
sc=kmat*x;srcol=kmat'*sc;
sr=col2im(srcol,[4 4],[N1 N2],'distinct');
subplot(121); imagesc(s)
colormap(gray); title('Original image')
subplot(122); imagesc(sr)
colormap(gray); title('Reconstructed image')
```

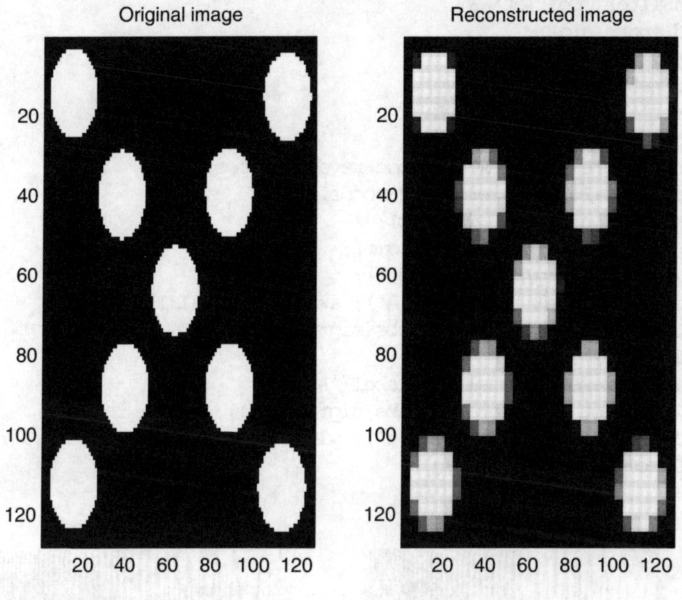

Figure 14.6. *Original and reconstructed image using the KLT*

The best compression rate is provided by the KLT since it leads to completely uncorrelated vector components in the projection space. It minimizes the mean square error minimization and performs the best mean power concentration in a minimum number of coefficients. However, the DCT compression is much faster, while its performance is close to that provided by the KLT.

EXERCISE 14.3.

Compare the compression performance of the DCT and the Wavelet Packet Decomposition (WPD) using the functions from the toolbox "Wavelab" (URL: http://www-stat.stanford.edu/~wavelab).

Generate a linearly frequency modulated signal, whose frequency varies between 1 kHz and 15 kHz, sampled at 100 kHz, on 512 points. Perform the compression using the two methods, by preserving only the largest 150 coefficients, and plot the obtained result. Repeat the compression in the case of 80 preserved coefficients.

```
fe=1e5; f1=1e3; f2=15e3; fn1=f1/fe; fn2=f2/fe;
sig=(vco(sawtooth(2*pi*[1:512]/512,1),[fn1 fn2],1))';
nocof=150; sigdct=dct(sig);
[sigdcts,idcts]=sort(abs(sigdct)); idcts=reverse(idcts);
sigdct(idcts(nocof+1:512))=zeros(512-nocof,1); sigrecdct=idct(sigdct);
```

```
qmf=MakeOnFilter('Coiflet',3);
[n,D]=dyadlength(sig);
wp=WPAnalysis(sig,D,qmf);
stree=CalcStatTree(wp,'Entropy');
[btree,vtree]=bestbasis(stree,D);
wpex=unpackbasiscoeff(btree,wp);
[wpexs,idpos]=sort(abs(wpex)); idpos=reverse(idpos);
wpex(idpos(nocof+1:512))=zeros(512-nocof,1);
wpnew=PackBasisCoeff(btree,wp,wpex);
sigrecdpo=WPSynthesis(btree,wpnew,qmf);
subplot(321); plot(sig/max(abs(sig)));
title('Original signal'); ylabel('A'); axis([0 512 -1 1]); grid
subplot(323); plot(sigrecdct/max(abs(sigrecdct)));
axis([0 512 -1 1]); grid
title('DCT compressed signal'); ylabel('A')
subplot(325); plot(sigrecdpo/max(abs(sigrecdpo)));
axis([0 512 -1 1]); grid; xlabel('t'); ylabel('A')
title('WPD compressed signal');
```

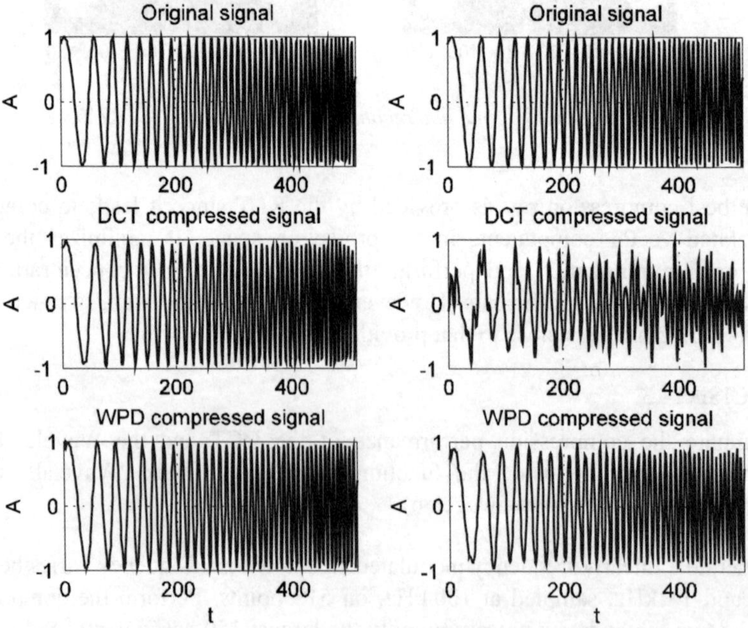

Figure 14.7. *DCT and WPD compression results*

The compression results are plotted on the left side column in the first case and on the right side column in the second case. Note the robustness of the WPD compression as the number of preserved coefficients decreases.

EXERCISE 14.4.

Repeat exercise 14.3 by preserving 99.5% of the original signal energy, for both DCT and WPD compression methods. Determine the number of required coefficients in the two cases and plot the compression result.

```
fe=1e5;f1=1e3;f2=15e3;fn1=f1/fe;fn2=f2/fe;
sig=(vco(sawtooth(2*pi*[1:512]/512,1),[fn1 fn2],1))';
 part =sum(sig.^2); enethreshold=0.995 ;
sigdct=dct(sig);
[sigdcts,ixdcts]=sort(abs(sigdct));
ixdcts=reverse(ixdcts);
sigdcts=reverse(sigdcts) ;
enedct=cumsum(sigdcts.^2);
ixdct=find(enedct>=enethreshold*energ);
nocoeffdct=ixdct(1);
sigdct(ixdcts(ixdct(1)+1:512))=zeros(512-ixdct(1),1);
sigrecdct=idct(sigdct);
qmf=MakeOnFilter('Coiflet',3);
[n,D]=dyadlength(sig);
wp=WPAnalysis(sig,D,qmf);
stree=CalcStatTree(wp,'Entropy');
[btree,vtree]=BestBasis(stree,D);
wpex=UnpackBasisCoeff(btree,wp);
[wpexs,ixdpos]=sort(abs(wpex));
ixdpos=reverse(ixdpos); wpexs=reverse(wpexs);
enedpo=cumsum(wpexs.^2);
ixdpo=find(enedpo>=enethreshold*energ);
nocoeffdpo=ixdpo(1);
wpex(ixdpos(ixdpo(1)+1:512))=zeros(512-ixdpo(1),1);
wpnew=PackBasisCoeff(btree,wp,wpex);
sigrecdpo=WPSynthesis(btree,wpnew,qmf);
subplot(311);plot(sig/max(abs(sig)));
title('Original signal');
ylabel('A');axis([0 512 -1 1]); grid
subplot(312); plot(sigrecdct/max(abs(sigrecdct)));
axis([0 512 -1 1]); grid
title('DCT compressed signal');ylabel('A')
subplot(313); plot(sigrecdpo/max(abs(sigrecdpo)));
axis([0 512 -1 1]); grid
title('WPD compressed signal');
xlabel('t');ylabel('A')
```

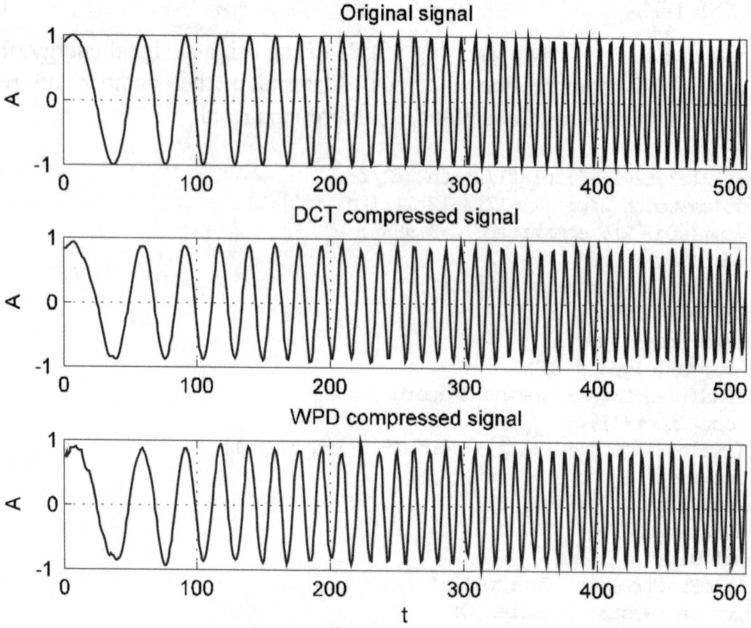

Figure 14.8. *Comparison of DCT and WPD compression methods for the same preserved energy level*

It can be seen that the compression results are very similar when the same preserved energy level is considered. However, 99.5% of the signal energy is confined into 140 coefficients in the case of the DCT, while the WPD requires only 83 coefficients for the same preserved energy percentage, thanks to its time-frequency localization property.

The variation of the cumulated energy of each decomposition coefficient is plotted in Figure 14.9.

```
enedct=cumsum(sigdcts.^2);
enedpo=cumsum(wpexs.^2);
figure; plot(enedct,'-');
hold on; plot(enedpo,'—r')
title('Variation of the cumulated energy of the decomposition coefficients')
ylabel('Energy');
xlabel('Number of preserved coefficients');
grid; axis([0 512 0 1.1*energ]);
legend('DCT','WPD',0)
```

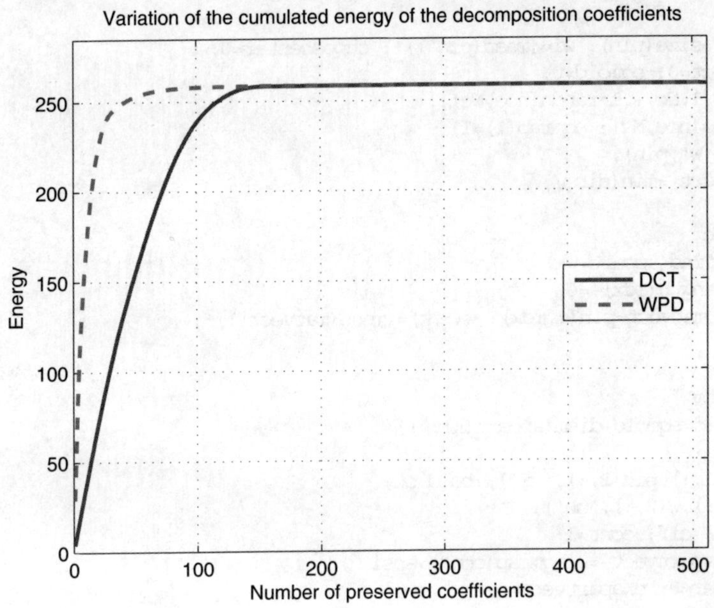

Figure 14.9. *Variation of the cumulated energy of the DCT and WPD coefficients*

Note that the higher the compression rate is, the more effective the use of the WPD is. It can be also very useful for the compression of signals which are wideband, highly irregular or contain discontinuities.

EXERCISE 14.5.

a. Generate 9 Gaussian clusters, containing 1,000 points, with variance 1, and centered on: (0,0), (0,5), (0,10), (5,0), (5,5), (5,10), (10,0), (10,5) and (10,10).

b. Use the K-means algorithm to find the code vectors corresponding to the 9 clusters.

a.

```
Nc=9; epsi=1e-3; nv=900; sigma=1;
centr=[0 0 0 5 5 5 10 10 10;0 5 10 0 5 10 0 5 10];
[noc,ncls]=size(centr); nvc=nv/ncls; pin=zeros(noc,nv);
for k1=1:ncls
    c=centr(:,k1); cc=c*ones(1,nvc); pinc=cc+sigma*randn(noc,nvc);
    pin(:,(k1-1)*nvc+1:k1*nvc)=pinc;
end
```

b.

```
[noc,nov]=size(pin); wi=(mean(pin'))'; chposcenter=0;
chapartvect=0; eqmold=0;
w=wi*ones(1,Nc)+.1*randn(noc,Nc);
wnew=zeros(noc,Nc); eqmrap(1)=1;
dism=dist(w',pin);
[valm,appart]=min(dism);
figure
for k=1:Nc
    idx=find(appart==k);
    if length(idx)~=0
        matvect=pin(:,idx); w(:,k)=(mean(matvect'))';
    end
end
for k=1:nov
    eqmold=eqmold+dism(appart(k),k);
end
plot(pin(1,:),pin(2,:),'xr'); hold on;
plot(w(1,:),w(2,:),'ob');
pause(.1); clf; cont=1;
while (chapartvect~=0|eqmrap(cont)>epsi)
chposcenter=0; chapartvect=0;
dism=dist(w',pin);
[valm,appart]=min(dism);
for k=1:Nc
    idx=find(appart==k);
    if length(idx)~=0
        matvect=pin(:,idx); w(:,k)=(mean(matvect'))';
    end
end
dism=dist(w',pin); [valm,appartnew]=min(dism); eqmnew=0;
for k=1:nov
    eqmnew=eqmnew+dism(appartnew(k),k);
end
cont=cont+1;
eqmrap(cont)=abs((eqmold-eqmnew)/eqmold); eqmold=eqmnew;
if appartnew~=appart
    chapartvect=chapartvect+1;
end
appart=appartnew; plot(pin(1,:),pin(2,:),'.r');
hold on; plot(w(1,:),w(2,:),'ob'); clf
end
subplot(211); plot(pin(1,:),pin(2,:),'xr');
hold on; plot(w(1,:),w(2,:),'ob'); axis([-5 15 -5 20])
legend('Original vectors','Code vectors',0)
subplot(212); plot(eqmrap);
title('Variation of the relative mean square error')
```

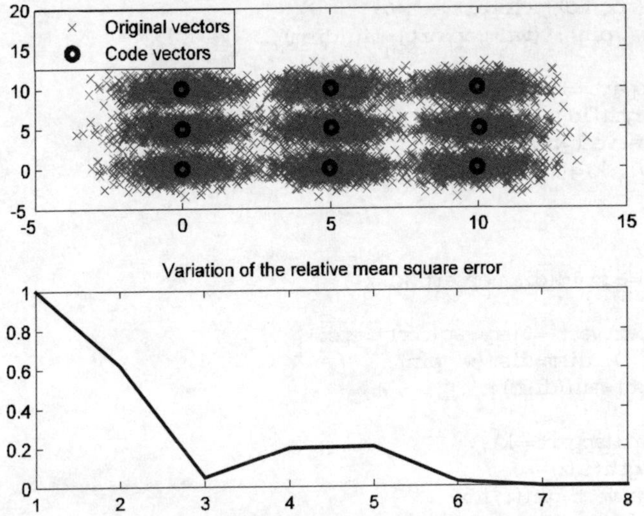

Figure 14.10. *Code vectors obtained using the K-means method*

The stop criterion for the K-means algorithm implemented above consists of two conditions which have to be met simultaneously: the training vectors have to be stable and the relative mean square error has to be under the appropriately chosen "epsi" threshold.

Note the fast convergence of this algorithm and the fact that the obtained solution is well suited to the distribution of the input vectors.

EXERCISE 14.6.

Compare the performance of the KLT, the VQ using the K-means algorithm and the QV using the Kohonen map for the compression of the image "*trees*", which can be found in the file "imdemos" of the image processing toolbox.

Cut up the image into blocks of 4×4 pixels. Keep 5 coefficients for each block in the case of the KLT and use 32 code vectors for the K-means algorithm and the Kohonen map. Consider a Manhattan distance based neighborhood system for the last case.

```
warning off; load imdemos trees; s=double(trees);
[N1,N2]=size(trees); pin=im2col(s,[4 4],'distinct');
Nc=32; epsi=1e-3;
[noc,nov]=size(pin); wi=(mean(pin'))';
w=wi*ones(1,Nc)+.1*randn(noc,Nc);
eqmold=0; eqmrap(1)=1; cont=1;
```

```
wnew=zeros(noc,Nc); chapartvect=0;
dism=dist(w',pin); [valm,appart]=min(dism);
for k=1:Nc
idx=find(appart==k);
     if length(idx)~=0
          matvect=pin(:,idx);
          w(:,k)=(mean(matvect'))';
     end
end
for k=1:nov
     eqmold=eqmold+dism(appart(k),k);
end
while (chapartvect~=0|eqmrap(cont)>epsi)
chapartvect=0; dism=dist(w',pin);
[valm,appart]=min(dism);
for k=1:Nc
     idx=find(appart==k);
     if length(idx)~=0
          matvect=pin(:,idx);
          w(:,k)=(mean(matvect'))';
     end
end
dism=dist(w',pin);
[valm,appartnew]=min(dism); eqmnew=0;
for k=1:nov
     eqmnew=eqmnew+dism(appartnew(k),k);
end
cont=cont+1;
eqmrap(cont)=abs((eqmold-eqmnew)/eqmold);
eqmold=eqmnew;
if appartnew~=appart
     chapartvect=chapartvect+1;
end
appart=appartnew;
end
smkam=zeros(size(pin));
for k=1:nov
     smkam(:,k)=w(:,appart(k));
end
pinm=min(pin'); pinM=max(pin'); pini=[pinm' pinM'];
wkoh=initsm(pini,Nc); dp=[50 5000 1];
mtvois=nbman([Nc Nc]);
wkoh=trainsm(wkoh,mtvois,pin,dp);
dism=dist(wkoh,pin);
[valm,appart]=min(dism); smkoh=zeros(size(pin));
for k=1:nov
     smkoh(:,k)=(wkoh(appart(k),:))';
end
srkam=col2im(smkam,[4 4],[N1 N2],'distinct');
srkoh=col2im(smkoh,[4 4],[N1 N2],'distinct');
```

```
subplot(221); imagesc(s);
colormap(gray); title('Original image')
subplot(222); imagesc(srtkl);
colormap(gray); title('KLT based compression')
subplot(223); imagesc(srkam);
colormap(gray); title('K-means based compression')
subplot(224); imagesc(srkoh); colormap(gray);
title('Kohonen map based compression')
```

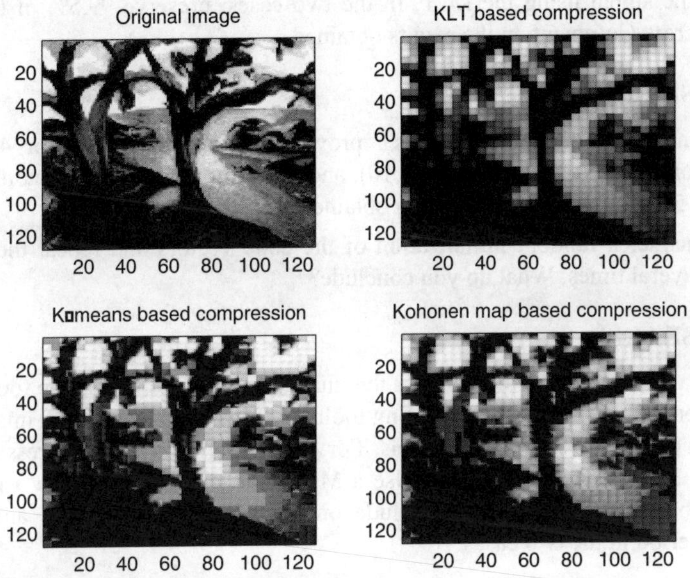

Figure 14.11. *Comparison of KLT, K-means and Kohonen map based compression methods*

Note the performance level of the two VQ based compression methods. The compression rate is higher than for any other method, since only one integer corresponding to the code vector index replaces a whole data block. However, the set of code vectors is supposed to be available for decompression.

14.3. Exercises

EXERCISE 14.7.

Compare the compression performed using the DCT and the KLT for the image "*flower*", which can be found in the file "imdemos" of the image processing toolbox. Use the MATLAB codes provided in the first two solved exercises. Consider successively image block partitions of 4×4, 8×8 and 16×16 pixels.

Comment on the effect of the block size on the compression effectiveness and calculation time.

EXERCISE 14.8.

Perform the wavelet packet based compression of a sinusoid having the frequency 1 kHz, sampled at 100 kHz, represented on 512 points and exhibiting random phase discontinuities at the zero-crossing instants. Perform the compression of the same signal using the DCT. In the two cases, preserve 99.5% of the initial signal energy. Comment on the results obtained.

EXERCISE 14.9.

a. Run the algorithm "K-means", provided in exercise 14.5, for 4 clusters centered on: (0,0), (0,10), (10,0), (10,10), and for a number of code vectors equal to 4, then to 5. Comment on the solution obtained.

b. Consider a random initialization of the code vectors and repeat the training process several times. What do you conclude?

EXERCISE 14.10.

Train a Kohonen map to compress the image "Saturn", which can be found in the file "imdemos" of the image processing toolbox. Consider successively image block partitions of 2×1, 2×2 and 4×4 pixels. For each case, plot the compression result using 16, 32 and 64 code vectors. Use a Manhattan distance and then a Euclidian distance based neighborhood. Conclude on the compression quality and on the training period in the two cases.

References

APPRIOU A., *Décision et reconnaissance des formes en signal*, Hermes, 2002.

ASSELIN DE BEAUVILLE J.P., KETTAF F.Z., *Bases théoriques pour l'apprentissage et la décision en reconnaissance des formes*, Cépaduès, 2005.

BELLANGER M., *Analyse des signaux et filtrage numérique adaptatif*, Masson, 1989.

BELLANGER M., *Traitement numérique du signal: théorie et pratique*, Masson, 1989.

BELLANGER M., *Filtres numériques: analyse et synthèse des filtres unidimensionnels*, Masson, 1990.

BISHOP C.M., *Neural Networks for Pattern Recognition*, Clarendon Press, Oxford, 1995.

CANDY J.V., *Signal Processing: The Model Based Approach*, Masson, 1989.

CHARBIT M., *Eléments de théorie du signal: les signaux aléatoires*, Ellipses, 1990.

COHEN L., *Time-Frequency Analysis*, Prentice Hall, 1995.

COULON F., *Théorie et traitement des signaux,* Dunod, 1987.

DUDA R.O., HART P.E., STORK D.G., *Pattern Classification*, Wiley, 2001

DUVAUT P., *Traitement du signal: concepts et applications*, Hermes, 1991.

FLANDRIN P., *Temps fréquence*, Hermes, 1998.

HAMMING R., *Digital Filters,* Prentice Hall, 1989.

JACKSON L., *Digital Filters and Signal Processing,* Kluwer, 1989.

KAY S.M., *Modern Spectral Estimation: Theory and Applications*, Prentice Hall, 1988.

KAY S.M., *Fundamentals of Statistical Signal Processing: Estimation Theory*, Prentice Hall, 1993.

KUNT M., *Traitement de l'information: techniques modernes de traitement numérique des signaux*, PPUR, 1991.

KUNT M., *Traitement numérique des signaux,* Dunod, 1986.

LACOUME J.L., *Théorie du signal,* P.U.F., 1983.

LE CHEVALIER F., *Principes de traitement des signaux radar et sonar*, Masson, 1989.

MARPLE S.L., *Digital Spectral Analysis with Applications*, Prentice Hall, 1987.

MATEESCU A., *Traitement numérique des signaux,* Editions Techniques, Bucarest, 1997.

MITRA S.K., *Handbook of Digital Signal Processing,* John Wiley and Sons, 1993.

OPPENHEIM A.V., *Advanced Topics in Signal Processing,* Prentice Hall, 1988.

OPPENHEIM A.V., *Discrete-Time Signal Processing,* Prentice Hall, 1989.

PAPOULIS A., *Signal Analysis,* McGraw-Hill, 1977.

PAPOULIS A., *Circuits and Systems: A Modern Approach,* Holt, Rinehart and Wiston, 1980.

PAPOULIS A., *Probability, Random Variables and Stochastic Processes*, McGraw-Hill, 1984.

PEEBLES P.Z., *Probability, Random Variables and Random Signal Principles*, McGraw-Hill, 1987.

PERSONNAZ L., *Réseaux de neurones formels pour la modélisation, la classification et la commande*, CNRS Editions, 2003

PICINBONO B., *Eléments de théorie du signal,* Dunod, 1986.

PICINBONO B., *Signaux aléatoires: probabilités et variables aléatoires, avec problèmes résolus*, Masson, 1989.

PORAT B., *Digital Processing of Random Signals: Theory & Methods*, Prentice Hall, 1994.

RANDALL R.B., *Frequency Analysis,* Bruel & Kjaer, 1987.

SCHARF L.L., *Statistical Signal Processing: Detection, Estimation and Time Series Analysis*, Addison Wesley, 1991.

STANKOVIC L., *Time-Frequency Signal Analysis,* Research monograph 1993-2003, 2004.

STOICA P., MOSES R., *Introduction to Spectral Analysis*, Prentice Hall, 1997.

THERRIEN C.W., *Discrete Random Signals and Statistical Signal Processing*, Prentice Hall, 1992.

VAIDYANATHAN P.P., *Multirate Systems and Filter Banks,* Prentice Hall, 1993.

VAN TREES H.L., *Detection, Estimation and Modulation Theory*, John Wiley, 1971.

List of Authors

Cornel IOANA
Institut national polytechnique
Grenoble
France

Ali MANSOUR
Ecole nationale supérieure des ingénieurs
UBO
Brest
France

André QUINQUIS
Ecole nationale supérieure des ingénieurs
UBO
Brest
France

Emanuel RADOI
Ecole nationale supérieure des ingénieurs
UBO
Brest
France

Index

P

R

S

T

U, V

W

Y, Z